PROBLEM-BASED LEARNING IN COMMUNICATION SYSTEMS USING MATLAB AND SIMULINK

PROBLEM-BASED LEARNING IN COMMUNICATION SYSTEMS USING MATLAB AND SIMULINK

KWONHUE CHOI

Yeungnam University, Gyeongsan, Korea

HUAPING LIU

Oregon State University, Corvallis

IEEE PRESS

WILEY

Library of Congress Cataloging-in-Publication Data is available.

ISBN: 978-1-119-06034-5

10 9 8 7 6 5 4 3 2 1

CONTENTS

PREFACE

THE CHALLENGES OF LEARNING AND TEACHING COMMUNICATIONS

Many digital communication topics taught in the traditional way require understanding mathematical expressions and algorithmic procedures to learn abstract concepts. The majority of existing textbooks facilitate teaching this way with systematic and thorough explanation of communication theories and concepts, mainly via mathematical models and algorithmic procedures. This is the natural outcome when computers and software were not so universally accessible decades ago as they are today. However, most students find such a way of learning digital communications ineffective and often frustrating. And even if they are able to follow the instructors in the classroom, their understanding of the concepts is often superficial. The accessibility of powerful software like MATLAB/Simulink and the Internet to students could be exploited to revolutionize the teaching of math intensive subjects such as digital communications. Through decades of classroom experience, we have learned that students' learning becomes significantly more effective if they are led to "construct" the system themselves and observe waveforms and statistics at various stages of the system or algorithm, a process called "active" learning here.

However, given the tools and texts available on the market to the instructors, implementing this active learning process is by no means easy. First, the majority of the textbooks are optimized for instruction in the traditional way. Some recent textbooks provide problems that involve the use of MATLAB/Simulink or similar software and codes or computer models to perform certain simulation. Readers can replicate these codes/models and conduct simulation, which would certainly reinforce some aspects they have learned. Such an approach is still far short of encouraging active learning by students. Second, there are some existing hardware training kits designed for educational purposes that can be used for labs/experiments of communications classes. However, these training kits are often expensive and cover only a limited number of topics of communications. Additionally, students need to learn hardware design skills such as DSP programming and VHDL to be able to use such a tool.

UNIQUE FEATURES OF THIS BOOK

This book is written to encourage active learning of communication theories and systems by its readers. Toward this goal, major communication concepts and algorithms are examined through carefully designed MATLAB/Simulink projects. Each project implements the simulation construction and execution steps or sequences that match how an actual communications system or algorithm works. These steps progressively explore the intermediate results between steps that students can "see" and comprehend what happens behind theories and mathematical expressions. The bulk of MATLAB simulation codes or Simulink models for these projects are provided. This ensures that students will be able to complete even complex projects such as Viterbi decoding, multiple-input multiple-output (MIMO) detection, and orthogonal frequency division multiplexing (OFDM) demodulation.

However, important parameters and codes lines or model blocks that are critical for learning the algorithm or communications process are left out for students to complete. This makes mechanically executing a certain completed code without understanding the technical details impossible. Step-by-step instructions are designed for each problem. Readers can conveniently check the results of each intermediate step and compare the various parameter choices and their effects and are thus led to actively figure out the intended answers and build up a complete system/algorithm.

Summarizing it, this book is written with the following three main goals in mind:

1. The framework of the codes/models provided in the book efficiently guides students through the simulation and actively engages students in learning the materials.
2. The codes/blocks provided minimize the amount of time students need to complete their simulations and ensure that they will be able to complete even complex projects without getting lost in the middle and giving up.
3. In completing the main algorithm/concept-specific incomplete parts, students will effectively be internalizing the theories.

In Chapters 4, 7, 9, 10, 11, 13, 20, 22, 23, and 30, students will learn how to convert constructed waveforms in simulations into electric signals and then to listen to those signals if they are audio signals, or observe the eye-patterns, scatter plots, or signal trajectories by using an oscilloscope for digitally modulated signals. In Chapters 13 and 30, students are encouraged to complete actual wireless communications in the band near-ultrasonic frequencies, requiring only a mobile phone and a PC with a microphone. We have found that all such present-day projects that embrace student interests can motivate them to explore more intensely how communication systems work.

Although, students are not required to know MATLAB/Simulink to use this book, Chapter 1 provides carefully designed projects that enable students to self-learn the MATLAB/Simulink skills needed for the rest of the projects in this book. All that a student needs are access to MATLAB, a headphone and an oscilloscope for some projects.

THE STRUCTURE OF THIS BOOK

The 30 chapters of this book cover MATLAB/Simulink basics (Chapter 1), basic signals and systems (Chapters 2–8), analog communications (Chapters 9–13), probability and random signals (Chapters 14–15), basics of digital communication techniques (Chapters 16–24), and wireless communication techniques (Chapters 25–30).

The majority of these chapters are structured as follows.

Aims: Summarize the topics and goals of the chapter.

Prelab: The theoretical background for the topic, if necessary; prerequisite problem sets for students to become familiar with the required MATLAB functions and features for the chapter.

Main lab: Problems for the main topic.

Further studies: Problems for advanced topics, if there are any.

A user's guide is provided at the beginning of the book, where the problem numbers corresponding to the prelab, main lab, and further studies of all chapters are tabulated.

To minimize the time students would otherwise have to spend on nonessential (in terms of learning core concepts and algorithms) but necessary and time-consuming tasks, MATLAB code script (incomplete m-files), Simulink models (incomplete .mdl/.slx files), and data files (.mat files) are provided so that students can easily access the core materials.

HOW TO EFFICIENTLY USE THIS BOOK

Teaching with this book:

1. As a supplementary textbook (mainly for assigning labs and projects) for undergraduate- and junior-level graduate communications and wireless communications classes as well as undergraduate signal and systems classes. A content-mapping table of the sections of this book with the sections of four widely adopted existing textbooks that cover essentially the same materials is provided.

2. As the main textbook for the aforementioned courses. While this book is not written to compete with existing communications theory and system textbooks, it is all-inclusive in that it covers, all major topics of communications.

With option 1, instructors can conveniently make lab assignments using the content-mapping table to choose appropriate projects from this book to reinforce student's learning experience. Because the projects in this book are designed to guide students step by step toward more complex projects, instructors need only spend minimal time and effort to cover all the material in class.

With option 2, instructors can use their own lecture notes to summarize the theory parts of the chapters/sections of this book that they plan to teach in class. For graduate classes, such class presentations may not be needed, since graduate students should be able to search for additional information, if needed. Students should nevertheless follow through the projects and write reports.

These uses of the book will reduce the amount of work that the instructors need to put into the class presentations, but the students still gain a thorough understanding of each concept through active learning. Instructors can customize the different chapters for different courses. For example, when this book is used for an undergraduate signals and systems class, Chapters 1–7 would be ideal, plus some materials on z-transform (for most curricula, students should have learned Laplace transform before taking signals and systems). In the first two to three weeks, students could complete Chapter 1 by themselves while the instructor focuses on basic signals and system properties. When the instructor is ready to start teaching signals and systems in both time and frequency domains, filter design, and sampling and reconstruction, students will then have all the MATLAB/Simulink skills needed to work on the corresponding projects. For an analog communications class, Chapters 1 and 8–13 should be covered. For a junior-level digital communications course, Chapters 1 and 14–24 may be covered. For a junior-level graduate wireless communications course (provided that students have taken digital communications), some or all of Chapters 1 and 25–30 can be covered.

SUPPLEMENTS

The following supplements are available from the companion website:

All MATLAB code or Simulink model samples and templates (incomplete m-files and incomplete .mdl/.slx models) and data files (.mat files).

Correction table for each edition if found.

Content-mapping table of the sections of this book with the sections of widely adopted existing textbooks if updated.

ACKNOWLEDGMENTS

This book has gone through many revisions over the past 12 years to make it a useful tool for instructors and effective guide for students learning communications systems. The writing of the book would have been impossible without the tremendous help from many of our colleagues and students. In particular, we thank Dr. Bong-seok Kim for checking every technical detail and Ms. Sahar Amini for proofreading the manuscript.

Our editor, Mary Hatcher, has very competently steered us through this project. We especially appreciate her steadfast support of our book and patience in guiding us through the publication process.

Huaping Liu is also extremely grateful to his wife Catherine and sons Frank, Ethan, Raymond, and Andrew for their endurance and not making demands on his time during the writing of this book. He also offers special thanks to two of his sons, Ethan and Raymond, for giving him many useful writing tips and for helping him revise the writing of chapters.

NOTATION AND LIST OF SYMBOLS

[WWW]: Sections or problems that require a data file or problems for which a script-file (m-file) is provided from the companion website (http://www.wiley.com/go/choi_problembasedlearning).

[T]: Theory-based sections or problems that do not require MATLAB or Simulink.

[A]: Advanced problems or materials.

m-file: MATLAB script-files

Terms using this style and font: MATLAB/Simulink-related terms, for example, variable/parameter/function/block/file name.

LIST OF ACRONYMS

AM	amplitude modulation
AWGN	additive white Gaussian noise
BER	bit error rate
CLT	central limit theory
CNR	carrier-to-noise ratio
CP	cyclic prefix
CSI	channel state information
DSB-LC	double side-band with a large carrier
DSB-SC	double side-band-suppressed carrier
EGC	equal gain combining
ESD	energy spectral density
FDM	frequency division multiplexing
ICI	inter-carrier interference
IFFT	inverse Fast Fourier transform
ISI	inter-symbol interference
LSSB	lower single-side band
MIMO	multiple input multiple output
ML	maximum likelihood
MLD	maximum likelihood detection (or decoding)
MPSK	M-ary phase shift keying
MRC	maximum ratio combining
NBFM	narrowband FM
NUS	near ultrasonic
OFDM	orthogonal frequency division multiplexing
OQPSK	offset QPSK

OSIC   ordered successive interference cancellation
PAM   pulse amplitude modulation
PD   phase detector
PDF   probability density function
PLL   phase locked loop
PSD   power spectral density
QAM   quadrature amplitude modulation
QM   quadrature multiplexing
QPSK   quadrature phase shift keying
SD   spatial diversity
SDC   selection diversity combining
SIC   successive interference cancellation
SM   spatial multiplexing
SRRC   square-root raised cosine
SSB   single-side band
STBC   space time block code
USSB   upper single-side band
VCO   voltage controlled oscillator
WSS   wide-sense stationary
ZF   zero forcing

CONTENT-MAPPING TABLE WITH MAJOR EXISTING TEXTBOOKS

NOTE: Mapping table for newer versions of the major textbooks will be updated on the companion website.

PART I. COMMUNICATION SYSTEM

	Corresponding Sections of Widely Adopted Existing Textbooks	
Chapter	**Introduction to Communication Systems** by Ferrell G. Stremler, 3rd ed. Addison Wesley, 1990.	**Introduction to Analog and Digital Communication** by S. Haykin and M. Moher, 2nd ed. John Wiley & Sons, 2007
2	2.5~2.7	–
3	2.12, 2.13, 2.15, 3.3, 3.9	2.1~2.3, 2.5
4	3.2, 3.5, 3.6, 3.15, 3.17	2.6
5	3.5~3.9	2.3
6	2.19, 3.11~3.13	2.7
7	3.15, 3.16	5.1~5.2
8	4.1~4.7.1	2.8
9	5.1, 5.2	3.1~3.3
10	5.3	3.5, 3.9
11	5.4	3.6, 3.8
12	6.1, 6.2	4.1~4.2, 4.4
13	6.7.2, 6.7.3	4.8

PART II. DIGITAL COMMUNICATION

	Corresponding Sections of Widely Adopted Existing Textbooks	
Chapter	**Digital Communications: Fundamentals and Applications** by B. Sklar, 2nd ed. PHIPE, 2002	**Digital Communications** by J. G. Proakis, 5th ed. McGraw-Hill, 2008
14	1.1~1.5	2.3
15	1.4~1.5.5	2.3, 2.7-1
16	3.1~3.2.1	2.3, 4.2-1
17, 18	3.1.3, 3.2.5.3, 4.2.6, 4.3.1	4.2, 5.1~5.1-1
19	4.3.2	2.2, 2.3, 4.2-2
20	3.2.3, 3.4.2	9.2~9.2-3
21	4.1~4.4.2, 4.7.1	3.2-2, 4.2-2
22	4.4.3~4.8.3, 9.8.1, 9.8.2.1	3.2-2, page 124 (OQPSK)
23	9.8.3, 9.5.1	3.2-3
24	7.1~7.4	8.1~8.1-1, 8.2~8.2-1, 8.3, 8.4
25	15.5.4	13.1, 13.4
26		11.2, 13.6
27		11.2, 13.6
28		15.4
29		15.1~15.2
30		11.2, 13.6

LAB CLASS ASSIGNMENT GUIDE

1. Prelab report: Mainly theoretical background and prerequisites.
2. Lab repot: To be completed during lab sessions.
3. Further study report: To be completed after lab. Mainly advanced problems.

Chapter	Prelab	Main lab	Further study
1	1.A~B, **1.C~1.E***, 5	**2.A, 3.A**, 3.B, **4.A**, 4.B, 4.C, **6**	2.B, 3.C, 3.D
2	**1.A~1.G**, 1.H~1.J, **2.A, 2.E-1**	**2.B, 2.C, 2.E-2~2.E-4**	2.D, 2.E-5, 2.F~G
3	**1, 2.A~2.C, 4.A**	2.D~F, 3, **4.B**, 4.C	
4	**1, 2, 3.A~3.C**	3.D, **4**	–
5	**1, 2.A-1~2, 2.G-1**	**2.A-3~4, 2.B~2.F, 2.A~2.F, 2.G-2, 4.A~D**	3.A~B, 4.E, **4.F**
6	**1, 2.A~2.C**	**2.D-1~2.D-9, 3**, 4.A~C	2.D-10~2.D-12, 4.D
7	**1, 2, 3.A, 3.B, 3.E-2**	**3.C, 3.D, 3.E-1, 3.E-3~3.E-9**	4
8	**1, 2.A, 2.E-1, 2.E-5, 2.G-1, 2.H, 2.1**, 2.I, 2.J-4, 3.A-1	**2.B~2.D-2, 2.E-2~2.E-4, 2.E-6, F, 2.G-2~2.G-8, 2.H-2~2.H-4,** 2.J-1~2.J-3, 2.J-5~2.J-9, 3.A-2, 3.A-3, 3.B, 3.C-1~3.C-3	2.D-3~4, 3.C-4
9	**1, 2.A-1, 2.C-1**	**2.A-2~2.B, 2.C-2~2.C-7**	3
10	**1, 3.B-1**	**2, 3.A, 3.B-2~3.B-6**	3.C

(*Continued*)

Chapter	Prelab	Main lab	Further study
11	**1, 4.B-1**	**2.A, 2.B-1~2.B-4, 3**	2.B-5~2.B-10, 4.A~4.D
12	**1.A~2.D, 3.A-1~3,** 3.A-4~5	**2.E,** 3.B~3.E	
13	**1**	**2.A, 2.C, 2.D-4**	2.D-5, 2.E
14	**1, 2, 3, 4.A**	**4.B~6**	7
15	**1, 2, 3.A,** 3.B~C, **3.D~E, 4.A~B,** 4.C, **4.D~E, 5, 7.B-8**	6.A~6.B, 6.E-1~6.E-3, **7.A-1~7.A-5, 7.B-1~7.B-7,7.C-1~7.C-3, 7.C-5**	6.C~6.D, 6.E-4, 6.E-5, 7.A-6, 7.A-7, 7.B-9~7.B-11, 7.C-4, 7.C-6, 7.C-7, 7.D
16	**1, 2.D-1~3**	**2, 3.A~3.B**	3.C~G
17	**1, 2**	3, 4	
18	**1.A~B**	**1.C, 2, 3**	
19	**1**	2	
20	**1 2.A~C**	**2.D~3**	4, 5 (Oscilloscope needed)
21	**1, 2, 3.A**	**3.B~4**	5
22	**1,2,3**	**4, 5.A~5.C-2, 5.C-4**	5.C-3, 5.C-5, 5.C-6, 6, 7
23	**1, 2, 3.B, 4.E, 4.F-2~3**	**3.C~3.F, 4.A~4.D**	4.F-1, 4.F-4, 5 (Oscilloscope needed)
24	**1, 2, 3.B**	**3.C, 4.A~4.B-2, 4.C-1**	3.D,3.E, 4.B-3~4.B-5, 4.C-2~4.C-4
25	**1.A, 1.B, 1.C-1~6,** 1.C-7, **1.D, 2.A~I,** 2.J, **3, 4.C-1**	**4 (except 4.C-6)**	4.C-6, 5
26	**1, 2, 3, 4.A**	**4.B~5**	
27	**1**	**2.A-1, 2.A-3~2.A-5, 2.B-2~2.B-4, 3.A~3.D (except 3.B-2)**	2.A-2, 2.A-6, 2.B-1, 3.B-2, 3.E
28	**1, 2.A**	**2.B, 3.A, 3.B-1~3.B-5**	3.B-6, 3.B-7, 4
29	**1, 2**	**3, 5, 6**	4, 7
30	**1**	**2, 3**	4

ABOUT THE COMPANION WEBSITE

This book is accompanied by a companion website:

www.wiley.com/go/choi_problembasedlearning

The website includes:

- All MATLAB code or Simulink model samples and templates (incomplete m-files and incomplete .mdl/.slx models) and data files (.mat files).
- Correction table for each edition if found.
- Content-mapping table of the sections of this book with the sections of widely adopted existing textbooks if updated.

1

MATLAB AND SIMULINK BASICS

- Arithmetic operators.
- Vector and matrix manipulation.
- Symbolic math.
- Script file (m-file) and user-defined functions.

1.1 OPERATING ON VARIABLES AND PLOTTING GRAPHS IN MATLAB

The fundamental MATLAB commands can be categorized into six groups, each of which is covered in one subsection. The first four subsections deal with the operations of different variable types and the last two subsections deal with the plotting commands that are frequently used in this book. On a PC that is installed with MATLAB, start MATLAB. A command window will appear where one can type in and execute MATLAB commands. Execute the set of commands/codes in the boxes and check the results. This self-study method is one of the fastest ways to master the basic MATLAB commands.

In a report, document what each command does. Focus on the specific actions and purposes, rather than the execution results. For commands that return an error message, document the reasons. Follow this guideline for all exercises in Section 1.1.

A sample report is available from the companion website. For information to access this website, refer to the guide at the beginning of this book.

Problem-Based Learning in Communication Systems Using MATLAB and Simulink, First Edition.
Kwonhue Choi and Huaping Liu.
© 2016 The Institute of Electrical and Electronics Engineers, Inc. Published 2016 by John Wiley & Sons, Inc.
Companion website: www.wiley.com/go/choi_problembasedlearning

1.A   Operation of scalar variables.

1.	X	6.	X*Y-X*3-Y	11.	X=12e6
2.	X=12	7.	X=Y^2	12.	clc
3.	X=X+2	8.	Z=sqrt(Y)	13.	x=rand
4.	Y=X+3	9.	X=2; Y=4; Z=X+Y	14.	x=rand
5.	Y*6	10.	Z=X^Y	15.	help rand

In addition, explain why the same command executed twice in item 13 and item 14 generates different results.

1.B   Operation of complex numbers.

1.	i	6.	Z=X*Y	11.	angle(Z)
2.	j	7.	real(Z)	12.	who
3.	X=1+3*j	8.	imag(Z)	13.	whos
4.	Y=-2+j ;	9.	conj(Z)	14.	clear
5.	Z=X+Y	10.	abs(Z)	15.	who

1.C   Operation of vectors.

1.	X=2 : 2: 10	12.	Y=[2; 1; 4; -3]	23.	Y=rand(1,5)
2.	Y=1 : 5	13.	Z=Y'	24.	Y=rand(4)
3.	Z=X + Y	14.	Z(1)	25.	Y=[7 3 -1 2]
4.	Z=X.*Y	15.	Z(2)	26.	mean(Y)
5.	Z=X*Y	16.	Z(1:3)	27.	var(Y)
6.	Z=X./Y	17.	Z(2:4)	28.	min(Y)
7.	Z=X/Y	18.	length(Z)	29.	max(Y)
8.	2*Y	19.	X=[2 4 8 16]	30.	[a b]=min(Y)
9.	Z=0 : 10	20.	Y=log2(X)	31.	sort(Y)
10.	sum (Z)	21.	Y^2	32.	Y=[Y 5]
11.	Y=[2 1 4 –3]	22.	Y.^2	33.	Z=[Y(3:4) X(1:2)]

1.D   Operation of matrices.

1.	X=[3 6 -2 -1; 0 5 2 1; 7 -1 4 8];	11.	Y(1, :)=X(2, :)	21.	Z=X.^2 +3*Y
2.	X(2,1)	12.	Y(2, :)=X(1, :)	22.	max(Z)
3.	X(2,3)	13.	Y(3, :)=[1 2 3 4]	23.	[T1 T2]=max(Z)
4.	X(1, :)	14.	Z=X – Y	24.	mean(Z)
5.	X(:, 2)	15.	Z=X*Y	25.	max(mean(Z))
6.	X(1:2,:)	16.	Z=X*Y'	26.	max(max(Z))
7.	X(: , 2:3)	17.	Z=X.* Y	27.	Z=rand(4)
8.	Y=[1 0 2; 3 2 1; 2 3 4]	18.	Z=X^2	28.	X=inv(Z)
9.	Y'	19.	Z=X.^2	29.	Y=X*Z
10.	Y=zeros(3,4);	20.	Z=2.^X	30.	size(Z)

1.E   Plotting some basic functions.

1.	x=0:0.1:10	9.	plot(x,y2)	17.	plot(x,y1)
2.	y1=sin(x)	10.	y3=exp(-x)	18.	subplot(3,1,2)
3.	y2=cos(x);	11.	plot(x,y3,'r')	19.	plot(x,y2)
4.	plot(x)	12.	legend('sin(x)','cos(x)','exp(-x)')	20.	subplot(3,1,3)
5.	plot(y1)	13.	axis([-5 15 -3 3])	21.	plot(x,y3)
6.	plot(x,y1) %Compare to 5	14.	axis([0 10 -2 2])	22.	semilogy (x,y3)
7.	grid	15.	figure	23.	help plot
8.	hold on	16.	subplot(3,1,1)	24.	help semilogy

1.F   Boolean operations and plotting graphs over a limited range of the x axis.

1.	A=[0 1 2 3 4];	7.	C=([1 0 1 1 1]==[1 0 1 0 0])	13.	y=(1<x)&(x<4);
2.	A<3	8.	C=([1 0 1 1 1]~=[1 0 1 0 0])	14.	plot(x,y); axis([0 10 -2 2]); grid;
3.	B=(A>2)	9.	x=0:0.01:10;	15.	y=1<x<4;
4.	C=([1 1 0 0] & [1 1 1 0])	10.	y=(x<3);	16.	plot(x,y); axis([0 10 –2 2]); grid;
5.	C=([1 1 0 0] I [1 1 1 0])	11.	figure		
6.	C=~[1 0 1 0 0]	12.	plot(x,y); axis([0 10 -2 2]); grid;		

In addition, explain the difference between items 13 and 15.

1.2   USING SYMBOLIC MATH

In the symbolic math in MATLAB, the characters (or words) such as **a**, **b**, and **temp** are treated as symbolic variables, not numeric variables. Mathematical expressions can be computed or manipulated in symbolic forms. Find out what else can be done using symbolic math in the following problems.

2.A   Write down what each of the lines in the following box does and capture the execution result.

```
>>syms a b c x t
>>y=sin(t);
>>diff(y)
>>int(y)
>>int(y, t, 0, pi)
>>z=int(x^2*exp(-x),x,1,3)
>>double(z)
>>limit(sin(t)/t,t,0)
>>symsum(x^2, x,1,4)
>>T=solve(a*x^2+b*x+c,x)
```

```
>>T2=solve(a*x^2+b*x+c,b)
>>a=1;b=2;c=3;
>>z=eval(T)
>>a=t;
>>z=eval(T)
```

2.B Verify the following quantities by using the symbolic math. Capture the calculation results.

$$\int_{-\infty}^{\infty} e^{-z^2} dz = \sqrt{\pi}, \tag{1.1}$$

$$\sum_{r=1}^{\infty} \left(\frac{1}{3}\right)^r = \left(\frac{1}{2}\right), \tag{1.2}$$

$$\lim_{x \to \infty} \left(1 + \frac{1}{x}\right)^x = e. \tag{1.3}$$

2.C Calculate the following integral by using the symbolic math. Be sure to perform **double(c)** after symbolic integration. Also explain why executing **double ( )** is needed to obtain the solution.

$$c = \int_1^2 \sin(z) e^{-z} dz. \tag{1.4}$$

1.3 CREATING AND USING A SCRIPT FILE (m-FILE)

The commands and functions we have covered so far are all executable directly in the command window. Using a "script file," which is also called an "m-file" in the earlier versions of MATLAB, users can execute various algorithms or can implement user-defined functions. In this book, the traditional term "m-file" will be used to refer to a MATLAB "script file."

3.A Follow the steps below and learn how to create and execute an m-file.

Step 1. Open a new script file editing window.

Step 2. [WWW]Shown in the box below is an m-file that plots $y = x \sin(ax)$, for the cases of $a = 0.1, 0.25, 0.3, 0.4, 0.5, 0.6, 0.7,$ and 0.8 over the range $0 < x < (10 + D)$, where D = the last digit of your student ID number. Write this m-file and save it as **CH1_3A.m**. The m-file name must begin with a letter; files with a name that begins with a number will not execute in MATLAB. Be sure not to use a space or mathematical operator (e.g., $+, -, /, *$) in the file name.

NOTE: For the parts with the superscript [WWW] prefixed, the companion website provides the supporting file or the required data. For information to access this website, refer to the guide section: "ABOUT THE COMPANION WEBSITE" at the beginning of this book.

```
clear
x=0:0.1:(10+The last digit of your student ID number);
for n=1:8
  a=n/10;
  if (a==0.2)
    a=0.25;
  end
  y(n,:)=x.*sin(a*x);
end
plot(x,y)
xlabel('x')
ylabel('y=x sin(ax)')
legend('a=0,1','a=0.25','a=0.3','a=0.4','a=0.5','a=0.6','a=0.7','a=0.8')
grid on
```

3.A-1   Add a comment to explain each line in the m-file. Capture the commented m-file.

3.A-2   Execute the m-file you have created. You can either click **'run'** button in the menu bar of the m-file editor or press the F5 key on the keyboard or type the m-file name in the command window as

```
>>CH1_3A
```

Capture the result. To capture a figure, you may navigate to **'Edit/Copy Figure'** in the menu of the figure and then paste it in your report.

3.A-3   Execute the following commands in the MATLAB command window. Based on the results, document the meanings of the two variables. Do not capture the execution results.

```
>>x
>>y
```

3.B   Let us write an m-file to plot sine waveforms of 10 different frequencies by properly modifying the m-file created in 3.A.

3.B-1   Consider 10 sine waveforms whose frequencies are 1, 2, 3, 4, ..., 10 Hz. In your code, calculate the smallest period (the highest frequency) among these waveforms and denote it by T. Then, overlay the 10 sine waveforms in the range of $-2T < t < 2T$, the range of the time axis (x axis) in the graph. Use **legend()** to label the 10 waveforms. Use a **'for'** loop as done in the m-file in 3.A.
  Capture the m-file and the execution result.

3.B-2 Use the command **mesh()** to plot the 10 sinusoids in a three-dimensional plot. Then click the axis rotation button in the menu of the figure to rotate the axis of the graph. Execute the m-file and capture the plot.

3.C The objective of this problem is to write an m-file to find the position of the maximum value in each column (or row) of a matrix and to calculate the mean of each column/row of the matrix.

3.C-1 [WWW]The m-file below, by using **rand()**, generates a 10 (row) × 9 (column) matrix **A**. The elements of the matrix are independent and identically and uniformly distributed between 0 and 3. By using **max( )** twice, the m-file finds the maximum elements of **A** and its row and column indexes.

```
%Do not append ';' at the end of the lines in this m-fie in order to see the result
of each line.
clc; clear
A=3*rand(9,10)

[B C]=max(A)
[D E]=max(B)

Max=D
Position=[C(E) E]
```

    (a) Add a comment for each line to explain what it does. Capture the m-file.
    (b) Execute the m-file. Capture all the execution results displayed in the command window. You may need to scroll up or down the command window to avoid missing any part of the execution results.
    (c) Is the execution result of each line what you expected to see?

3.C-2 In the m-file of 3.C-1, add the part that computes the mean (use the function **mean( )**) of each row of the matrix. Also add the function to find the row with the largest mean value. Capture the m-file and the execution result.

3.D [WWW]The m-file below plots the discontinuous function $y(t)$ given in equation (1.5) using logical operators.

$$y(t) = \begin{cases} \sin\left(2\pi \times 5t + \dfrac{\pi}{3}\right) & 1 \le t \le 2, \\ 0 & 0 \le t < 1 \text{ or } 2 < t \le 5. \end{cases} \qquad (1.5)$$

```
clear;
t=0:0.01:5;
x=(1<=t)&(t<=2);
x2=sin(2*pi*5*t+pi/3);
y=x.*x2;
plot(t, y); axis([-1 6 -2 2])
```

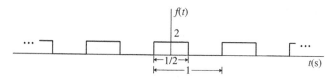

FIGURE 1.1   Periodic function $f(t)$.

3.D-1   (a) Add your explanation to each line as a comment and capture the m-file. (b) Execute this m-file and capture the result.

3.D-2   Write an m-file to plot $f(t)$ in Fig. 1.1 using **sin( )** (or **cos( )**) and Boolean operators (==, >, < , <=, >=). Capture your m-file and the execution result.

1.4   [A]USER-DEFINED MATLAB FUNCTION

Similar to many other programming languages, MATLAB also supports the use of user-defined functions to avoid repeatedly editing the main body of a code. User-defined functions are similar to the built-in MATLAB commands or functions; they follow certain syntax and are normally saved in the same folder where the main m-file is located in, but it can be saved in a different folder. Through the following problem you will learn how to write and to use user-defined MATLAB functions.

4.A   Let us write a MATLAB function that converts a number in linear scale into dB scale. In MATLAB, open the script file editor and write the following m-file. Save the m-file as **lin2dB.m** (if you click "save," the default file name will be **lin2dB.m**).

```
function xdB=lin2dB(x)
xdB=10*log10(x);
```

4.A-1   Execute **lin2dB(100)** in the command window. Capture the result.

4.A-2   Execute **lin2dB([1 2 10 20 1/10])** in the command window. Capture the result and check whether or not the results are correct.

4.B   Let us write a MATLAB function that plots the Gaussian probability density function.

4.B-1   Write the following m-file and save it. Add your comments explaining what each line does or means.

```
function plot_gaussian(m, v)
x=m+sqrt(v)*(-5:0.01:5);
fx=1/sqrt(2*pi*v)*exp(-(x-m).^2/(2*v));
plot(x,fx)
```

4.B-2   Execute **plot_gaussian(0, 1)** in the command window and capture the result.

4.B-3   Try a few arbitrary values for the arguments (i.e., the mean **m** and variance **v**) of **plot_gaussian()**. Capture your results.

4.C   Write a user-defined function **swap(A,row0col1,c,d)** that swaps two rows (or columns) of a matrix. If **row0col1** is 0, then **swap(A, row0col1,c,d)** swaps the **c**-th row and the **d**-th row of a matrix **A** and returns the swapped matrix. If **row0col1** is 1, then **swap(A,row0col1,c,d)** swaps the **c**-th column and the **d**-th column of a matrix A and returns the swapped matrix.

4.C-1   [WWW]An incomplete version of **swap.m** is provided below. Complete all parts marked by '**?**' and add a comment for each line you are completing.

```
function e=swap(A,row0col1,c,d)
e=A;
if row0col1==0
   e(d,:)=A(c,:);
   e(?,:)=A(?,:);
end

if row0col1==1
   ??;
   ??'
End
```

4.C-2   Execute the following command lines and capture the results. Check whether or not your swap function works correctly.

```
>>x=rand(4,5)
>>y=swap(x,0,2,4)
>>z=swap(y,1,5,1)
```

1.5   DESIGNING A SIMPLE SIMULINK FILE

Complete all of the following steps but document only the results of Step 5.F-2 and Step 5.G-6 in a report.

5.A   Creating a new Simulink design file.

Step 5.A-1   Start MATLAB.

Step 5.A-2   In the command window, execute **simulink** to start Simulink as shown below. You can also start Simulink from the menu bar, which might be different for different MATLAB versions.

```
>> simulink
```

Step 5.A-3   Press 'Cntrl+N' keys when the **Simulink Library browser** window is active. A Simulink design window, which we call "design window" in short hereafter, will open. Alternatively, you can use the shortcut icon in the menu bar, which may be different for different Simulink versions.

5.B   Adding blocks to the Simulink design window.

In the design window, you can import and add various functional blocks from the Simulink library.

Step 5.B-1   The left side of **Simulink Library browser** window provides a list of the function blocks.

Let us add a block that generates a sine waveform in the new Simulink model. The sine waveform generator is one of the Simulink sources. A click on **Simulink/Sources** will show all the blocks in the source directory. In order to get familiar with Simulink, you might navigate to different categories such as **Math Operations** and **Logic and Bit Operations** to check out the blocks in these directories.

Step 5.B-2   Click the block **Sine Wave** in **Simulink/Sources** and drag it into the empty design window created in Step 5.A-3. This can also be done by right-clicking the block and then choose 'Add...'.

Step 5.B-3   Browse through the **Simulink/Sinks** category and add the **Scope** block in your design window as shown in Fig. 1.2.

If you are not sure in which directory (category) your desired block is, you can search for it by entering the block name in the search input field in the menu of **Simulink Library Browser** window.

5.C   Connecting the blocks.

In order to get a desired system function, we must properly connect the output of each block to the input of another block. Let us connect the output of the **Sine Wave** block to the input of the **Scope** block. This can be done by simply clicking and dragging the output port of the **Sine Wave** block to the input port of the **Scope**. One can click on the source block and then 'Cntrl+click' on the destination block.

5.D   Setting block parameters and simulation time.

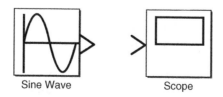

Sine Wave   Scope

FIGURE 1.2   Adding blocks to a new design.

The Simulink blocks typically have their default parameters. Double-click the block to open the parameter setting window where a description of that block is also provided.

NOTE: The same block may have different names, parameter names, and procedures to set its parameters in different Simulink versions. If the instructions do not work for your Simulink version, you may use the completed Simulink design files uploaded on the companion website.

Step 5.D-1   Open the parameter setting window of the **Sine Wave** block. Check all the parameters and try to understand what each of these parameters means.

Step 5.D-2   In this tutorial, we consider an example to generate $\sin(4\pi t)$. Read the description of the **Sine Wave** block and properly set **Amplitude, Bias, Frequency (rad/s), Phase (rad)** to generate $\sin(4\pi t)$. Note that in MATLAB pi is a reserved variable equal to π.

Set the parameter Sample time of **Sine Wave** block to 1/100. The note below provides some details about the parameter **Sample time** that is required for most of the blocks to be used later.

NOTE: All the signals generated in Simulink blocks have their own **Sample time** parameter. The Sample time parameter sets the sample time interval of the signal generated by the block. Typically, **Sample time** should be set much smaller than the inverse of the Nyquist rate. Such setting will make the sampled signal look like a continuous signal when plotted. On the contrary, too small a value for **Sample time** will increase simulation time. Note that for blocks with input port(s), **Sample time** of -1 simply copies (inherits) the **Sample time** of the input signal(s).

Step 5.D-3   Open the **Scope** display window by double-clicking the **Scope** block. Then, in the menu bar of the Scope display window, click the icon named **Configuration Properties** (or **Parameters** in some old versions) to open the **Scope** parameter setting window.

(a) The parameter **Number of ports (Number of axes** or simply **Axes** in some old versions) determines the number of input ports of the **Scope** block. Set it as 1, since only one **Sine Wave** block's output will be monitored.

(b) Click the **Logging (History** or **Data History** in some old versions) tab and unselect the check box **Limit data points to last.**

Be aware that the graphical user interface such as the menu bar and the parameter input fields might be different for different versions of Simulink.

Step 5.D-4   There is one input field in the menu bar of the design window. That input field is for a parameter **Simulation stop time**. The number typed in that field determines the execution time of the simulation, that is, the time up to which point the signal is generated and processed, not the actual time required for running the simulation. In this tutorial, we want to see 20 cycles of the output waveform of the **Sine Wave** block set in Step 5.D-2, that is, $\sin(4\pi t)$. Thus, we set the **Simulation stop time** to $20 \times (2\pi/4\pi) = 10$ seconds. Type in 10 in that input field.

5.E   Saving the files.

By using '**File/Save as**' in the menu bar, save your design (currently **untitled***). In the Simulink versions before R2012a, the file extension is ***.mdl**. For R2012a and newer versions, the file extension is ***.slx** by default, but the extension ***.mdl** is still supported. You can save your design in any folder of your choice. Save your design file as a new file.

5.F   Running the simulation and observing the output waveforms.

Step 5.F-1   On the left side of the **Simulation stop time** input field, there is a play button. Click it to run the simulation.

Step 5.F-2   If the simulation is complete, double-click the Scope block to open the **Scope** display window. Capture the **Scope** display window. Examine whether the waveform displayed in the **Scope** display window displays 20 cycles of the desired sine waveform, that is, $\sin(4\pi t)$.

Step 5.F-3   Change the parameters of the **Sine Wave** block to generate a different sine waveform and capture your result. Examine whether the waveform is generated as you set.

5.G   Adding more blocks and observing multiple waveforms.

Before proceeding to the following steps, be sure to restore the parameters of **Sine Wave** to those set in Step 5.D-2 to generate $\sin(4\pi t)$.

Step 5.G-1   If more than one block of the same function are needed for the design, you can copy and paste the one configured by right-clicking it and selecting **copy** in the pop-up menu and then right-clicking anywhere else in the design and selecting **paste**. You can also copy and paste the block by 'Cntrl+C' and 'Cntrl+V'. Add one more Sine Wave block using copy and paste. By default, the pasted block will be named **Sine Wave1**.

Step 5.G-2   Search for the block **Add** (or **Sum**) in the **Simulink library browser** and add it to the slx (or mdl) file.

Step 5.G-3   Referring Step 5.D-3, set **Number of ports** of the **Scope** block to 3. Then, set **Layout** to **3×1** (no need in some old versions). Now the **Scope** block should display three input ports.

NOTE: Throughout this book, be sure to properly set **Layout** dimension of the **Scope** blocks to separately display the input signals as done here.

Step 5.G-4   Change the parameter **Amplitude of Sine Wave** into 2 to generate $2\sin(4\pi t)$ and set the parameters of **Sine Wave1** to generate $\sin(5.2\pi t)$.

Step 5.G-5   Connect the blocks as shown in Fig. 1.3. To connect an output of a block to the inputs of multiple destination blocks, left click and drag for connecting to the first destination block. Then, right-click and drag for connecting to the rest of the destination blocks.

Step 5.G-6   Run the simulation. Capture the **Scope** display window.

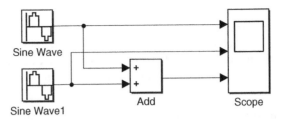

FIGURE 1.3   A test design for sine waveform generation and observation.

Step 5.G-7   You can change the viewing ranges of the x axis (time axis) and y axis in the **Scope** display window using the zoom icons in the menu bar. Locate the corresponding icons for **Zoom** (to zoom in on data in both the x and y directions), **'Zoom X-axis'**, **'Zoom Y-axis'**, and **Autoscale**. **Autoscale** displays the whole graph. Selecting any of the other three allows you to use the cursor to specify any viewing range.

1.6   CREATING A SUBSYSTEM BLOCK

In a Simulink model, right-clicking any component will pop-up a menu that allows the user to **'Create Subsystem from Selection'**, among many other functions. This feature allows us to group certain parts, for example, the frequently used parts of a design, into a single subsystem. The subsystem can be saved as a "user-defined" block to enrich the library Simulink provides. For a complex design with large number of components, creating subsystems will make the design a lot easier to read and to understand.

In this section, we design two user-defined blocks, a **Sound Source** and a **Spectrum Viewer**, that will be used frequently later in other chapters. Complete all of the following steps but document only the results of 6.C-1 and 6.C-2 in a report.

6.A   Creating the Sound Source subsystem block.

Step 6.A-1   [WWW]Download **sound.mat** from companion website and save it in your work directory. Design a new Simulink model as shown in Fig. 1.4.

Set the parameters of each block as follows. Do not change other parameters not mentioned here.

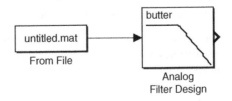

FIGURE 1.4   Design for the subsystem named **Sound Source**.

FIGURE 1.5   Creating a subsystem **Sound Source**.

1. **From File**
   - **File name = sound.mat**
2. **Analog Filter Design**
   - **Passband edge frequency[rads/s] = 2*pi*4e3**

Step 6.A-2   Select both blocks. This can be done either by pressing and holding your primary mouse button (typically the left button) while dragging the cursor to box in all components you want to select or by holding down the 'Shift' key while selecting the individual components one by one. To select all components in the design, you can simply use 'Cntrl+A'.

Then right-click one of the selected blocks to activate a pop-up menu and select '**Create Subsystem from Selection**', or simply press 'Ctrl+G'. Change the default subsystem name, **Subsystem,** into **Sound Source** as shown in Fig. 1.5. Save the current design as **Sound_Source.mdl/slx** in a directory.

Step 6.A-3   Double-click the **Sound Source** block to see the internal design. Capture the internal design window.

6.B   Creating the **Spectrum Viewer** subsystem block

Step 6.B-1   Open a new design window and design a new Simulink model as shown in Fig. 1.6. Note that the **Spectrum Analyzer** was named **Spectrum Scope** in earlier versions of Simulink. Be sure to use the **Signal Specification** block in the **Simulink/Signal Attribute** category and use the **Spectrum Analyzer** block in **DSP System Toolbox/Sinks**.

Step 6.B-2   For old Simulink versions that provide **Spectrum Scope**, instead of **Spectrum Analyzer**, set the parameters of **Spectrum Scope** as follows. Do not change any parameters not mentioned here.

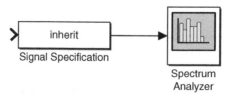

FIGURE 1.6   Design for the subsystem named **Spectrum Viewer**.

1. **Scope Properties** tab
   - **Spectrum Units = dBm** (only for the versions that have this parameter)
   - **Buffer input** : Select (check the box)
   - **Buffer size** = 1024
   - **Number of spectral averages** = 200
2. **Axis Properties** tab
   - **Frequency range = [-Fs/2 ... Fs/2]** (only for the versions that have this parameter)
   - **Minimum Y-limit** = -40
   - **Maximum Y-limit** = 25

For Simulink versions that provide **Spectrum Analyzer**, instead of **Spectrum Scope**, set the parameters of **Spectrum Analyzer** as follows.

1. Open the **Spectrum Analyzer** display window and browse 'View/Spectrum Settings' from the menu bar or click the icon named **Spectrum Settings** on the toolbar. Then, set the parameters as shown below. Do not change any other parts not mentioned here.
   - **Main options/Type = Power**
   - In **Main options**, change **RBW(Hz),** which is default selection into **Window length** and set **Window length** = 1024.
   - **Windows options/Overlap (%)** = 6.25
   - **Trace options/Units** = dBm
   - **Trace options/Average** = 200
2. Browse **View/Configuration Properties...** from the menu bar or click the icon named **Configuration Properties** on the toolbar. Set the parameters as follows.
   - **Y-limits (Minimum)** = -40
   - **Y-limits (Maximum)** = 25

The details of the parameter settings above have been tested in several Simulink versions. For some other Simulink versions or future versions, you may need to investigate a bit more, but the process will be pretty similar.

Step 6.B-3   Set the parameters of the **Signal Specification** block as follows. Do not change other parameters not mentioned here.

   - **Sample time** = 1/16e4

Step 6.B-4   As done in Step 6.A-2, select all and create the subsystem. Change the subsystem name from **Subsystem** into **Spectrum Viewer**. Save the current design as **Spectrum_Viewer.mdl/.slx.**

6.C   Testing the subsystems created.

   In this section, we observe the output spectrum of the **Sound Source** user-defined block created in 6.A using the **Spectrum Viewer** user-defined block created in 6.B.

6.C-1   Design a new mdl/.slx as shown in Fig. 1.7. To import **Sound Source** and **Spectrum Viewer** to your new design window, open **Sound_Source.mdl/.slx** and

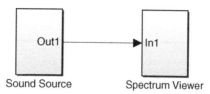

Sound Source   Spectrum Viewer

FIGURE 1.7   Design for testing the user-defined blocks **Sound Source** and **Spectrum Viewer**.

Spectrum_Viewer.mdl/.slx that you have saved as mentioned in in 6.A and 6.B and copy and paste them.

Capture the competed design window.

6.C-2   Set **Simulation stop time** to 3 seconds and run the simulation. After simulation is finished, capture the **Spectrum analyzer** display window. Follow the guidelines in the note below for capturing the window.

NOTE: Before capturing the **Spectrum Analyzer** display window, be sure to decrease the height of the window to get a width:height ratio of about 7:1 for the graph portion as shown in Fig. 4.4 in Chapter 4. Also do not **autoscale** or change the axis limits unless you are instructed to do so. Follow this guideline throughout all the problems in this book that require the **Spectrum Analyzer** display window.

2

NUMERICAL INTEGRATION AND ORTHOGONAL EXPANSION

- Calculate definite integrals using numerical integration.
- Express an arbitrary function as a linear combination of orthogonal basis functions.

2.1 SIMPLE NUMERICAL INTEGRATION

In MATLAB, numerical integration of an arbitrary function can be done by using the so-called "Riemann sum" [1, 2]. The numerical integration method introduced in this chapter will be used extensively in this book. An m-file example that calculates the definite integral $\int_{x=3}^{12} 2xe^{-2x}dx$ using the simplest version of "Riemann sum" is shown below:

```
clear
a=3;
b=12;
xstep=0.01;
x=a:xstep:b ;
y=2*x.*exp(-2*x);
S=sum(y)*xstep
```

Problem-Based Learning in Communication Systems Using MATLAB and Simulink, First Edition.
Kwonhue Choi and Huaping Liu.
© 2016 The Institute of Electrical and Electronics Engineers, Inc. Published 2016 by John Wiley & Sons, Inc.
Companion website: www.wiley.com/go/choi_problembasedlearning

1.A   Write the m-file above and add a comment for each line explaining what it does. (a) Capture the commented m-file. (b) Execute the m-file and capture the result.

1.B   The function **sum(x)** computes the sum of the elements of vector **x**. Hence, the last line of the above code, **sum(y)*xstep**, can also be written as **xstep* y(1)+ xstep*y(2)+ xstep*y(3) + ...**, where the **n**-th term, that is, **xstep* y(n)**, is equal to the area of a rectangle with a width **xstep** and height **y(n)**. Explain why **sum(y)*xstep** is an approximate to the definite integral $\int_{x=3}^{12} 2xe^{-2x}dx$.

1.C   The value of this integral approximately equals 0.0088. Change the step size of variable **x** to 0.05, 0.1, 0.5, and 1 and execute the m-file for each case. Show the results and explain why the error, that is, the difference between the numerical integration result and the actual value, becomes larger as **xstep** increases.

1.D   Write an m-file that calculates $\int_{x=-1}^{5} x^2 e^x dx$ using numerical integration. Capture the m-file and execution result.

1.E   Write an m-file that calculates $\int_{x=5}^{x=5.01} x^3 e^{-x}dx$ using numerical integration. Capture the m-file and execution result.

1.F   Write an m-file that calculates $\int_{x=0}^{250} \frac{e^{-x}}{2x+1}dx$ using numerical integration. Capture the m-file and execution result.

1.G   Calculate the integrals in 1.D~1.F by using symbolic math discussed in Section 1.2 of Chapter 1. Be sure to perform **double( )** after symbolic integration to get the real values.

1.G-1   Compare the results obtained by using symbolic math with the numerical integration result. Use the normalized error defined below to compute the difference of these results:

$$\text{Normalized error} = \left| \frac{\text{Exact value} - \text{Numerical integration result}}{\text{Exact value}} \right| \times 100\%. \quad (2.1)$$

1.G-2   If the normalized error is large, say, greater than 1%, explain why.

1.G-3   Make proper changes to your numerical integration code until the normalized error is small. Capture the results.

1.H   [A]Consider the two functions $f(x) = 2x$, $g(x) = e^{-x}$. Write an m-file that calculates $\int_0^t f(\tau)g(t - \tau)d\tau$ for $t = 1$ by using numerical integration. Capture the m-file and execution result.

1.I   [A]Write an m-file that calculates $\int_0^{10} 2t \times e^{j\omega t}dt$ for $\omega = -1, 0, 2, 1e3$ by using numerical integration. Capture the m-file and execution result.

1.J   [A]Calculate the integrals given in 1.H and 1.I using symbolic math. (a) Capture the m-file and execution result. (b) Compare the results with the numerical integration results.

2.2   ORTHOGONAL EXPANSION

2.A   Set the variable $T = 8.XXX$, where XXX = the last three digits of your student ID number. For example, if your student ID is 20840258, then $T = 8.258$. Let $s_1(t)$ be a sine waveform with period $2T$ and $s_n(t)$ be a sine waveform with a frequency that equals n times of the frequency of $s_1(t)$.

The goal is to check whether or not the elements of a signal set $\{s_1(t), s_2(t), s_3(t),...\}$ are mutually orthogonal over the range $0 \le t \le T$ using numerical integration and symbolic math methods.

2.A-1   [T]Mathematically verify that $\{s_1(t), s_2(t), s_3(t),...\}$ is an orthogonal set 1. The signal $s_1(t)$ defined above can be expressed as $s_1(t) = \sin(2\pi f_1 t)$ with a period $2T$. Thus, $f_1 = 1/(2T)$. Similarly, $s_n(t)$ defined above can be expressed as $s_n(t) = \sin(2\pi n f_1 t)$. Show that the inner product $\int_0^T s_l(t)s_k^*(t)dt$ equals 0 for any positive integers l and k by substituting $s_l(t)$ and $s_k(t)$ into this expression.

2.A-2   [WWW]Complete the following m-file, which generates the sample vectors for $s_1(t)$ and $s_2(t)$ over the range $0 \le t \le T$ and calculates the energy of $s_1(t)$ and $s_2(t)$ using numerical integration. Document the complete m-file and the execution results.

```
clear
T=8.XXX ; % The last three digits of your student ID number
t_step=1e-3;
t=0:t_step:T;
f1=1/(2*T); % Frequency of s₁(t)
s1t=sin(2*pi*f1*t); % Sampled vector of s₁(t)
s2t=sin(??); % Sampled vector of s₂(t)
E1=sum(abs(s1t).^2)*t_step % Create the energy E₁=∫₀ᵀ |s₁(t)|²dt with numerical
integration
E2=??
%Do not attach ';' to the last two lines calculating E1 and E2.
```

2.A-3   [WWW]From the signal set $\{s_1(t), s_2(t), s_3(t),...\}$, select $s_l(t)$ and $s_k(t)$, where l equals the last digit of your student ID plus 1 and k equals the second to the last digit of your student ID plus 1. The following is an incomplete m-file, which generates the sample vectors of $s_l(t)$ and $s_k(t)$ and verifies that the inner product of the two vectors equals 0 with numerical integration.

(a) Complete this m-file, that is, determine all the quantities marked by '**?**'. Capture the complete m-file and the execution result.
(b) Determine whether or not the two signals are orthogonal. Note that with numerical methods, there is always a residual error.

```
clear
T=8.XXX ; % XXX=The last three digits of your student ID number.
t_step=1e-3;
t=0:t_step:T;
f1=1/(2*T); % Frequency of s₁(t)

l=A+1; % A = The last digit of your student ID number
k=B+1; % B= The second last digit of your student ID number.

slt=sin(2*pi*f1*l*t); % Sampled vector of s_l(t).
skt=??; % Sampled vector of s_k(t).

InnerProduct=sum(slt.*conj(?))*?        %Do not attach ';' to see the result.
```

2.A-4 [WWW]Now, we verify that the inner product is 0 by using symbolic math, where **t** is a symbolic variable, rather than a sampled time vector. Complete the following m-file and show the execution result.

```
clear
T=8.XXX ; % The last three digits of your student ID number
syms t
f1=1/(2*T);

l=A+1; % A = The last digit of your student ID number.
k=B+1; % B= The second last digit of your student ID number.

slt=sin(2*pi*l*f1*t);
skt=??;

InnerProduct=int(?*conj(?),t,?,?)
double(InnerProduct)
```

2.B [WWW]Consider the function $f(t) = t^3 e^{-t}\cos(t)$ over the range of $0 \leq t \leq T$, where T takes the same value as given in 2.A. We will approximate this function as $f(t) \cong \sum_{n=1}^{N} f_n s_n(t)$ using the orthogonal set $\{s_1(t), s_2(t), s_3(t), \ldots\}$ given in 2.A. Note that f_n, which minimizes the approximation error between $f(t)$ and $\sum_{n=1}^{N} f_n s_n(t)$, is derived as a function of $f(t)$ and $s_n(t)$ as

$$f_n = \frac{\int_0^T f(t) s_n^*(t)\,dt}{\int_0^T s_n(t) s_n^*(t)\,dt}, \tag{2.2}$$

where $s_n^*(t)$ is the complex conjugate of $s_n(t)$.

The incomplete m-file below calculates $f_1, f_2, ..., f_n$ using numerical integration and plots the curves for both $f(t)$ and $\sum_{n=1}^{N} f_n s_n(t)$, with $N = 3$ as an example.

```
clear
N=3;
T=8.XXX ; % XXX= The last three digits of your student ID number
t_step=1e-3;
t=0:t_step:T;
ft=(t.^3).*exp(-t).*cos(t) ; % Sampled vector of f(t)=t³e⁻ᵗcos(t)
f1=1/(2*T);
ft_approx= zeros(1, length(t)); %Generate an all-zero vector of the same length as the
sampled time vector 't'
for n=1:N
    snt=sin(2*pi*n*f1*t); % Sampled vector of sₙ(t).
    f_n=(sum(?.*conj(?))*t_step)/(sum(?.*conj(?))*t_step);
    %Generate fₙ in (2.2) with numerical integration.
    ft_approx = ft_approx + f_n*snt ; % Calculate the partial sum  ∑ fₙsₙ(t).
                                        n = 1
end
figure
plot(t, ft)
hold on
plot(t, ft_approx, 'r')
legend( 'ft', 'ft_{approx}' )
```

2.B-1 Complete the m-file above (fill in the places marked by '?' with appropriate MATLAB variables) and capture the completed m-file.

2.B-2 Complete Table 2.1; that is, determine the mathematical expression that corresponds to each the MATLAB variables in the m-file above.

2.B-3 Execute the m-file and capture the resulting graph.

2.C Execute the m-file of 2.B for the cases of $N = 5$, 7, and 15. Check whether $\sum_{n=1}^{N} f_n s_n(t)$ approximates $f(t)$ better as N increases. Analyze the resulting figure to validate your observation.

TABLE 2.1 Variables and Corresponding Expression in the m-file in 2.B.

Variable	Expression
ft	
Snt	$\sin(2\pi n f_1 t)$
f_n	
ft_approx	

2.D [A]Consider the case in which the period of the sine waveform $s_1(t)$ is T, rather than $2T$. The frequency of $s_n(t)$, which equals n times of the frequency of $s_1(t)$, is different from those in 2.A-1 as well.

2.D-1 Modify the line 'f1=1/(2*T)' into 'f1=1/T' in the m-file created in 2.B. Set $N = 35$ in the modified m-file and execute it. Capture the execution result.

2.D-2 Compare the plot obtained in 2.D-1 with the plots obtained in 2.C for which the period was set to $2T$. Check whether or not the set $\{s_1(t), s_2(t), s_3(t),...\}$ in 2.D-1 is a "complete" orthogonal set based on the note below.

NOTE: A property of a "complete" orthogonal set: An orthogonal set $\{s_1(t), s_2(t), s_3(t),...\}$ is "complete" if the difference between the original function $f(t)$ and the approximated functions $\sum_{n=1}^{N} f_n s_n(t)$ converges to 0 as N approaches infinity [3–6].

2.E Consider a periodic function $f_T(t)$ with a period $T = t_2 - t_1$ generated by periodically extending a time limited function $f(t)$ defined over $t_1 \le t \le t_2$. The exponential Fourier series coefficient of $f_T(t)$ is calculated as [7, 8]

$$
F_n = \frac{\int_{t_1}^{t_2} f(t)e^{-jn\omega_0 t}dt}{\int_{t_1}^{t_2} e^{jn\omega_0 t}e^{-jn\omega_0 t}dt}
$$

$$
= \frac{1}{t_2 - t_1}\int_{t_1}^{t_2} f(t)e^{-jn\omega_0 t}dt. \tag{2.3}
$$

2.E-1 (a) [T]Complete the Fourier series coefficient expression in equation (2.3) by expressing ω_0 as the function of t_1 and t_2. (b) Prove that with the ω_0 obtained in (a), the elements of the set of complex sinusoids $\{..., e^{-j2\omega_0 t}, e^{-j\omega_0 t}, e^{j0\times\omega_0 t}, e^{j\omega_0 t}, e^{j2\omega_0 t}, ...\}$ are mutually orthogonal over $t_1 \le t \le t_2$.

2.E-2 [WWW]Calculate the exponential Fourier series coefficients F_n of the periodic function $f_T(t)$, for which the time limited function $f(t)$ is given in 2.B. Perform the orthogonal expansion over the interval $0 \le t \le T$ by using the complex sinusoids set $\{..., e^{-j2\omega_0 t}, e^{-j\omega_0 t}, e^{j0\times\omega_0 t}, e^{j\omega_0 t}, e^{j2\omega_0 t}, ...\}$. The following m-file calculates the exponential Fourier series coefficient F_n using the numerical integration method and plots $f(t)$ and $\sum_{n=-3}^{3} F_n e^{jn\omega_0 t}$ over the interval $0 \le t \le T$. Execute the m-file below and show the execution result.

```
clear
N=3;
T=8.XXX ; % XXX=The last three digits of your student ID number.

t1=0;
t2=T;
t_step=1e-3;
t=t1:t_step:t2;
```

```
ft=(t.^3).*exp(-t).*cos(t) ;

w0=2*pi/(t2-t1); % For this problem, we may directly set w0= 2*pi/T.

ft_approx=zeros(1, length(t));
for n=-N:N
  nth_exp=exp(j*n*w0*t);
  f_n=(sum(ft.*conj(nth_exp))*t_step)/(sum(nth_exp.*conj(nth_exp))*t_step);%or
  (sum(ft.*conj(nth_exp))*t_step)/T
  ft_approx = ft_approx + f_n*nth_exp;
end

figure
plot(t, ft)
hold on
plot(t, ft_approx, 'r')
legend('ft' , 'ft_{approx}')
```

2.E-3   Modify the m-file to plot $\sum_{n=-7}^{7} F_n e^{jn\omega_0 t}$ and $\sum_{n=-15}^{15} F_n e^{jn\omega_0 t}$. Execute the modified m-file and capture the results.

2.E-4   Based on the result in 2.E-3, check whether or not the orthogonal function set in 2.E-2, that is, $\{\ldots, e^{-j2\omega_0 t}, e^{-j\omega_0 t}, e^{j0 \times \omega_0 t}, e^{j\omega_0 t}, e^{j2\omega_0 t}, \ldots\}$, with a proper ω_0 is "complete."

2.E-5   [A]Explain why the orthogonal function set in 2.D is not "complete" and the orthogonal function set in 2.E-2 is "complete" although the smallest frequency (the fundamental frequency or frequency spacing) used in 2.D and 2.E-2 is the same.

2.E-6   [A]Explain why the orthogonal function set used in 2.A to 2.C is "complete" although it consists of only real-valued sine functions just like those used in 2.D, which is "incomplete."

2.F   [A]Perform the orthogonal expansion of an arbitrary time-limited function; that is, express an arbitrary function as a linear combination of orthogonal functions. Equations (2.4)–(2.6) illustrate three examples of the time-limited functions.

$$f(t) = t^2 + 2t + 4, \quad -1 \le t \le 3, \tag{2.4}$$

$$f(t) = \frac{\log(5 + t)e^{\sin(x)}}{t^2 + 1}, \quad 2 \le t \le 10, \tag{2.5}$$

$$f(t) = \frac{\sin(5t) + 3}{e^{-t} + 0.1} + 0.05t^2, \quad 0.5 \le t \le 5. \tag{2.6}$$

2.F-1   Choose one of your own continuous-time function $f(t)$ defined over the period of $t_1 \leq t \leq t_2$ with your own choices of t_1 and t_2. You may choose one of the examples given in equations (2.4)–(2.6). Write your chosen $f(t)$.

2.F-2   Write an m-file to plot your chosen $f(t)$ over the time interval $t_1 \leq t \leq t_2$. Be sure to set **t_step** sufficiently small so that your graph looks smooth. Show your graph.

2.F-3   Properly modify the m-file completed in 2.E-2 to generate $\sum_{n=-15}^{15} F_n e^{jn\omega_0 t}$ for your chosen $f(t)$. Overlay the resulting curve on the curve of $f(t)$ over the period $t_1 \leq t \leq t_2$. Check whether $\sum_{n=-15}^{15} F_n e^{jn\omega_0 t}$ approximates $f(t)$ well over the time period $t_1 \leq t \leq t_2$.

REFERENCES

[1] G. E. Shilov and B. L. Gurevich, *Integral, Measure, and Derivative: A Unified Approach*, Mineola, New York: Dover, 1978.

[2] T. Apostol, *Mathematical Analysis*, North Reading, MA: Addison-Wesley, 1974.

[3] D. C. Lay, *Linear Algebra and Its Applications*, 3rd ed., Boston, Columbus, North Reading, MA: Addison-Wesley, 2006.

[4] G. Strang, *Linear Algebra and Its Applications,* 4th ed., Belmont, CA: Brooks/Cole, 2006.

[5] S. Axler, *Linear Algebra Done Right,* 2nd ed., New York: Springer, 2002.

[6] W. Rudin, *Real and Complex Analysis*, New York: McGraw-Hill, 1987.

[7] W. Rudin, *Principles of Mathematical Analysis,* 3rd ed., New York: McGraw-Hill, Inc., 1976.

[8] A. Zygmund, *Trigonometric Series,* 3rd ed., Cambridge, UK: Cambridge University Press, 2002.

3

FOURIER SERIES AND FREQUENCY TRANSFER FUNCTION

- Obtain the Fourier series of periodic signals and conduct simulation in Simulink.
- Obtain the frequency transfer function of a linear system and simulate it in Simulink.

3.1 DESIGNING THE EXTENDED FOURIER SERIES SYSTEM

The periodic square wave $f_T(t)$ with period T can be represented by the Fourier series [1, 2] as

$$f_T(t) = \frac{4}{\pi} \left(\cos \omega_0 t - \frac{1}{3} \cos 3\omega_0 t + \frac{1}{5} \cos 5\omega_0 t - \cdots \right), \quad \text{where } \omega_0 = \frac{2\pi}{T}. \quad (3.1)$$

The goal of this section is to check, in Simulink, whether the right-hand side of (3.1) converges to $f_T(t)$. Since the right-hand side of equation (3.1) is the sum of infinite number of scaled cosine waveforms of different frequencies, it is impossible to generate it exactly in Simulink. Thus, we will perform a partial-sum approximation.

1.A In the following steps, set T to the last two digits of your student ID number.

1.A-1 Determine the value of T; for example, if your student ID is 20123247, then $T = 47$.

1.A-2 Calculate the fundamental frequency ω_0 for the T value obtained in 1.A-1.

Problem-Based Learning in Communication Systems Using MATLAB and Simulink, First Edition.
Kwonhue Choi and Huaping Liu.
© 2016 The Institute of Electrical and Electronics Engineers, Inc. Published 2016 by John Wiley & Sons, Inc.
Companion website: www.wiley.com/go/choi_problembasedlearning

1.B   Design a slx/mdl file as shown in Fig. 3.1.

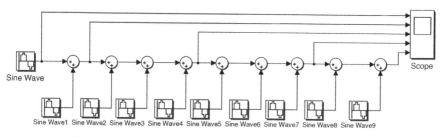

FIGURE 3.1   Design for a partial-sum approximation of $f_T(t)$.

Be sure to properly arrange the **Sine Wave** blocks so that their names occur in numerical order as shown in Fig. 3.1. Set the parameters of the **Sine Wave** blocks as follows:

- **Phase** = pi/2
- **Sample time** = 0.1

1.B-1   Determine the trigonometric formula and explain the reason for setting the **Phase** of all **Sine Wave** blocks to $\pi/2$ in order to approximate the right-hand side of equation (3.1).

1.B-2   Set **Amplitude** and **Frequency (rad/s)** for all **Sine Wave** blocks according to equation (3.1) given the value of T set for your design. For example, if $T = 47$, then set **Amplitude** = 4/pi*(−1/7) and **Frequency (rad/s)** = 7*(2*pi/47) for **Sine Wave3**.

1.B-3   Run the simulation for a duration of 20 T. For example, if $T = 47$, then the simulation stop time = 47*20. Make sure that **Layout** of the **Scope** display window is properly set to separately display the input signals and then capture the **Scope** display window.

1.B-4   Document your observations on the Fourier series approximation of the periodic square wave as the number of sinusoids increases in the sum. Assess both the shape and the period of the approximated waveform.

3.2   FREQUENCY TRANSFER FUNCTION OF LINEAR SYSTEMS

If a complex sinusoid of frequency ω rad/s—that is, $\exp(j\omega t)$—is applied as the input of a linear system, the steady-state output is also a complex sinusoid of the same frequency ω rad/s. The amplitude and phase of output complex sinusoid are determined by the system frequency response. In other words, the output $g(t)$ can be expressed as $g(t) = H(\omega) \exp(j\omega t)$, where $H(\omega)$ is a complex-valued function of the input frequency ω. Rewrite $g(t)$ as $g(t) = H(\omega) \exp(j\omega t) = |H(\omega)| e^{j\angle H(\omega)} \exp(j\omega t) = |H(\omega)| \exp(j\omega t + j\angle H(\omega))$. Clearly, the magnitude and the phase of the output complex sinusoid $g(t)$ are equal to $|H(\omega)|$ and $\angle H(\omega)$, respectively. The function $H(\omega)$,

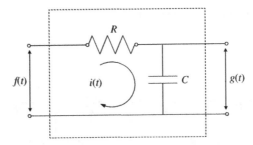

FIGURE 3.2   RC low pass filter.

which is the Fourier transform of the system impulse response, is called the frequency transfer function [3,4]. We will explain this concept using a simple RC circuit [5,6].

2.A [T]Let $f(t)$ and $g(t)$ denote, respectively, the input and the output voltages of the RC circuit shown in Fig. 3.2. Prove that the input and the output relationship of this circuit can be written as a linear differential equation expressed as

$$f(t) = RC\frac{dg(t)}{dt} + g(t). \tag{3.2}$$

2.B [T]To find $H(\omega)$, substitute the input $f(t) = \exp(j\omega t)$ and the output $g(t) = H(\omega)\exp(j\omega t)$ into the differential equation derived in 2.A and then rearrange the equation. Show that $H(\omega)$ can be written as $H(\omega) = 1/(1 + j\omega RC)$.

2.C [T]Assume $R = 0.5\ k\Omega$ and $C = (1000 +$ the last three digits of your student ID number) μF. For example, if your student ID is 20123465, the set $C = 1465e\text{-}6$ F. For each of the input frequencies ω in the first column of the following table, calculate $|H(\omega)|$ and $\angle H(\omega)$ for $H(\omega)$ given in 2.B and complete Table 3.1.

TABLE 3.1   Output Amplitude and Phase of an RC Low Pass Filter Shown in Fig. 3.2.

Input frequency ω [rad/s]	Amplitude of output $|H(\omega)|$	Phase of output, $\angle H(\omega)$, where $\angle H(\omega) = \arctan\left(\frac{Im(H(\omega))}{Re(H(\omega))}\right)$ rad
−120		
−40		
−10		
−5		
0		
5		
10		
40		
120		

2.D [WWW]Complete the m-file below that plots $|H(\omega)|$ as a function of ω in the range of $\omega = [-120 \sim 120]$ using the following steps:

Step 1. Define a vector **w** that represents the discrete version of ω, and set its initial value at -120 and final value of 120 with a step size of 0.1.

Step 2. Calculate $|H(\omega)|$ for each element of ω. Avoid using the '**for**' loop since the results for all elements of the vector can be calculated using a single line of MATLAB code via vector operation.

Step 3. Plot $|H(\omega)|$ versus ω using the **plot( )** command.

```
clear
w=-120:0.1:120;
R=0.5e3;
C= 1XXXe-6; % XXX is the last three digits of your student ID number
Hw=??; % Be sure to use ./, instead of /, if you divide element by element.
plot(?, abs(?))
```

2.E Modify the m-file completed in 2.D to plot $\angle H(\omega)$ in the range of $\omega = [-120 \sim 120]$ and capture your modified m-file.

2.F Examine the execution results of the m-files completed in 2.D and 2.E. Also explain why this RC circuit is called a "low pass filter" based on the simulation results.

3.3   VERIFICATION OF THE FREQUENCY TRANSFER FUNCTION OF LINEAR SYSTEMS IN SIMULINK

In this section, we verify the answers to the problem in Section 3.2 in Simulink. Make sure that your MATLAB/Simulink copy has Simulink toolbox **SimPowerSystems**. If you are not sure whether your MATLAB copy has this toolbox or whether you cannot locate this toolbox from the library list in the **Simulink Library Browser,** you can type in **SimPowerSystems** in the search input field of the **Simulink Library Browser** to search for it.

3.A [WWW]First, model the RC circuit in Fig. 3.2 in Simulink. Design an mdl/.slx file as shown in Fig. 3.3. Search and add **powergui, Sine Wave, Controlled Voltage Source,** two **Series RLC Branch, Voltage Measurement,** and **Ground.** The correct Simulink library path for **Ground** here is **Sim power systems\Elements.** To rotate any of the blocks, right-click the block to pop-up a menu.

Next, set the parameters of each block as follows. Do not change the parameters not mentioned below.

1. **Sine Wave**
   • **Sample time** = 1e-4

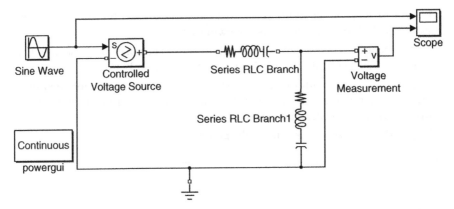

FIGURE 3.3 Simulink design for RC low pass filter shown in Fig. 3.2.

2. **Series RLC Branch**
   - **Branch type** = R
   - **Resistance (Ohm)** = 0.5e3
3. **Series RLC Branch1**. Note that the **Series RLC Branch 1** shown in Fig. 3.3 is rotated clockwise by 90°.
   - **Branch type** = C
   - **Capacitance (F)** = 1XXXe-6, where XXX = the last three digits of your student ID number.
4. **Scope**: Open Scope display window and then select the Parameters icon.
   - Unselect **Limit data points to last** in the **Logging** tab or **Data Logging** tab in some Simulink versions.

3.A-1 Capture the Simulink design window of the mdl/.slx file you designed.

3.A-2 Determine which port of which block corresponds to the input $f(t)$ and output $g(t)$ in Fig. 3.2, respectively.

3.B In the following steps, change the input frequency in the mdl/.slx file completed in 3.A and measure the corresponding amplitude and the frequency of the output.

3.B-1 Set the input frequency to 120 rad/s, that is, set the parameter **Frequency (rad/s)** of the **Sine Wave** block to 120 and then run the simulation for 10 seconds. After the simulation is completed, **Autoscale** the **Scope** display window. Capture the **Scope** display window.

3.B-2 Measure and record the output signal frequency from the window captured in 3.B-1. Compare the measured output frequency with the input frequency.

3.B-3 Measure and record the output signal amplitude after the transient response disappears. Also assess whether it matches the answer in the table completed in 2.C.

TABLE 3.2   Output Frequency and Amplitude of the RC Low Pass Filter in Section 3.A.

Input frequency ω (rad/s)	Output frequency	Amplitude of output in steady state
−120		
−40		
−10		
−5		
0 (Set **Phase** of the **Sine Wave** to pi/2)		
5		
10		
40		
120		

3.B-4   Repeat 3.B-1 to 3.B-3 for each of the frequencies in Table 3.2. Measure the frequency and amplitude of the output and complete the table.

Note that a negative value is not allowed for the parameter **Frequency (rad/s)** of the **Sine Wave** block. However, using the identity $\sin(-\omega t) = -\sin(\omega t)$, a negative frequency can be realized by setting **Amplitude** to −1 with a positive value for **Frequency (rad/s)**.

3.B-5   Compare your theoretical results of the output amplitudes in the second column of the Table 3.1 and the simulation results in 3.B-4.

3.4   STEADY-STATE RESPONSE OF A LINEAR FILTER TO A PERIODIC INPUT SIGNAL

In this section, we determine the output signal $g(t)$ when a nonsinusoidal, periodic signal $f(t)$ is applied as the input to the RC circuit shown in Fig. 3.2. First, we derive the output expression using the linearity property of the filter. Then, we simulate the system in Simulink to verify the theoretical results.

Suppose that $f(t)$ is a periodic square function of period 2 expressed as

$$f(t) = \begin{cases} 1 & 0 \le t < 1 \\ -1 & 1 \le t < 2 \end{cases} \quad \text{and} \quad f(t) = f(t+2) \quad \forall t. \tag{3.3}$$

Recall that the periodic function $f(t)$ can be represented by the Fourier series as

$$f(t) = \sum_{n=-\infty}^{\infty} F_n e^{jn\omega_0 t}, \quad \text{where} \quad \omega_0 = \frac{2\pi}{T(=2)} = \pi \quad \text{and} \quad F_n = \begin{cases} \dfrac{2}{jn\pi}, & n = \text{odd}, \\ 0, & n = \text{even}. \end{cases}$$

$$\tag{3.4}$$

Equation (3.4) shows that the periodic square wave is represented as a weighted superposition of complex sinusoids $(\dots,\; e^{j(n-1)\omega_0 t},\; e^{jn\omega_0 t},\; e^{j(n+1)\omega_0 t},\; \dots)$. Based on the linearity property of linear systems (the RC circuit in this case) and the concept of frequency transfer function, the system output $g(t)$ can also be written as a weighted superposition of complex sinusoids. The following projects explore this problem.

4.A [T]Let $H(\omega)$ be the system frequency transfer function. For the special input signal $e^{j\omega t}$, the output is expressed as $H(\omega)\,e^{j\omega t}$.

4.A-1   The frequency of $e^{jn\omega_0 t}$ is $n\omega_0$ rad/s. Determine the output for the input signal $e^{jn\omega_0 t}$.

4.A-2   Determine the output for the input signal $F_n e^{jn\omega_0 t}$. Explain how the linearity property is applied in reaching this output signal.

4.A-3   Determine the output expression for the input $F_{-2}e^{-j2\omega_0 t} + F_{-1}e^{-j\omega_0 t} + F_0 e^{j0\omega_0 t} + F_1 e^{j\omega_0 t} + F_2 e^{j2\omega_0 t}$. Again, explain how the linearity property is used in deriving this output signal.

4.A-4   Explain why the output $g(t)$ can be expressed as $g(t) = \sum\limits_{n=-\infty}^{\infty} G_n e^{jn\omega_0 t}$ with $G_n = H(n\omega_0)\,F_n$ if the input $f(t)$ is a periodic function, which can be expressed as $\sum\limits_{n=-\infty}^{\infty} F_n e^{jn\omega_0 t}$.

4.A-5   Based on the result in 4.A-4, explain why the output $g(t)$ is also periodic if the input $f(t)$ is a periodic signal.

4.B   In this subsection, we plot the output $g(t) = \sum\limits_{n=-\infty}^{\infty} G_n e^{jn\omega_0 t}$ with G_n obtained in 4.A-4.

4.B-1   [WWW]The exact signal $g(t)$ requires infinite number of terms in the Fourier series expansion. The following m-file plots the partial-sum (**gt_approx** in the m-file) approximation of $g(t)$ with n taking on the integers from -100 to 100.

Complete the places marked by '?' in the m-file below. Add a comment for each line of the m-file to explain what it does. Especially for the lines with the sign '=', explain what the variables on the left-hand side represent and the reason (why and how) the right-hand side expression is appropriately constructed accordingly.

Capture the completed m-file.

```
clear
R=0.5e3;
C=1XXXe-6; % XXX is the last three digits of your student ID.
t=0:(1/1000):20;
T=2;
w0=(2*pi)/?;
gt_approx=0;
```

```
for n=-100:100
  if mod(n,2)==0
    Fn=0; %  Equation (3.4)
  else
    Fn=?; %  Equation (3.4)
  end
  w=n*w0;
  Hw=1/(1+j*R*C*w);
  Gn=Hw*?; % 4.A-4
  gt_approx=gt_approx + Gn*exp(j*n*w0*t);
end
figure
plot(t,gt_approx);
grid
```

4.B-2   Execute the m-file above. Capture the plot.

4.C   The theoretically output waveform derived in 4.B-2 can be verified in Simulink. In the mdl/.slx file designed in 3.A, replace the **Sine Wave** block with a **Signal Generator** block (you may search for it in the **Simulink Library Browser**).
Set the parameters of the **Signal Generator** block as follows.

- **Wave form = Square**
- **Amplitude = −1**
- **Frequency = 1/2**

4.C-1   Run the simulation for 20 seconds. After the simulation is completed, select **Autoscale** on the **Scope** display window and then capture the result.

4.C-2   Assess whether the input waveform in 4.C-1 is consistent with equation (3.3).

4.C-3   Determine whether the simulated output waveform in 4.C-1 is the same as the theoretical output waveform derived and plotted in 4.B-2. The (x, y) coordinates of the local peaks are good checkpoints for this assessment. Also, in comparing the waveforms, exclude the initial portion of the simulated output waveform since it takes the system some time to reach the steady state.

REFERENCES

[1] W. Rudin, *Principles of Mathematical Analysis*, 3rd ed., New York: McGraw-Hill, Inc. 1976.
[2] A. Zygmund, *Trigonometric Series*, 3rd ed., Cambridge, UK: Cambridge University Press, 2002.

[3] C. L. Phillips, J. M. Parr, and E. A. Riskin, *Signals, Systems and Transforms*, Upper Saddle River, NJ: Prentice Hall, 2007.

[4] J. P. Hespanha, *Linear System Theory*, Princeton, NJ: Princeton University Press, 2009.

[5] A. Agarwal, J. H. Lang, *Foundations of Analog and Digital Electronic Circuits*, Cambridge, MA: Morgan Kaufmann, 2005.

[6] J. D. Irwin, *Basic Engineering Circuit Analysis*, River Street, Hoboken, NJ: Wiley, 2006.

4

FOURIER TRANSFORM

- Investigate signal spectra in Simulink.
- Perform the Fourier transform of the sampled audio signals.

4.1 THE SPECTRUM OF SINUSOIDAL SIGNALS

1.A [T]Determine the inverse Fourier transforms of the following three signals expressed in the frequency domain using the defining equation $f(t) = \frac{1}{2\pi} \int_{-\infty}^{\infty} F(\omega)e^{jt\omega}d\omega$ [1–3]. For simplicity, we will use the term "spectrum" to refer to the Fourier transform of a time function when it does not cause confusion.

1.A-1 $2\pi\delta(\omega - \omega_0)$.

1.A-2 $\pi\left[\delta(\omega - \omega_0) + \delta(\omega + \omega_0)\right]$.

1.A-3 $-j\pi\left[\delta(\omega - \omega_0) - \delta(\omega + \omega_0)\right]$.

1.B [WWW]Start a new Simulink design as shown in Fig. 4.1. For the multiple copies of the same block (the two **Sine Wave** blocks and three **Spectrum Viewer** subsystem blocks), make sure that numbers in the names of these blocks occur in numerical orders from left to right in the model, as shown in Fig. 4.1.

Problem-Based Learning in Communication Systems Using MATLAB and Simulink, First Edition.
Kwonhue Choi and Huaping Liu.
© 2016 The Institute of Electrical and Electronics Engineers, Inc. Published 2016 by John Wiley & Sons, Inc.
Companion website: www.wiley.com/go/choi_problembasedlearning

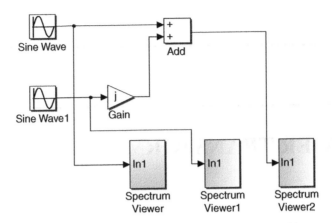

FIGURE 4.1   System to observe the spectra of sinusoids.

Set the parameters of each block as follows. Do not change the parameters not mentioned below.

1. **Sine Wave**
   - **Frequency (rad/s)** = 2*pi*(20+Last digit of your student ID number)*1000
   - **Phase** = pi/2
2. **Sine Wave1**
   - **Frequency (rad/s)** = 2*pi*(20+Last digit of your student ID number)*1000
   - **Phase** = 0
3. **Spectrum Viewer, Spectrum Viewer1,** and **Spectrum Viewer2**

Download **Spectrum_Viewer_only_for_CH4.mdl/.slx** from the companion website. Copy the subsystem **Spectrum Viewer** saved in that file and paste it in your design if your Simulink version is compatible (i.e., it opens the .mdl/.slx file without a problem).

If your Simulink version is not compatible, then copy and paste the subsystem **Spectrum Viewer** created in 6.B of Chapter 1 and revise the parameter of **Spectrum Analyzer** (previously **Spectrum Scope**) inside **Spectrum Viewer**.

- For versions 2013 and 2014b, **View/Spectrum Setting/Trace options/Unit = Watts**
- For version 2009, **Spectrum Unit** of **Scope properties** tab = **Watts**
- For version 2007, **Amplitude scaling** of **Axis properties** tab = **Squared magnitude**

In these versions, the setting typically displays the spectrum in linear scale, not dB scale.

For convenience, set the parameter for the first **Spectrum Viewer** only and then make two copies of this subsystem to maintain the same set of parameters. Do not modify the original copy of **Spectrum Viewer** in **Spectrum_Viewer.mdl/slx** created in 6.B of Chapter 1. We will open this file and use this original copy in other chapters in order to observe the spectrum in dB scale.

4. **Gain**
   - **Gain** $= j$

1.B-1   Complete the output mathematical expressions of the following blocks:

(a) **Sine Wave1**: ? (The answer will be $\sin(2\pi \times 25000t)$ if the last digit of your student ID is 5)
(b) **Sine Wave**: ? (Use the identity $\sin(x + \pi/2) = \cos(x)$)
(c) **Gain**: ?
(d) **Add**: ?

1.B-2   Run the simulation for 10 seconds. After the simulation is completed, capture the scope windows of the three **Spectrum Viewers**. Since we have not changed the scope window location setting, the three scope windows will overlap. Move the windows away from one another so that they do not overlap. Follow the guideline at the end of Chapter 1 for capturing the scope window.

1.B-3   Again, note that the scope window of **Spectrum Viewer** shows the absolute value of the spectrum (Fourier transform) of the input signal. We should be able to clearly see the line spectra in all three **Spectrum Viewer** scope windows. Determine the following items for each window:

(a) Number of spectral lines
(b) Position of the spectral lines (the x axis values and provide the unit)
(c) Whether the number of spectral lines and the positions of the spectral lines are consistent with your answers to Problem A

1.B-4   Change the **Frequency (rad/s)** parameter of the **Sine Wave1** block to 2*pi*3e3, and run the simulation again. Capture the **Spectrum Viewer1** scope window. Determine the positions of the spectral lines and check whether they are consistent with your answer to 1.A.

1.B-5   Set the **Phase** of the **Sine Wave1** block to any real number and run the simulation again. Capture the **Spectrum Viewer1** scope window. Assess whether the position of line spectrum depends on the phase. Why?

4.2   THE SPECTRUM OF ANY GENERAL PERIODIC FUNCTIONS

We now analyze the spectrum of any general periodic functions. The Fourier transform $F\left[f_T(t)\right]$ of the periodic function $f_T(t)$ with period T is given as

$$F\left[f_T(t)\right] = \sum_{n=-\infty}^{\infty} 2\pi F_n \delta\left(\omega - n\omega_0\right), \tag{4.1}$$

where $\omega_0 = 2\pi/T$ is the fundamental frequency and F_n is the Fourier series coefficient obtained as [1–3]

$$F_n = \frac{1}{T}\int_0^T f_T(t)e^{-jn\omega_0 t}dt \quad \left(\text{or } \int_{t_0}^{t_0+T} f_T(t)e^{-jn\omega_0 t}dt\right), \tag{4.2}$$

where t_0 is an arbitrary real number.

2.A   [T]Prove equation (4.1) through the following steps.

2.A-1   Complete the expression of the Fourier series expansion of $f_T(t)$, that is, $f_T(t) = \sum_{n=-\infty}^{\infty} ? \times e^{jn\omega_0 t}$ (determine the quantity marked by '**?**').

2.A-2   Take the Fourier transform of both sides of the equation completed in 2.A-1 $F\left[f_T(t)\right] = F\left[\sum_{n=-\infty}^{\infty} ? \times e^{jn\omega_0 t}\right]$. Then, by using the linearity property of the Fourier transform and the Fourier transform of complex exponential functions, that is, $F\left[e^{jn\omega_0 t}\right] = 2\pi\delta\left(\omega - ?\right)$, prove equation (4.1). You might refer to 1.A-1.

2.B   [T]Determine the Fourier transform of the periodic signal shown in Fig. 4.2 using the following two steps:

2.B-1   Derive F_n by using equation (4.2).

2.B-2   Substitute F_n into equation (4.1) and simplify the equation.

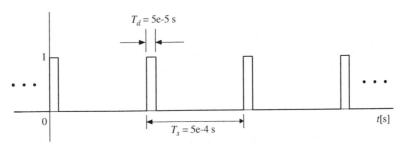

FIGURE 4.2   Periodic signal $f_T(t)$.

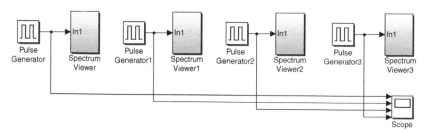

FIGURE 4.3   System to observe the waveforms and spectra of the periodic signals.

4.3   ANALYSIS AND TEST OF THE SPECTRA OF PERIODIC FUNCTIONS

3.A   Start a new Simulink design as shown in Fig. 4.3.

Set the parameters of each block as follows. Do not change any parameters not mentioned below.

1. **Pulse Generator**
   - **Period** = 5e-4
   - **Pulse Width(% of period)** = 5
2. **Pulse Generator1**
   - **Period** = 5e-4
   - **Pulse Width(% of period)** = 10
3. **Pulse Generator2**
   - **Period** = 2.5e-4
   - **Pulse Width(% of period)** = 10
4. **Pulse Generator3**
   - **Period** = 2.5e-4
   - **Pulse Width(% of period)** = 20
5. **Spectrum Viewer, Spectrum Viewer1, Spectrum Viewer2,** and **Spectrum Viewer3**: Copy and paste **Spectrum Viewer** in the design file of 1.B.

3.A-1   Set **Simulation stop time** to 2e-3 and run the simulation for 2 microseconds. Properly resize the **Scope** block's display window and capture it. Prior to capturing, right-click option **Axes Properties** on all three signal waveforms to set **Y_min** to −0.5 for better presentation.

3.A-2   From the **Scope** block's display window, measure the parameters below. You might need to properly enlarge the x axis of the **Scope** block's display window to obtain an accurate value.

(a) Output of **Pulse Generator**: frequency($=1$/period) $= 1/0.5\text{e-}3 = 2$ kHz, pulse width $= 2.5\text{e-}5$ s
(b) Output of **Pulse Generator1**: frequency $= ?$ kHz, pulse width $= ?$ s
(c) Output of **Pulse Generator2**: frequency $= ?$ kHz, pulse width $= ?$ s
(d) Output of **Pulse Generator3**: frequency $= ?$ kHz, pulse width $= ?$ s

3.A-3   Now run the simulation for 3 seconds. Reposition the four scope windows of the **Spectrum Viewer** blocks so that they do not overlap. **Autoscale** all figures and capture them. Note that the method to **Autoscale** a figure might be different depending on Simulink version.

3.B   Let us examine the spectral characteristics of the periodic function.

3.B-1   The spectra of the four signals captured in 3.A-3 commonly have the shape of a line spectrum. Is this consistent with the Fourier transform expression of periodic functions given in equation (4.1)? From equation (4.1), discuss what causes the line spectrum.

3.B-2   Measure and record the line spectrum intervals in kHz of the four signals captured in 3.A-3, and check whether they are consistent with equation (4.1).

3.C   We now analyze the relationship between the envelope of the line spectrum and the pulse width.

3.C-1   The envelope of the line spectrum has the shape of a sinc function, or, more exactly, the absolute value of the sinc function, since the scope shows the "magnitude" spectrum. Explain why this spectrum shape makes sense.

3.C-2   We can see that the envelopes of the line spectrum of **Pulse Generator** and **Pulse Generator2** have their first null points (the point where $y = 0$) at 40 kHz. However, the envelopes of the line spectrum of **Pulse Generator1** and **Pulse Generator3** have their first null points at 20 kHz. From this observation, answer the following questions: (1) Does the frequency or the pulse width of the periodic signal determine the shape of line spectrum envelope? (2) From this observation, what is the general relationship between the pulse shape and the line spectrum envelope?

3.C-3   We want to create a period signal that has the line spectrum as shown in Fig. 4.4. The interval of spectral lines is ? kHz, and the first null of the sinc-shaped envelope occurs at 80 kHz. Find the frequency (=1/period) and pulse width of this periodic signal.

  (a) Frequency = ? Hz
  (b) Pulse width = ? second

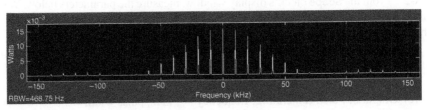

FIGURE 4.4   Spectrum of a desired periodic signal.

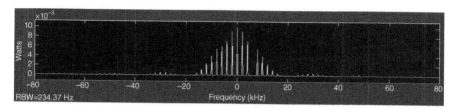

FIGURE 4.5 Desired line spectrum.

3.C-4 Properly set the parameters of **Pulse Generator3** so that the output frequency and the pulse width equal the values given in C-3. Only for this problem, set **Spectrum Viewer3/Signal specification/Sample time** = 1/32e4. Again, do not modify the parameters of the original block saved in **Spectrum_Viewer.mdl/.slx**.

Run the simulation for 3 seconds. Capture the **Spectrum Viewer3** display window. If the result is not the same as the spectrum in 3.C-3, it means that the parameters were not set correctly. If this happens, correct the parameter settings.

3.D [A]In this subsection, we wish to generate a periodic signal that has the same line spectrum as **Pulse Generator1** but does not have the line component located at 6 kHz. That is, we wish to generate a periodic signal that has the line spectrum as shown in Fig. 4.5.

3.D-1 Denote the output of **Pulse Generator1** as $f_T(t)$. Consider a signal $g(t) = f_T(t) - F_3 e^{j3\omega_0 t}$, where ω_0 is the frequency (=1/period) of $f_T(t)$ and F_n is the nth Fourier series coefficient of $f_T(t)$. Then $g(t)$ will have the line spectrum as shown in Fig. 4.5. Explain why.

3.D-2 From the answer in 3.A-2, the output of **Pulse Generator1** is same as $f_T(t)$ shown in Fig. 4.2. By substituting $f_T(t)$ in Fig. 4.2 into equation (4.2) and with some manipulations (can use symbolic math), show that the Fourier series coefficient F_3 of $f_T(t)$ is approximately equal to $0.0505 - 0.0694j$. You may use the answer to 2.B-1 for this calculation.

3.D-3 The mdl/slx file in Fig. 4.6 generates $g(t)$ in 3.D-1 to observe its spectrum. Properly set the parameters of each block to complete the design.

The following notes might be helpful for completing this design:

(a) **Pulse generator1** is the same one as the **Pulse generator1** in the design of the Section 3.A. You can copy it from that design.

(b) Use Euler's identity to express $e^{j3\omega_0 t}$ as the sum of a sine signal and a cosine signal, and set the parameters of **Sine Wave, Sine Wave1,** and **Gain** so that the output of the **Add** block equals $e^{j3\omega_0 t}$. Set **Frequency** = 3*(2*pi)*(1/5e-4) and **Phase** = pi/2 for the **Sine Wave** block. Also properly set the parameters of **Sine Wave1** and **Gain**.

(c) Set **Gain1** to be the complex number $-F_3$ so that the output of the **Add1** block equals $g(t) = f_T(t) - F_3 e^{j3\omega_0 t}$.

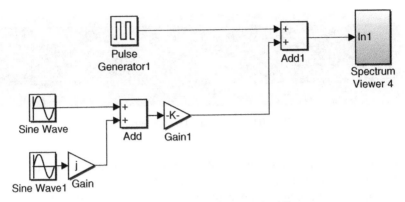

FIGURE 4.6   System for generating a desired line spectrum.

(d) **Spectrum Viewer4** in this design is the same one as **Spectrum Viewer** in the design of 1.B or 3.A. Copy and paste it. The suffix number 4 is appended just to distinguish **Spectrum Viewer** in this design from those in the design of Fig. 4.3.

Capture the completed design.

3.D-4   Execute the completed mdl/slx file for 3 seconds. **Autoscale** the scope window of **Spectrum Viewer4** and capture the window. Check whether the spectrum line located at 6 kHz is removed.

3.D-5   Modify the mdl/slx file to remove the spectrum line element located at −2 kHz. Capture the scope window of **Spectrum Viewer4** to verify that your design is correct.

4.4   SPECTRUM OF A NONPERIODIC AUDIO SIGNAL

4.A [WWW]Download **sampled_ft.mat** from the companion website into your MATLAB work folder. Your MATLAB work folder path is specified in the menu bar of the MATLAB main window. Execute the following lines of code to import the variables saved in **sampled_ft.mat** into the workspace and determine the name and the size of the variables imported:

```
>>load sampled_ft.mat
>>whos
```

4.B The variable **ft_vector** in **sampled_ft.mat** is a sampled vector of an audio signal $f(t)$, and the vector **t_vector** contains the corresponding sampling time instants.

4.B-1   Execute the following in the command window to see the sampling interval. Is it equal to 1/8192?

```
>>t_vector(2) - t_vector(1)
```

4.B-2 Execute the following in the command window to observe the waveform. Capture the resulting figure.

```
>>plot(t_vector, ft_vector)
```

4.C Examine the figure captured in 4.B-2 by zooming into various parts of it and determine whether **ft_vector** is a nonperiodic signal.

4.D Let $F(\omega)$ be the Fourier Transform of $f(t)$. Note that **ft_vector** is the sampled version of $f(t)$. In this subsection, we calculate $F(\omega)$ at three different frequencies, $F(30)$, $F(70)$, and $F(200)$ using numerical integration. Refer to Section 2.1 of Chapter 2 for numerical integration.

4.D-1 [T]Express $F(\omega)$ as an integral of $f(t)$ using the defining equation of the Fourier transform. Suppose that $f(t) = 0$ outside of the sampling interval $0 \leq t \leq 2$. Under this assumption, the integration interval of the Fourier transform can be set as [0~2] seconds, rather than $[-\infty \sim \infty]$.

4.D-2 [WWW]The following m-file computes $F(30)$, $F(70)$, and $F(200)$ via numerical integration. Complete the quantities marked by '**?**' in the m-file and add a comment for each line of the m-file to explain what it does. Especially for the lines with the sign '=', explain what the variables on the left-hand side represents and the reason (why and how) the right-hand side expression is appropriately constructed accordingly. The comment for the line in bold shows an example. Capture the completed m-file.

```
clear
load sampled_ft.mat %Note that there are 'ft_vector' and 't_vector' saved in sampled_ft.mat.
t_step= t_vector(2) - t_vector(1);

Fw30=sum(?.*exp(-j*30*t_vector))*t_step %Left: F(30), Right: numerical calculation
of ∫_{-∞}^{∞} f(t)e^{-j30t}dt.
Fw70=?
Fw200=?
```

4.E [WWW]The following m-file calculates $F(\omega)$ for $\omega = -25000 : 50 : 25000$. This sampled version of $F(\omega)$ is denoted by **Fw_vector**. This m-file also plots the magnitude spectrum.

```
clear
load sampled_ft.mat %Note that there are 'ft_vector' and 't_vector' saved in sam-
pled_ft.mat.
t_step= t_vector(2) - t_vector(1);

Fw_vector=[];
w_vector=[];
for w=-25000:50:25000

  Fw=sum(?.*exp(-j*?*t_vector))*t_step;

  w_vector=[w_vector w];
  Fw_vector=[Fw_vector Fw];

end
plot(w_vector,abs(Fw_vector))
xlabel('Frequency [rad/sec]')
grid
```

4.E-1   Complete the places marked by '**?**' in the m-file and add a comment for each line of the m-file to explain what it does. Especially for the lines with the sign '=', explain what the variables on the left-hand side represents and the reason (why and how) the right-hand side expression is appropriately constructed accordingly. Capture the completed m-file.

4.E-2   In the m-file shown above, the line in bold computes the Fourier transform numerically. Explain why it is repeated in the '**for**' loop.

4.F   Complete the following steps:

4.F-1   Execute the m-file completed in 4.E-1 and capture the magnitude spectrum.

4.F-2.   Determine the approximate frequency range in kHz of the main spectral lobe of $f(t)$. Be sure to properly convert the frequency unit, since the x axis of the captured spectrum is in rad/s.

4.G   After executing the m-file, execute the following in the command window to determine the frequency where the spectrum of $f(t)$ reaches the maximum value. Capture the execution result.

```
>> [T1 T2]=max(abs(Fw_vector)) % T2 is the index of the largest element of Fw_vector.
>>abs(w_vector(T2))% Read/display the frequency where the absolute value of
Fw_vector reaches the peak.
```

NOTE: The magnitude spectrum in 4.F-1 is symmetric to zero frequency. Thus the command **max()** in the first line may find the maximum point in the negative frequency range. This is why we take **abs()** in the second line **abs(w_vector(T2))**.

4.H   From books or Internet materials, determine the typical audio signal frequency range. Based on your answers to 4.F and 4.G above, determine whether $f(t)$ could be an audio signal.

4.I   Connect a speaker or headphone to the audio output of your PC. Execute the following line in the command window to play **ft_vector**, that is, the sampled vector of $f(t)$. Describe what you hear.

```
>> soundsc(ft_vector) % Execute '>>help sound' for details about sound() or soundsc()
```

4.J   In this problem, we interpret differentiation from linear filter perspectives. To this end, we plot the spectrum of $df(t)/dt$, the derivative of the audio signal $f(t)$, and listen to the resulting signal $df(t)/dt$.

4.J-1   Recall that **t_step** denotes the sample interval of **ft_vector**. Explain why the differentiation (slope) at the nth sample of the vector **ft_vector** can be approximated as **(ft_vector(n+1)-ft_vector(n))/t_step**.

4.J-2   If we create a new vector **diffout** by performing 'diffout= (ft_vector(2:L)-ft_vector(1: (L-1))/t_step', where **L** denotes the length of **ft_vector**, then according to 4.J-1 above, the vector **diffout** approximates the sampled version of $df(t)/dt$. In the following m-file, a vector **Dw_vector** denotes a sampled version of the Fourier transform of $df(t)/dt$. The m-file generates **Dw_vector** by numerically performing the Fourier transform on the vector **diffout** and then plots the magnitude spectrum.

Complete the places marked by '**?**' in the m-file and add a comment for each line of the m-file to explain what it does. Especially for the lines with the sign '=', explain what the variables on the left-hand side represent and the reason (why and how) the right-hand side expression is appropriately constructed accordingly.

(a) Capture the completed m-file.
(b) Execute the m-file and capture the magnitude spectrum of $df(t)/dt$.

```
clear
load sampled_ft.mat
t_step= t_vector(2) - t_vector(1);
L=length(ft_vector);
diff_out=(ft_vector(2:L)-ft_vector(1:(L-1)))/t_step;
t_vector=t_vector(1:(L-1)); %To delete the last time instance for new 't_vector' of diff_out
whose length is L-1 not L.

Dw_vector=[];
w_vector=[];
```

```
for w=-25000:50:25000
  w_vector=[w_vector w];

  Dw=sum(?.*exp(-j*?*?))*t_step; % To numerically calculate the Fourier transform
  of df(t)/dt at ω=w.
  Dw_vector=[Dw_vector Dw];
end
plot(w_vector,abs(Dw_vector));
xlabel('Frequency [rad/sec]');
grid
```

4.J-3   Solve the following problems:

(a) [T]Prove that $|D(\omega)| = |\omega| \times |F(\omega)|$, where $F(\omega)$ and $D(\omega)$ are the Fourier transforms of $f(t)$ and $df(t)/dt$, respectively.

(b) From the equation $|D(\omega)| = |\omega| \times |F(\omega)|$, describe how the frequency changes the difference between the shapes of $|D(\omega)|$ and $|F(\omega)|$.

(c) Check whether $|F(\omega)|$ captured in 4.F-1 and $|D(\omega)|$ captured in 4.J-2(b) are consistent with your description in (b). Be sure to compare the overall shapes of the two spectra, rather than the exact values of the spectra along the y axis. Focus on the differences in the shape of the two spectra.

4.J-4   Execute the command **soundsc(ft_vector)** again to play $f(t)$ in the command window. Carefully listen to the sound and notice that there is a bass guitar sound of very slow beat in the background. You are advised to use an earphone or headphone because a regular PC speaker does not produce the bass well. Next execute **soundsc(diff_out)** to play $df(t)/dt$.

(a) Compare the volumes of the bass guitar sounds in $f(t)$ and in $df(t)/dt$.

(b) Explain what has caused the sound differences on the basis of your answer to 4.J-3(b).

4.J-5   Explain (a) why differentiation is considered as a linear system and (b) why it is considered sort of a **high pass** filter.

REFERENCES

[1] R. Bracewell, *The Fourier Transform and Its Applications*, New York: McGraw-Hill, 1978.

[2] J. Duoandikoetxea, *Fourier Analysis*, Providence, RI: American Mathematical Society, 2001.

[3] H. Dym and H. McKean, *Fourier Series and Integrals*, Waltham, MA: Academic Press, 1985.

5

CONVOLUTION

- Generate the sampled time-limited functions and process them.
- Calculate the convolution of two arbitrary time functions using numerical integration.
- Verify the properties of convolution with the impulse function.
- Investigate the frequency domain properties for convolution in the time domain.

5.1 SAMPLED TIME-LIMITED FUNCTIONS

1.A We can create a sampled vector of a time-limited signal using Boolean operations.

1.A-1 Execute the following m-file and capture the result.

```
clear all;
t=-5:0.01:5;
y=(-2<t)&(t<-1);
plot(t, y)
axis([-5 5 -1 2]);grid on
```

Problem-Based Learning in Communication Systems Using MATLAB and Simulink, First Edition.
Kwonhue Choi and Huaping Liu.
© 2016 The Institute of Electrical and Electronics Engineers, Inc. Published 2016 by John Wiley & Sons, Inc.
Companion website: www.wiley.com/go/choi_problembasedlearning

FIGURE 5.1   A sample function $f_1(t)$.

1.A-2   Explain what each line does. Also comment on the shape of the graph generated in 1.A-1.

1.B   [WWW] The m-file below creates two vectors **f1** and **f2**, which are the sampled versions of $f_1(t)$ and $f_2(t)$ shown in Figs. 5.1 and 5.2, respectively. Complete the m-file (determine the quantities marked by '**?**') and capture the completed m-file.

```
clear all;
tstep=0.01;
t=-5:tstep:5;

f1=0.5*( (t>?)&(t<?) );
f2=((t>-1)&(t<0)) - ((t>0)&(t<1)) ;
```

1.C   Execute the m-file above. Then execute the following lines of code to plot $f_1(t)$. Capture the figure and verify that **f1** is correctly generated.

```
>>figure
>>plot(t,f1);
>>axis([-5 5 -5 5]);grid on
```

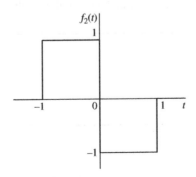

FIGURE 5.2   A sample function $f_2(t)$.

1.D Execute the following lines of code to plot $f_2(t)$. Capture the figure and verify that **f2** is correctly generated.

```
>>figure
>>plot(?,?);
>>axis([-5 5 -5 5]);grid on
```

1.E [WWW]We know that $f_1(-\tau)$ is the reflected signal of $f_1(\tau)$ about the y axis. The m-file below creates the sampled version (vector name **f1mirror**) of $f_1(-\tau)$ and plots it over the range of $-5 \leq \tau \leq 5$. (a) Execute the m-file and capture the result. (b) Check whether it generates $f_1(-\tau)$.

```
clear
tau_step=0.01;
tau=-5:tau_step:5;
f1mirror=0.5*((tau>-1)&(tau<0));
plot(tau,f1mirror); axis([-5 5 -5 5]);grid on;
```

1.F Once we have the sampled version of a certain function $x(t)$, the sampled version of function $x(t - t_0)$ can be created by using the command **circshift()**.

1.F-1 Execute the following lines of codes and capture the results. Note that the notation **a'** denotes complex conjugate transpose of matrix (or vector) **a**; it simply does transposition for a real quantity **a**. Explain what each line does.

```
>>rand(1, XXXX); % XXXX= The last four digits of your student ID. Do not explain
   this line.
>>temp=rand(1,10)
>>temp2=circshift(temp',1)'
>>temp2=circshift(temp',3)'
>>temp2=circshift(temp',-1)'
```

1.F-2 Execute '**temp=circshift(f1mirror',100); plot(tau, temp); axis([-5 5 -5 5]); grid on;**' in the command line and capture the resulting plot. If you have cleared the vector **f1mirror** in MATLAB workspace, then repeat 1.E to regenerate **f1mirror** before starting this problem.

1.F-3 The plot in 1.F-2 shows **f1mirror** right-shifted by 1 second. Provide a discussion on how the 1 second of time-shift is introduced.

5.2   TIME-DOMAIN VIEW OF CONVOLUTION

In this problem, we investigate convolution in the time domain [1], mainly by visualizing the intermediate process of convolution.

2.A   Graphical interpretation of $f_1(a - \tau)$

2.A-1   [T]By how much and in which direction along the τ axis should $f_1(-\tau)$ be shifted to generate $f_1(a - \tau)$?

2.A-2   [T] For the signal $f_1(t)$ shown in Fig. 5.1, sketch $f_1(a - \tau)$ as a function of τ for the cases $a = -1$, 0, and 2.5.

2.A-3   [WWW]The m-file below creates and plots the sampled versions (vectors) of $f_1(a - \tau)$ by using **circshift()** for $a = -1$, -0.5,0, 0.5, 1, 2.5, and 3 over the range $-5 \leq \tau \leq 5$. For instance, in the first line in bold, the variable **f1mirror_delayed** corresponds to the sampled vector of $f_1((-1) - \tau)$; in the second line in bold, the variable **f1mirror_delayed** corresponds to the sampled vector of $f_1((-0.5) - \tau)$. The seven graphs will be placed vertically in one window using **subplot(7,1,1)**, ..., **subplot(7,1,7)**.
    Add the portion for $a = 0.5$, 1, 2.5, and 3 to complete the m-file. Execute the completed m-file. Stretch the figure window vertically for a better view and capture it.

```
clear
tau_step=0.01;
tau=-5:tau_step:5;
f1mirror=0.5*((tau>-1)&(t<0));
figure

a=-1;
delay_samples=round(a/(tau_step));
f1mirror_delayed=circshift(f1mirror',delay_samples)';
subplot(7,1,1)
plot(tau,f1mirror_delayed); axis([-5 5 -0.5 1]);grid on

a=-0.5;
delay_samples=round(a/(tau_step));
f1mirror_delayed=circshift(f1mirror', delay_samples)';
subplot(7,1,2)
plot(tau,f1mirror_delayed); axis([-5 5 -0.5 1]); grid on

%Repeat the lines of code above for a=0, 0.5, 1, 2.5 and 3.
...
...
```

2.A-4   Examine the plots in the figure to assess whether your sketches in 2.A-2 are correct.

2.B [WWW]To the m-file in 2.A-3, insert the line '**f2=((tau>-1)&(tau<0))-((tau>0)& (tau<1));**', which generates the sampled vector of $f_2(\tau)$. Further modify the m-file to create the sampled versions (vectors) of $f_2(\tau) \times f_1(a-\tau)$ for $a = -1, -0.5, 0, 0.5,$ 1, 2.5, and 3 over the range $-5 \leq \tau \leq 5$. Then plot the corresponding seven curves as a function of τ and display them vertically in one figure. Execute the m-file and capture the figure. Stretch the figure window vertically for a better view.

2.C   From the captured graph in 2.B, explain why for the functions $f_1(t)$ and $f_2(t)$ being considered, the indefinite integral $\int_{-\infty}^{\infty} f_2(\tau) \times f_1(a-\tau) d\tau$ can be replaced by the definite integral $\int_{-5}^{5} f_2(\tau) \times f_1(a-\tau) d\tau$ regardless of the value of the parameter a.

2.D   Computing $\int_{-\infty}^{\infty} f_2(\tau) \times f_1(a-\tau) d\tau$ using numerical integration.

2.D-1 [WWW]The m-file below calculates $\int_{-\infty}^{\infty} f_2(\tau) \times f_1(t-\tau) d\tau$ for the case of $t = 0.2$, using numerical integration. The last line of the m-file displayed in bold computes the integral $\int_{-\infty}^{\infty} f_2(\tau) \times f_1(0.2-\tau) d\tau$ numerically. Refer to Section 2.1 of Chapter 2 for the numerical integration processes.
   Execute the m-file and capture the result.

```
clear
tau_step=0.01;
tau=-5:tau_step:5;
f2= ((tau>-1)&(tau<0)) -((tau>0)&(tau<1)) ;
f1mirror=0.5*((tau>-1)&(tau<0));

t=0.2;
delay_samples=round(t/tau_step);
f1mirror_delayed=circshift(f1mirror', delay_samples)';

sum(f2.*f1mirror_delayed)*tau_step
```

2.D-2   Manually calculate $\int_{-\infty}^{\infty} f_2(\tau) \times f_1(t-\tau) d\tau$ for $t = 0.2$ and assess whether it matches the numerical integration result.

2.D-3   You can calculate $\int_{-\infty}^{\infty} f_2(\tau) \times f_1(t-\tau) d\tau$ for other values of t by changing the line '**t=0.2;**'. Execute the m-file with several different values of **t** in the range of $(-1, 2)$. Compare the numerical results with the manual calculation results.

2.E [WWW]The m-file below numerically calculates $\int_{-\infty}^{\infty} f_2(\tau) \times f_1(t-\tau) d\tau$ for t from -3 to 4 with a step of 0.05 and generates a vector **f2convf1**.

```
clear
tau_step=0.01;
tau=-5:tau_step:5;
f1mirror=0.5*((tau>-1)&(tau<0));
f2= ((tau>-1)&(tau<0)) -((tau>0)&(tau<1)) ;
t_vector=[];
f2convf1=[];
for t=?:0.05:?

  delay_samples=round(t/tau_step);
  f1mirror_delayed=circshift(f1mirror', delay_samples)';
  f2convf1_at_t=sum(f2.*f1mirror_delayed)*tau_step;

  t_vector=[t_vector t];
  f2convf1=[f2convf1 f2convf1_at_t];

end
figure
plot(t_vector, f2convf1); grid on
```

2.E-1 Complete this m-file (determine the quantities marked by '**?**') and then capture it.

2.E-2 The line '**f2convf1_at_t = sum(f2.*f1mirror_delayed)*tau_step;**' implements the convolution expression $\int_{-\infty}^{\infty} f_2(\tau) \times f_1(t-\tau) d\tau$ by using numerical integration and stores the result in the variable **f2convf1_at_t**. Explain why this numerical integration needs to be performed repeatedly inside the '**for**' loop.

2.E-3 Identify the differences between the two variables, **f2convf1_at_t** and **f2convf1**.

2.F Execute the completed m-file in 2.E-1 and capture the convolution result graph.

2.G Manual calculation of convolution.

2.G-1 [T]Obtain the expression of $f_1(t) * f_2(t)$ for $f_1(t)$ and $f_2(t)$ given, respectively, in Figs. 5.1 and 5.2, and sketch $f_1(t) * f_2(t)$.

2.G-2. Check whether your sketch in 2.G-1 is consistent with the graph in 2.F.

5.3 CONVOLUTION WITH THE IMPULSE FUNCTION

3.A Convolution with an impulse located at the origin.

3.A-1 [WWW]The m-file below plots another example of $f_1(t)$ that is different from the one in Fig. 5.1. Execute the m-file and the capture the resulting figure.

```
clear
t_step=0.01;
t=-5:t_step:5;
f1=1/t_step*(t==0);
figure
plot(t, f1); axis([-5 5 -5 10]);grid on
```

3.A-2   The graph of $f_1(t)$ generated in 3.A-1 should look like a well-known function. What is this function?

3.A-3   In the m-file of 2.E, recall that **f1mirror** is the sampled version of $f_1(-\tau)$. For the function $f_1(t)$ in 3.A-2, modify the line '**f1mirror = 0.5 * ((tau>-1)&(tau<0))**' of the m-file in 2.E into '**f1mirror = 1/tau_step*(tau==0);**'. Execute the modified m-file and capture the convolution result graph.

3.A-4   Compare the convolution result graph with the graph of $f_2(t)$. From this comparison, summarize the properties of convolution with the function observed in 3.A-2.

3.B.   Convolution with the shifted impulse.

3.B-1   In the m-file in 3.A-1, modify the line '**f1=1/t_step*(t==0);**' into '**f1=1/t_step*(t==1.5);**'. Execute the modified m-file and capture the resulting figure. Identify the difference between the two examples of $f_1(t)$ before and after this modification.

3.B-2   For the $f_1(t)$ in 3.B-1, modify the line '**f1mirror = 1/tau_step*(tau==0);**' of the m-file in 3.A-3 into '**f1mirror = 1/tau_step*(tau==-1.5);**'. Execute the modified m-file and capture the convolution result graph.

3.B-3   Compare the graph of $f_2(t)$ and the convolution result graph in 3.B-2. Based on the comparison, generalize the properties you summarized in 3.A-4.

5.4   FREQUENCY-DOMAIN VIEW OF CONVOLUTION

4.A   The m-file below calculates the Fourier transforms of $f_1(t)$ in Fig. 5.1 and $f_2(t)$ in Fig. 5.2 numerically. It also plots the magnitude and phase spectra of the two signals. We have used the two vectors **f1** and **f2** to represent the time-domain sampled versions of $f_1(t)$ and $f_2(t)$. Let two other vectors **Fw1** and **Fw2** represent the frequency-domain sampled vectors of the Fourier transforms of $f_1(t)$ and $f_2(t)$, respectively. The numerical approach to compute the Fourier transform is described in Section 4.E of Chapter 4.

```
clear
tstep=0.01;
t=-5:tstep:5;
f1=0.5*((t>0)&(t<1));
f2= ((t>-1)&(t<0)) -((t>0)&(t<1));

Fw1=[];
Fw2=[];
w_vector=[];
w_step=2*pi*0.01;
for w=(2*pi*?):w_step:(2*pi*?) % Note that frequency 'w' has the unit of rad/sec not Hz.

  Fw1_at_w=sum(f1.*exp(-j*w*t))*tstep;
  Fw2_at_w=?;

  w_vector=[w_vector w];

  Fw1=[Fw1  Fw1_at_w];
  Fw2=[Fw2  Fw2_at_w];
end
save Fw1_Fw2.mat Fw1 Fw2
figure(1)
subplot(2,1,1)
plot(w_vector, abs(Fw1)) % Amplitude spectrum of f1(t)
subplot(2,1,2)
plot(w_vector, angle(Fw1)) % Phase spectrum of f1(t)
figure(2)
subplot(2,1,1)
plot(w_vector, abs(?))% Amplitude spectrum of f2(t)
subplot(2,1,2)
plot(w_vector, angle(?)) % Phase spectrum of f2(t)
```

4.A-1 This m-file is incomplete; quantities to be determined are marked by '?'. Complete this m-file. To this end, set the frequency range as $[-10 \sim 10]$ Hz and the step size **Fw1** (and **Fw2** as well) as 0.01 Hz. Execute the completed m-file and capture the execution result.

4.A-2. The line '**Fw1_at_w=sum(f1.*exp(-j*w*t))*tstep;**' in the m-file computes the Fourier transform $\int_{-\infty}^{\infty} f_1(t)e^{-j\omega t}dt$ numerically. Explain why it should be calculated repeatedly in the '**for**' loop.

4.B The incomplete m-file below numerically calculates the convolution of $f_1(t)$ and $f_2(t)$, $f_2(t)* f_1(t)$, and the Fourier transform of $f_2(t)* f_1(t)$. Note that the vector **f2convf1** is the time-domain sampled version of $f_2(t)* f_1(t)$, and **Fourier_f2convf1** is the frequency-domain sampled vector of the Fourier transform of $f_2(t)* f_1(t)$.

Complete this m-file and then execute it. Capture the amplitude and phase spectra of $f_2(t)*f_1(t)$. Do not clear the vector **Fourier_f2convf1** in the workspace, since it will be needed for the next problem.

```
clear
tau_step=0.01;
tau=-5:tau_step:5;
f1mirror=0.5*((tau>=-1)&(tau<0));
f2= ((tau>=-1)&(tau<=0)) -((tau>=0)&(tau<=1)) ;

t_vector=[]; f2convf1=[];
w_vector=[]; Fourier_f2convf1=[];
tstep=0.05

for t=-3:tstep:4
    delay_samples=round(t/tau_step);
    f1mirror_delayed=circshift(f1mirror', delay_samples)';
    f2convf1_at_t=sum(f2.*f1mirror_delayed)*tau_step; % Convolution of f2 and f1

    t_vector=[t_vector t];
    f2convf1=[f2convf1 f2convf1_at_t];
end

for w=(2*pi*-10):(2*pi*0.01):(2*pi*10)
    Fourier_f2convf1_at_w=sum(?.*exp(-j*?*t_vector))*tstep;
    w_vector=[w_vector w];
    Fourier_f2convf1=[Fourier_f2convf1 Fourier_f2convf1_at_w];
end
figure
subplot(2,1,1)
plot(w_vector, abs(Fourier_f2convf1))% The amplitude spectrum of $f_2(t) * f_1(t)$.
subplot(2,1,2)
plot(w_vector, ?) % Phase spectrum of $f_2(t) * f_1(t)$.
```

4.C   Note that in the m-file completed in 4.A-1, **Fw1** and **Fw2** have been generated and saved in **Fw1_Fw2.mat**. Execute the following in the command window and capture the plot.

```
>>load Fw1_Fw2.mat
>>figure
>>plot(w_vector, abs(Fw1.*Fw2 - Fourier_f2convf1));
>>axis([-10 10 -2 2])
```

4.D Examine the plot in 4.C and determine whether **Fourier_f2convf1** is identical to **Fw1.*Fw2**. This verifies a well-known property of convolution. Provide a summary of this property.

4.E If $f(t)$ is applied to a linear system with impulse response $h(t)$, then the output $g(t)$ equals the convolution of $f(t)$ and $h(t)$. However, by the well-known property verified in 4.D, it is possible to determine the output $g(t)$ without resorting to convolution in the time domain. Develop a detailed procedure to accomplish this.

4.F [T]Consider the two time functions $f_1(t) = e^{-(k+1)t}u(t)$ and $f_2(t) = e^{-t}u(t)$, where k is the last digit of your student ID. Solve the following problems. You may use symbolic math. Note that the unit step function $u(t)$ is not defined in symbolic math. However, we can still employ symbolic math by taking into account the fact that any form of $u(t)$ in the integrand is equivalent to limiting the integration boundary without $u(t)$ appearing in the integrand.

4.F-1   Determine the convolution of the two functions, $f_1(t) * f_2(t)$.

4.F-2   Determine the Fourier transform of $f_1(t) * f_2(t)$ obtained in 4.F-1.

4.F-3   Determine the Fourier transform of $f_1(t)$, $F_1(\omega)$.

4.F-4   Determine the Fourier transform of $f_2(t)$, $F_2(\omega)$.

4.F-5   Verify that the answer to 4.F-2 equals $F_1(\omega) \times F_2(\omega)$.

REFERENCE

[1] S. Damelin and W. Miller, *The Mathematics of Signal Processing*, Cambridge, UK: Cambridge University Press, 2011.

6

LOW PASS FILTER AND BAND PASS FILTER DESIGN

- Design low pass filters (LPF) and band pass filters (BPF).
- Use LPF or BPF to extract desired spectral components.
- Study the frequency characteristics and impulse responses of LPF and BPF.

6.1   [T]ANALYSIS OF THE SPECTRUM OF SAMPLE AUDIO SIGNALS

1.A   Execute **'help audiovideo'** in the MATLAB command window to see the audio data files provided in MATLAB. Write down the names of all the audio data files.

1.B   Select any one of the audio data files and execute **'load** selected_file_name' (e.g., **load gong**) in the command window to load the selected audio file into the workspace. Then execute **whos** and capture the execution result. The variables **y** and **Fs** are generated in the workspace. The variable **y** is the audio sample vector and the variable **Fs** is the sampling frequency of the sampled vector **y**.

1.B-1   Type in **Fs** in the command window to see the sampling frequency. Write down the sampling frequency and calculate the sample interval.

1.B-2   Determine the length (in seconds) of the selected audio sample, that is, the size of the audio sample vector **y** and its sampling interval.

1.C   [T]Execute **soundsc(y)** in the command window to play the audio signal. Load the six audio sample files provided in MATLAB one by one and play all

Problem-Based Learning in Communication Systems Using MATLAB and Simulink, First Edition.
Kwonhue Choi and Huaping Liu.
© 2016 The Institute of Electrical and Electronics Engineers, Inc. Published 2016 by John Wiley & Sons, Inc.
Companion website: www.wiley.com/go/choi_problembasedlearning

of them. Describe the sound you heard for each of these signals (what it is, e.g., train whistle).

1.D   The MATLAB command **pwelch()** calculates and plots the power spectral density (PSD) of the input (sampled vector). Here, PSD is equivalent to the magnitude square of the Fourier transform (spectrum). The MATLAB function **pwelch(ht,[],[],[],Fs)**, where **ht** is the sampled version of a time function $h(t)$ and **Fs** is the sampling frequency, generates the PSD plot of $h(t)$, that is, $|H(\omega)|^2$ in dB-scale.

1.D-1   Execute the following in the command window to load and to plot, using **subplot**, the PSD of each of the six audio data samples provided in MATLAB in one figure. Stretch the figure vertically to clearly show all PSDs before capturing the window.

```
>> figure
>> subplot(6,1,1);load chirp;pwelch(y,[],[],[],Fs)
>> subplot(6,1,2);load gong;pwelch(y,[],[],[],Fs)
>> subplot(6,1,3);load handel;pwelch(y,[],[],[],Fs)
...
...
...
```

1.D-2   Based on the sound, which one of six audio data samples has the highest frequency (pitch)? Explain whether or not the PSD plots captured in 1.D-1 are consistent with what you heard. Do not focus on the absolute level of each PSD; instead, evaluate where the majority of the frequency components of each signal are located at in the frequency domain.

1.E   The PSD plots in 1.D show the signal spectra in the positive frequency range only. This is because the magnitude spectrum of a real-valued function $h(t)$ is an even function, that is, $|H(\omega)| = |H(-\omega)|$, where $H(\omega)$ is the Fourier transform of $h(t)$. Prove that $|H(\omega)| = |H(-\omega)|$ if $h(t)$ is a real-valued signal.

1.F   [WWW]The m-file below creates an audio sampled vector **y_plus_tone** and plots its PSD. The vector **y_plus_tone** is the sum of the audio data vector **y** in the audio sample file **handel** and the sampled vector of a 3.? kHz sine wave named tone, where '**?**' should be set equal to the last digit of your student ID number.

```
clear
load handel
t_step=1/Fs;
t=0:t_step:(length(y)-1)*t_step;
tone=sin(2*pi*3.?e3*t); %?=the last digit of the student ID number, e.g.,
      sin(2*pi*3.5e3*t) if the last digit is 5.
```

```
y_plus_tone=y'+tone;
figure
pwelch(y_plus_tone,[],[],[],Fs)
```

1.F-1   This m-file is incomplete and quantities to be completed are marked by '?'. Complete this m-file. Then execute it and capture the PSD plot.

1.F-2   Determine whether the PSD plot in F-1 is what you expect and why.

1.F-3   Execute **soundsc(y_pluse_tone)** in the command window to play the sound and describe how **y_pluse_tone** sounds. Explain the reason why it sounds so.

6.2   LOW PASS FILTER DESIGN

2.A   [T]Suppose that the frequency transfer function of a linear system is $H(\omega)$.

2.A-1   If the Fourier transform of the input signal is $X(\omega)$, express the output Fourier transform $Y(\omega)$ in terms of $H(\omega)$ and $X(\omega)$.

2.A-2   The impulse response $h(t)$ of this system can be obtained from its frequency transfer function $H(\omega)$. Express $h(t)$ in terms of $H(\omega)$.

2.A-3   From the answers to 2.A-1 and 2.A-2, we can show that the output $y(t)$ given input $x(t)$ can be obtained as

$$y(t) = h(t) * x(t), \text{ where } * \text{ denotes convolution.} \qquad (6.1)$$

Derive this equation. To this end, start from the answer to 2.A-1 and use the fact that the multiplication in frequency domain is equivalent to the convolution in time domain.

2.B   [T]Consider a linear system with the following frequency transfer function:

$$H(\omega) = \begin{cases} 1 & \text{if } |\omega| \leq 2\pi B \text{ [rad/s]}, \\ 0 & \text{if } |\omega| > 2\pi B \text{ [rad/s]}. \end{cases} \qquad (6.2)$$

2.B-1   Explain why this system is called a low pass filter [1] using the answer to 2.A-1.

2.B-2   What is the bandwidth of this LPF in Hz?

2.B-3   Determine $H(\omega)$ in the equation (6.2) as B approaches infinity. Using this result and the answer to 2.A-1, explain why $Y(\omega) = X(\omega)$ if the filter has an infinite bandwidth.

2.C   In this problem, we determine the impulse response $h(t)$ of a linear time invariant system from its frequency transfer function $H(\omega)$.

2.C-1 [T]Using the answer to 2.A-2, show that the impulse response of the LPF given in equation (6.2) can be derived as

$$h(t) = 2B \text{ sinc}(2Bt), \quad \text{where sinc}(x) \overset{\Delta}{=} \frac{\sin(\pi x)}{\pi x}. \qquad (6.3)$$

NOTE: In some of the existing textbooks, sinc(x) is defined as sin(x)/x. In this book, we adopted the definition in equation (6.3), which is also what the MATLAB function **sinc(x)** implements. In some literature, sinc(x) is written as Sa(x).

2.C-2 [WWW]The m-file below creates and plots the sampled and truncated version of $h(t)$ expressed in equation (6.3), assuming that $B = 200$ Hz in the range of $t = [-0.02 \; 0.02]$. The sampling interval is set to 1/8192, which is equal to the sampling interval of **y_plus_tone** created in 1.F-1.

```
clear
B=200;
t=-0.02:(1/8192):0.02;
ht=2*B*sinc(2*B*t);
plot(t,ht);
grid
```

Execute this m-file and capture the resulting figure.

2.C-3 Execute the m-file above for the cases of B equaling 100 Hz and 400 Hz and capture the figure for each case. From the captured plots, summarize the relationship between the bandwidth of the LPF and the length of its impulse response. The first zero crossing point in the impulse response is a good metric to quantify the impulse response length.

2.C-4 If B increases to infinity, what kind of function do you expect $h(t)$ to converge to?

2.C-5 Determine $y(t)$ in equation (6.1) by substituting the answer in 2.C-4 into $h(t)$ in equation (6.1).

2.C-6 Is the result in 2.C-5 consistent with the answer to 2.B-3?

2.C-7 Explain why filters with the impulse responses as shown in 2.C-2 and 2.C-3 are impractical to implement?

2.D Consider an LPF with a causal impulse response expressed as

$$h(t) = \begin{cases} 2B \text{ sinc}(2B(t - t_d)), & 0 \le t \le 2t_d, \\ 0 & \text{elsewhere.} \end{cases} \qquad (6.4)$$

2.D-1   [WWW]The m-file below creates and plots the sampled vector of $h(t)$ expressed in equation (6.4) for $B = 200$ Hz and $t_d = 0.02$. Execute the m-file and capture the result.

```
clear
B=200;
td=0.02
t=0:(1/8192):2*td;
ht=2*B*sinc(2*B*(t-td));
plot(t,ht);
grid
```

2.D-2   Explain why the linear system with the impulse response shown in 2.D-1 is practical as opposed to the system with the impulse responses shown in 2.C-2 or 2.C-3?

2.D-3   [WWW]Since the function $h(t)$ in equation (6.4) is different from the $h(t)$ in equation (6.3), the transfer functions corresponding to equations (6.3) and (6.4) are also different. We can obtain $H(\omega)$ of $h(t)$ expressed in equation (6.4) by taking its Fourier transform. However, it is mathematically cumbersome to calculate the Fourier transform of a truncated sinc function. In this problem, we resort to the numerical integration method to obtain $H(\omega)$, more precisely, the sampled version of $H(\omega)$ in the frequency domain. Numerical integration method was discussed in Section 2.1 of Chapter 2.

The MATLAB code fragment below generates a vector **Hw_vector**, the sampled version of $H(\omega)$. Each element of **Hw_vector** will be calculated through a separate numerical integration of **ht**, the sampled version of $h(t)$. The magnitude spectrum $|H(\omega)|$ and phase spectrum $H(\omega)$ are also plotted.

Some useful information for completing this m-file:

- Outside of the time period $0 \leq t \leq 0.04$, $h(t) = 0$; thus the integration boundary is [0, 0.04] and the Fourier transform of $h(t)$ can be written as $\int_{-\infty}^{\infty} h(t)e^{-j\omega t}dt = \int_{0}^{0.04} h(t)e^{-j\omega t}dt.$
- For each value of $\omega = -20000 : 10 : 20000$ rad/s, $H(\omega)$ is calculated via a separate numerical integration.

Recall that the vector **ht** is the sampled version of $h(t)$. The line 'Hw=sum(?.*exp(-j*?*t))*t_step;' numerically implements $\int_{0}^{0.04} h(t)e^{-j\omega t}dt$ for each specific value of ω from -20000 rad/s to 20000 rad/s in a 'for' loop. Complete the two places marked by '?' and execute the completed m-file. Capture the execution result.

```
clear
B=2000; %bandwidth, currently set to 2 KHz.
td=0.02; %delay
```

```
t_step=1/8192;
t=0:t_step:0.04;

ht=2*B*sinc(2*B*(t-td)); % Equation (6.4)

Hw_vector=[];
w_vector=[];
for w=-20000:10:20000
   w_vector=[w_vector w];
   Hw=sum(?.*exp(-j*w*?))*t_step;
   Hw_vector=[Hw_vector Hw];
end

figure
subplot(3,1,1)
plot(t,ht);xlabel('t [sec]');ylabel('h(t)');grid
subplot(3,1,2)
plot(w_vector,abs(Hw_vector))
xlabel('w [rad/sec]');ylabel('lH(w)l');grid
subplot(3,1,3)
plot(w_vector,angle(Hw_vector));
xlabel('w [rad/sec]');ylabel('\angle H(w)');grid;axis([-50 50 -1 1])
```

2.D-4   From the magnitude spectrum, measure as accurate as possible the 3dB bandwidth of the LPF in Hz.

2.D-5   Measure the slope of the phase spectrum.

2.D-6   Repeat the above experiment for **td**=0.01 and **td**=0.03. Capture the execution result for each case.

2.D-7   The three plots, one captured in 2.D-3 and two captured in 2.D-6, show $h(t-t_d)$ (top subplot) and its magnitude spectrum (middle subplot) and phase spectrum (bottom subplot) for three different delay values ($t_d = 0.02, 0.01$, and 0.03), respectively. Measure the delays of each of the three top subplots and the slopes of each of the three phase spectra and record them in Table 6.1.

2.D-8   Based on the results in Table 6.1, determine the effect of the delay, that is, the center of symmetry of $h(t - t_d)$ (which equals t_d in equation (6.4)), on its phase

TABLE 6.1   Time Delay and the Slope of the Phase Spectrum.

Delay	Measured delay (center of $h(t-t_d)$)	Measured slope of the phase spectrum
$t_d = 0.01$		
$t_d = 0.02$		Replicate the answer to 2.D-5 here
$t_d = 0.03$		

spectrum $H(\omega)$. Determine the functional relationship between t_d and the slope of $H(\omega)$.

2.D-9   Based on the plots captured in 2.D-3 and 2.D-6, summarize the effect of delay of the signal $h(t)$ on its magnitude spectrum $|H(\omega)|$.

2.D-10   [A]Execute the m-file for two more cases: **B** = 250 and **B** = 1000. Measure the bandwidth from the magnitude spectrum and check whether it is equal to **B**.

2.D-11   [A]In a time invariant linear system like the LPFs we have designed, the input signal undergoes the same amount of delay as the delay of $h(t)$. This is not desirable for real-time systems. In order to reduce the output delay of the LPF, reduce **td** to 0.001 in the m-file and execute it again. Capture the resulting plot. Based on the plot, comment on the penalty one has to pay in terms of the magnitude spectrum in order to reduce the output delay.

2.D-12   [A]Explain why reducing the delay causes the problem observed in 2.D-11.

6.3   LPF OPERATION

Suppose that $c(t)$ denotes the convolution of $a(t)$ and $b(t)$, that is, $c(t) = a(t) * b(t)$. Let the sampled vectors of $a(t)$ and $b(t)$ be **at** and **bt**, respectively. Then the sampled vector of $c(t)$, **ct**, can be simply created by using the command '**ct=conv(ht ,gt);**'. We will use this approach to perform low pass filtering in the following problems.

3.A   [WWW]In the following m-file, we input the audio sample vector saved in **handel.mat** to an LPF with a bandwidth of 2 kHz. The LPF outputs the vector **yt**.

```
clear
B=2000; %bandwidth, set to 2 kHz currently
td=0.02; % delay (= center of symmetry of the delayed sinc pulse ht)

t_step=1/8192;
t=0:t_step:0.04;
ht=2*B*sinc(2*B*(t-td)); % ht is the sampled vector of h(t), i.e., impulse response for
    LPF.

load handel
xt=y';
%xt is a variable declared for the sampled vector of x(t), input signal to LPF.
%In the command conv(a,b), the argument vectors a and b need to be row vectors. So,
    we need to change the column vector y (=audio sample vector in handel) into the row
    vector xt by using transpose operator ( ').

yt=conv(?,xt); % Use a theory about the relationship among input, output, and
    impulse response of linear system.
```

3.A-1   Determine the uncompleted variable in the m-file (marked by '**?**').

3.A-2   Execute the completed m-file. Then execute the following in the command window to listen to the input as well as the output sounds of the LPF. Compare the input and output sounds and summarize their differences.

```
>>soundsc(xt)
Execute the following when playing is completed.
>>soundsc(yt)
```

3.A-3   Change the bandwidth B of the LPF to 500 Hz in the m-file above and execute the completed m-file. Then listen to the sound of **yt** again. Repeat this experiment for **B** = 4000 Hz. Describe how the sound changes as the bandwidth of the LPF changes.

3.A-4   Execute the following lines in the command window and capture the execution result.

```
>> t_axis=1/8192*(1:length(yt));
>> plot(t_axis, yt)
>> axis([0 0.1 -8000 8000])
>> grid
```

3.A-5   From the graph in 3.A-4, determine the starting time, the point where the LPF output signal **yt** starts to rise to a visually noticeable level. Explain why the starting time is roughly equal to the delay of the impulse response of the LPF.

3.B   The goal of this problem is to recover the audio signal interfered by a large sinusoidal signal by using a properly designed LPF.

3.B-1   [WWW]The following m-file creates a sampled audio signal **xt** by adding a sine wave to the sampled sound vector **y** in **handel.mat**. It also plots the PSD of **xt**. Execute the m-file and capture the resulting PSD.

```
clear
load handel
rand(1, XXXX); % XXXX= the last four digits of your student ID number. Be sure to
      include this line.
f=3250+500*rand;
tone=sin(2*pi*f*(1:length(y))*1/Fs);
xt=y'+ tone;
pwelch(xt,[],[],[],Fs)
clear f;
```

3.B-2   Based on the PSD plot of **xt**, estimate the frequency of the added sine wave as accurately as possible. Properly enlarge the PSD plot for accurate reading.

3.B-3   Execute the following line in the command window to play **xt**. Describe how **xt** sounds.

```
>>soundsc(xt)
```

3.B-4   [WWW]By passing **xt** through an LPF, we can remove the beeping sound (caused by interference) without significantly distorting the original audio signal **y** in **handel.mat**. In the following m-file, we will generate the output of the LPF **yt** so that it sounds almost the same as the original audio signal **y**.

Do not include the command '**clear**' in the m-file because the variable **xt** created in 3.B-1 will be needed. Read the comments for each line first and then complete the three places marked by '**?**'. Capture the completed m-file.

```
B=?; % Determine the LPF bandwidth (constant) to filter out(eliminate) the sine wave
       included in xt.
td=0.02; %delay time
t_step=1/8192;
t=0:t_step:0.04;
ht=2*B*sinc(2*B*(t-td)); % The sampled vector of h(t), i.e., the impulse response of
       LPF.
yt=?(?,xt); % Pass xt through LPF to create the output yt. The first ? is the function name.
       The second ? is the variable name.
```

3.B-5   Justify your choice of the value of **B** set in the m-file above.

3.B-6   Execute the m-file in 3.B-4 and then execute the following line in the command window to play **yt**. Describe how **yt** sounds. Has the beeping sound been removed without distorting the sound of the original signal? If there is still a beeping sound or the original sound is noticeably distorted, go back to 3.B-4 and properly change B until you get the desired sound.

```
>>soundsc(yt)
```

6.4   [A]BAND PASS FILTER DESIGN

4.A   Execute the following lines in the command window to generate the impulse response of the LPF with a bandwidth of 300 Hz and plot its PSD. Capture the PSD plot.

```
>>t=0: (1/8192):0.04;
>>B=300;td=0.02;
>>ht=2*B*sinc(2*B*(t-td));
>>pwelch(ht,[],[],[], 8192);
```

4.B [T]Denote the Fourier transform of $x(t)$ by $X(\omega)$. Then prove that the Fourier transform of $x(t)\cos(\omega_0 t)$ is $\frac{1}{2}[X(\omega + \omega_0) + X(\omega - \omega_0)]$.

4.C The following problems provide an intuitive approach on how to design a BPF [1].

4.C-1   After executing the lines of code in 4.A, continue to execute the following lines. The first line creates a 3 kHz cosine waveform vector **cos3000** of the same length as that of **ht** created in 4.A. The second line multiplies **cos3000** by **ht** to create **ht_times_cos**, which is the sampled vector of a frequency up-converted signal $h(t)\cos(2\pi \times 3000t)$. The third line plots the PSD of **ht_times_cos**. Capture the resulting PSD plot.

```
>>cos3000=cos(2*pi*3000*t);
>>ht_times_cos=ht.*cos3000;
>>pwelch(ht_times_cos,[],[],[], 8192)
```

4.C-2   Using the frequency-shift formula derived in 4.B, validate the PSD result generated in 4.C-1. Note that **pwelch(a,[],[],[],b)** shows only the positive frequency components.

4.C-3   Consider an arbitrary signal $x(t)$ whose spectrum spreads over 0 to 4.5 kHz and denote its sampled vector by **xt**. Describe the difference of the PSD shape of **conv(xt,ht_times_cos)** in comparison with the PSD of **xt**. In your description, use the PSD shape of **ht_times_cos** and the frequency-domain view of convolution in the time domain (see Section 5.4 of Chapter 5).

4.D Fig. 6.1 shows the frequency transfer function of an ideal band pass filter that passes only the spectral components of the input signal in the frequency range of $[B_L \ B_H]$.

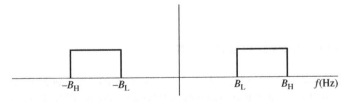

FIGURE 6.1   Frequency response of an ideal band pass filter.

4.D-1   Now, we denote the sampled impulse response of the BPF shown in Fig. 6.1 by **ht**. Based on the discussions in **4.A–4.C**, describe the steps to create **ht** in MATLAB.

4.D-2   [WWW]The following m-file performs the BPF operation to extract only the beeping sound from **xt** created in 3.B-1. Determine the appropriate values for the two places marked by '**?**' and justify your answer.

```
B=?; %B is used to generate ht below.
td=0.02;
t_step=1/8192;
t=0:t_step:0.04;
ht=2*B*sinc(2*B*(t-td)).*cos(2*pi*?*t); % Sample vector impulse response h(t) for BPF
yt = conv(xt,ht);
```

4.D-3   Execute the m-file above and then execute **soundsc(yt)** in the command window to play the BPF output **yt**. Describe how **yt** sounds. In case **xt** created in 3.B-1 has been cleared, execute the m-file in 3.B-1 first before executing the m-file above.

REFERENCE

[1] A. S. Sedra and K. C. Smith, *Microelectronic Circuits*, 3rd ed., Philadelphia: Saunders, 1991.

7

SAMPLING AND RECONSTRUCTION

- Study how the sampling changes the signal spectrum.
- Reconstruct a signal from its sampled version using low pass filtering.
- Implement frequency up-conversion using sampling and a band pass filter.

7.1 CUSTOMIZING THE ANALOG FILTER DESIGN BLOCK TO DESIGN AN LPF

1.A Start Simulink and open a new mdl/slx file and add the blocks as shown in Fig. 7.1. These blocks can be easily identified by searching for them in the Simulink library using block names.

Set the internal variables of the **Sine Wave** block and **Analog Filter Design** block as follows.

1. **Sine Wave**
    - **Sample time** = 1/3e4
2. **Analog Filter Design**
    - **Design method** = **Chevyshev II**
    - **Filter type** = **Low pass**
    - **Filter order** = **32**
    - **Stop band edge frequency** = **2*pi*1000**
    - **Stop attenuation in dB** = 40

Problem-Based Learning in Communication Systems Using MATLAB and Simulink, First Edition.
Kwonhue Choi and Huaping Liu.
© 2016 The Institute of Electrical and Electronics Engineers, Inc. Published 2016 by John Wiley & Sons, Inc.
Companion website: www.wiley.com/go/choi_problembasedlearning

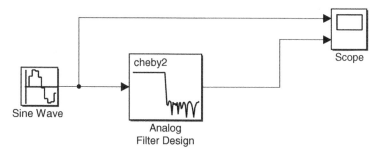

FIGURE 7.1   Test system for **Analog Filter Design** block.

1.A-1   The **Analog Filter Design** block is customized as an LPF with the above parameter setting. Determine the bandwidth of this LPF in Hz.

1.A-2   Set **Simulation Stop Time** to 0.05. Then set the parameter **Frequency (rad/s)** of the **Sine Wave** block as specified in Table 7.1. Run the simulation for each of these cases and measure the corresponding amplitudes of the LPF output and record them in the table.

1.A-3   Based on the simulation result, determine whether the LPF is designed correctly and why?

7.2   STORING AND PLAYING SOUND DATA

2.A   Open **Sound_Source.mdl** (or **Sound_Source.slx**) designed in 6.A of Chapter 1. If you do not have this file, go through 6.A of Chapter 1.

Add the **To file** block from the Simulink library to the design and connect it to the **Sound Source** subsystem as shown in Fig. 7.2. Save the design as a new file and do not overwrite **Sound_Source.mdl/slx,** since it will be needed for other projects later.

Capture your design window.

TABLE 7.1   **Test Inputs to LPF and the Output Amplitude.**

Frequency (rad/s) of the **Sine Wave** block	Amplitude of the LPF output
2*pi*(400 + XX)	
XX = Last two digits of your student ID number	
2*pi*900	
2*pi*950	
2*pi*1000	
2*pi*1050	
2*pi*1500	
2*pi*3000	

FIGURE 7.2   Test system for the subsystem **Sound Source**.

2.B   Set the block parameters of the design in 2.A through the following steps.

2.B-1   [WWW]Download **sound_CH7.mat** from the companion website to your MATLAB work folder. Your MATLAB work folder path is specified in the menu bar of the MATLAB main window.

Execute the following in the command window to verify whether downloading is successful.

```
>> ls *.mat
```

2.B-2   Set the parameter **File name** of the **From File** block in the subsystem **Sound Source** to **sound_CH7.mat**. Capture your parameter setting window.

2.B-3   Set the parameters of the **To File** block as follows. Capture your parameter setting window.

- **Save format = Array** (not required for old versions)
- **File name = signal.mat**
- **Sample time** = 1/8192

2.C   Set **Simulation stop time** to 20 seconds and run the simulation. After the simulation is finished, execute the following in the command window. Describe the sound you heard.

```
>> clear; load signal.mat; soundsc(ans(2,:))
```

7.3   SAMPLING AND SIGNAL RECONSTRUCTION SYSTEMS

Fig. 7.3 is a simple block diagram of a sampling [1, 2] and reconstruction system to restore the original signal from its sampled version by using an LPF.

3.A   In Fig. 7.3, $x(t)$ (=output of the **Sound Source** block) is an audio signal whose bandwidth B equals 4 kHz. For simplicity, we assume that the Fourier transform of $x(t)$, $X(\omega)$, has a triangular shape with a one-sided bandwidth of 4 kHz, as shown in Fig. 7.4.

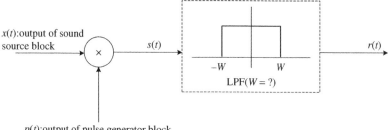

$x(t)$:output of sound source block

$s(t)$

$r(t)$

LPF($W = ?$)

$p(t)$:output of pulse generator block

FIGURE 7.3   Sampling and signal reconstruction system.

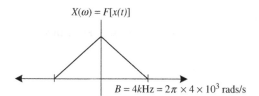

$X(\omega) = F[x(t)]$

$B = 4kHz = 2\pi \times 4 \times 10^3$ rads/s

FIGURE 7.4   Spectrum of the sound signal $x(t)$.

3.A-1   [T]What is the minimum sampling frequency F_s of the periodic sampling pulse signal $p(t)$ in Fig. 7.3 so that $x(t)$ can be reconstructed from its sampled version without distortion? Justify your answer.

3.A-2   [T]Suppose that the sampling pulse signal $p(t)$ is as shown in Fig. 7.5, with a sampling frequency $F_s = 2B = 8$ kHz, pulse amplitude 1, and pulse width equaling 1/10 of the period.

The Fourier transform of the sampled signal $s(t)$ ($=x(t)p(t)$) can be derived considering the following facts:

1. Since $p(t)$ is a periodic signal, it exhibits a line spectrum, and the magnitude of nth spectral line is $2\pi P_n$, where P_n denotes the Fourier series coefficient of $p(t)$ (refer to Section 4.2 of Chapter 4).

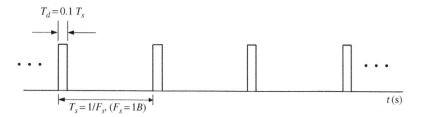

$T_d = 0.1\, T_s$

$T_s = 1/F_s,\ (F_s = 1B)$

t (s)

FIGURE 7.5   Sampling signal $p(t)$.

2. The Fourier transform of the product of two functions in the time domain corresponds to the convolution of the Fourier transforms of the two respective signals in the frequency domain (refer to Section 5.4 of Chapter 5).

Now for the sampling signal $p(t)$ in Fig. 7.5, determine the two quantities marked by '**?**' in the following equation:

$$S(\omega) = F[s(t)] = \sum_{n=-\infty}^{\infty} ? \times X(\omega - n \times ?), \qquad (7.1)$$

where $F[s(t)]$ denotes the Fourier transform of $s(t)$. From this equation, sketch $S(\omega)$.

3.B In this problem, we design the system shown in Fig. 7.3 in Simulink. We first sample the signal $s(t)$. This is done by sampling the output of the **Sound Source** block. Then we reconstruct $x(t)$ from its sampled version using an LPF. We use the **Pulse generator** block, **Product** block, and **Analog Filter Design** block to realize these signal processing steps.

First, modify the mdl/slx file in Section 7.2 as shown in Fig. 7.6:

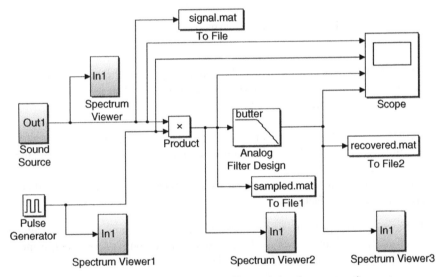

FIGURE 7.6 Simulink design of a sampling and signal reconstruction system.

Then, set the parameters of each block as follows. Do not change the parameters not mentioned here.

1. **Product**
   - **Sample time** = 1/(16e4)
2. **Pulse Generator**
   - **Period** = 1/(8e3)
   - **Pulse Width (% of Period)** = 10

3. **To File1**

   - **Save format = Array** (not required in old versions)
   - **File name = sampled.mat**
   - **Sample time = 1/8192**

4. **To File 2**

   - **Save format = Array** (not required in old versions)
   - **File name = recovered.mat**
   - **Sample time = 1/8192**

5. **Spectrum Viewer, Spectrum Viewer1, Spectrum Viewer2, … :** Use the subsystem block created in Section 1.6.B of Chapter 1. In case you do not have it, go through Section 1.6.B of Chapter 1 to create it. Make sure that the numbers in the names of these blocks occur in numerical order as shown in Fig. 7.6.

3.B-1   Capture the design window of your completed mdl/slx file.

3.B-2   Set **Simulation stop time** to 5e-4 seconds and run the simulation. Upon completing the simulation, properly enlarge the sampling signal $p(t)$ (=output of **Pulse generator**) in the **Scope** display window to measure its frequency and pulse width. Record the measured values together with the captured waveform. Is the captured waveform the same as shown in Fig. 7.5? NOTE: Prior to capturing the waveform, right-click the figure to set **Y-min** to −0.5 and **Y-max** to 1.5.

3.B-3   Set **Simulation stop time** to 1e-2 seconds and run the simulation. Upon completing the simulation, right-click to **Autoscale** $x(t)$ and $s(t)$ in the **Scope** display window. Then zoom into the range of [0.006, 0.01] along the x axis. Capture the **Scope** display window.

3.B-4   From the captured window in B-3, determine whether the sampled signal $s(t)$ is correctly created. That is, is it equal to $x(t)\,p(t)$?

3.B-5   Set the **Simulation stop time** to 4 seconds and run the simulation. Executing the following line of code in the command window will play which of the four signals $\{x(t), p(t), s(t), r(t)\}$ defined in Fig. 7.3?

```
>>load sampled.mat; soundsc(ans(2,:))
```

3.B-6   Execute the command above and describe the sound you hear. Does the signal sound right and why?

3.B-7.   Capture the internal **Spectrum Analyzer** (**Spectrum Scope** in some old MATLAB versions) display windows of **Spectrum Viewer, Spectrum Viewer1,** and **Spectrum Viewer2** (except **Spectrum Viewer3**). Before capturing the **Spectrum Analyzer** display windows, decrease the height of the window to get a width:height ratio of about 7:1 for the graph portion as shown in Fig 4.4 in Chapter 4. Do not **Autoscale**. Follow this guideline throughout all the problems in this book that require the **Spectrum Analyzer** display window.

The line spectrum should appear in the **Spectrum Viewer1/Spectrum Analyzer** display window.

(a) Explain the reasons that cause the line spectrum.
(b) Determine the frequency interval of the line spectrum.
(c) Based on the signals chosen for the system, determine the minimum frequency interval between the spectrum lines and compare it with the observed value in (b).

3.B-8   Is the spectrum in the **Spectrum Viewer2/Spectrum Analyzer** display window consistent with your sketch in 3.A-2? The spectrum will change as time goes by. In making this assessment, focus on (1) the overall spectral shape such as spectral envelope and (2) the frequency interval between the spectrum lines.

3.B-9   Determine how the frequency interval between the spectrum lines changes if the parameter **Period** of the **Pulse Generator** block is changed to 1/(32e3).

3.B-10   Change the parameter **Period** of the **Pulse Generator** block to 1/(32e3) and run the simulation again. Observe the spectrum of $s(t)$ in the **Spectrum Viewer2/Spectrum Analyzer** display window. Upon completing the simulation, capture the **Spectrum Viewer2/Spectrum Analyzer** display window. Is the captured spectrum consistent with the answer to 3.B-9?

3.B-11   Restore the parameter **Period** of the **Pulse Generator** block back to 1/(8e3). If we reduce **Pulse Width (% of Period)** to 1, will the spectrum in the **Spectrum Viewer2/Spectrum Analyzer** display window be different from the one captured in 3.B-7? Assess the difference in terms of the spectral envelope that connects the peaks of the replicas (called "harmonics"). Which one has a wider envelope? Provide a mathematical justification for your assessment.

3.B-12   Change the **Pulse Width (% of Period)** of the **Pulse Generator** block to 1 and run the simulation. Capture the **Spectrum Viewer2/Spectrum Analyzer** display window. Is the result consistent with your answer to 3.B-11?

3.C   The aim of this problem is to reconstruct the original signal $x(t)$ from the sampled signal $s(t)$. Set **Period** = 1/(8e3), **Pulse Width (% of Period)**=10 of **Pulse Generator**.

3.C-1   If the goal is to transmit or to store the information contained in the original signal $x(t)$, it is more convenient to use $s(t)$, instead of $x(t)$, as long as there is no information loss by using $s(t)$. Summarize the advantages of using $s(t)$ in terms signal processing and storage requirements.

3.C-2   In 3.B-6, we have checked that $s(t)$ sounds completely differently from $x(t)$. Hence we first reconstruct the original signal $x(t)$ from $s(t)$.

In the **Scope** display window captured in B-3, $s(t)$ is equal to $x(t)$ only for 10% of time and is 0 during the remaining 90% of time. In other words, $s(t)$ carries only 10% of the waveform of $x(t)$. If we further reduce the pulse width of the sampling pulse $p(t)$ for the advantages discussed in 3.C-1, the portion of time that $s(t) = x(t)$ will be reduced accordingly. Recovering the original signal from the sampled signal requires recovering the lost part of the waveform. Intuitively, from the captured waveform of $s(t)$ in 3.B-3, would it be possible to recover the lost part, say, 90% of the waveform of $x(t)$?

3.C-3   Based on $S(\omega)$ obtained in A-2 (or the display window of **Spectrum Viewer2/Spectrum Analyzer** captured in 3.B-7), intuitively $X(\omega)$ could certainly be recovered from $S(\omega)$. Therefore $x(t)$ could also be recovered from $s(t)$. Design a scheme that employs an LPF with a bandwidth $B = 4$ kHz to recover $x(t)$.

3.C-4   In the mdl/slx design simulated in Section 3.B, determine a proper value for the parameter **Passband edge frequency** of the **Analog Filter Design** block so that it generates the reconstructed signal. Note that the unit is rad/s.

3.C-5   Set **Passband edge frequency** of the **Analog Filter Design** block as obtained in 3.C-4. Set **Simulation stop time** to 5e-2 seconds and run the simulation. After the simulation is completed, **Autoscale** (right click) all the waveforms in the **Scope** display window and then capture the display window.

3.C-6   Check whether or not the shape of the restored signal $r(t)$, that is, the output of the **Analog Filter Design** block, is same as $x(t)$. Focus on the shapes not the signal magnitude.

3.C-7   If the design is correct, then $r(t)$ should be nearly identical to $x(t)$, except a scaling factor of 1/10. Analytically explain this. You may refer to the answer to 3.A-2.

3.C-8   Set **Simulation stop time** to 4 seconds and run the simulation. You may set it larger if the computing power of your PC can handle it. If we run the following line of code in the command window after the simulation, which one of the four signals $\{x(t), p(t), s(t), r(t)\}$ defined in Fig. 7.3 will be played?

```
>>load recovered.mat;soundsc(ans(2,:))
```

3.C-9   Execute the command above. Compare the played sound with the original sound (**signal.mat**) you heard in 2.C.

3.D   In this problem, we simulate a system that approximates the ideal impulse sampling (sampling pulse width equals 0).

3.D-1   To approximate the impulse sampling, set **Pulse Width (% of Period)** of **Pulse Generator** to 1 in the mdl/slx file. With this setting, what percentage of the waveform of $x(t)$ is directly carried through to $s(t)$?

3.D-2   Run the simulation for 4 seconds. Upon completing the simulation, execute the following in the command window to play $r(t)$ (= **Analog Filter Design** output signal). Does it sound the same as the original signal $x(t)$?

```
>>load recovered.mat;soundsc(ans(2,:))
```

3.E   The goal of this problem is to investigate the relationship between the sampling frequency and the feasibility to reconstruct the original signal from its sampled version.

3.E-1 . Another method to reduce the nonzero portion of the sampled signal $s(t)$ is to increase the Period of **Pulse Generator**. Suppose that the pulse width is fixed. Should the parameter **Period** be increased or decreased in order to reduce the nonzero portion of the sampled signal $s(t)$?

3.E-2   [T]At this point, the sampling signal $p(t)$; that is, the output of **Pulse Generator** is shown in Fig. 7.5, where the sampling frequency F_s is set to twice of the bandwidth of the original signal $x(t)$. If the sampling period is doubled, that is, 1/4e3, which is equivalent to reducing the sampling frequency F_s to B, then the answers to the two quantities marked by '**?**' in equation (7.1) should be modified accordingly. Determine the second quantity (for this problem, the first quantity is not of our concern). Modify the sketch of $S(\omega)$ completed in 3.A-2 according to the new quantity.

3.E-3   If the parameter **Period** of the **Pulse Generator** block is doubled to 1/4e3, **Pulse Width (% of Period)** should be decreased from 10% to 5% in order to make the actual pulse width remain unchanged.

 Set **Period**= 1/(4e3) and **Pulse Width (% of Period)**= 5 and run the simulation for 4 seconds. Observe the spectrum of $s(t)$ in the display window of **Spectrum Viewer2/Spectrum Analyzer** while the simulation is in progress. After the simulation is completed, run the simulation again and capture the display window of **Spectrum Viewer2/Spectrum Analyzer**.

3.E-4   Is the spectrum captured in 3.E-3 consistent with your sketch in 3.E-2? The magnitude of the instantaneous spectrum might change. Thus, in comparing the results in 3.E-3 and 3.E-2, focus only on their overall shapes.

3.E-5   What is the main difference between the spectra captured in 3.E-3 and in the display window of **Spectrum Viewer2/Spectrum Analyzer** of 3.B-7. From this observation, explain why it is impossible to reconstruct $x(t)$ from $s(t)$ if $F_s = B = 4$ kHz.

 NOTE: A phenomenon whereby the spectrum replicas in the sampled signal overlap with one another due to an insufficient sampling frequency is called "aliasing."

3.E-6   Run the simulation for 4 seconds. After the simulation is complete, execute the following to play the recovered signal $r(t)$, the output of the **Analog Filter Design** block. Judged from the sound of the recovered signal, is the original signal recovered from its sampled version?

```
>>load recovered.mat;soundsc(ans(2,:))
```

3.E-7   Repeat the simulation for each of the following sampling periods: $1/(2e3)$, $1/(5e3)$, $1/(6.4e3)$, and $1/(10e3)$. Describe how the spectrum of $s(t)$, $S(\omega)$, displayed by **Spectrum Viewer2/Spectrum Analyzer** differs with the difference sampling frequencies. After each simulation, play the recovered signal $r(t)$ and describe what it sounds like.

3.E-8   From the simulation results so far, can you develop a condition on the minimum sampling frequency F_s with which $x(t)$ can be recovered from $s(t)$ without distortion?

3.E-9   Quantitatively justify your conclusion made in 3.E-8.

7.4 FREQUENCY UP-CONVERSION WITHOUT RESORTING TO MIXING WITH A SINUSOID

Consider the scenario that in Fig. 7.3, the LPF is replaced by a BPF with a bandwidth 2B centered at the sampling frequency F_S. For the following problems, we assume that the sampling signal $p(t)$ is the one shown in Fig. 7.5 and denote the BPF output by $z(t)$.

4.A   Based on equation (7.1) and sketch of $S(\omega)$ completed in 3.A-2 or the display window of **Spectrum Viewer2/Spectrum Analyzer** captured in 3.B-7, it is straightforward to express $Z(\omega)$, the output spectrum of the BPF, by using two Fourier series coefficients of $p(t)$ and the $X(\omega)$. Write the expression of $Z(\omega)$ assuming that the BPF is ideal and the delay is negligible.

4.B   [T]The time-domain output of the BPF $z(t)$ can be written as

$$z(t) = x(t) \times A \cos(\omega t + \theta). \tag{7.2}$$

Derive A, ω, and θ.

4.C   We verify the answer to 4.B through simulation. Revisit the mdl/slx file designed in Section 7.3 and make sure that **Period** of the **Pulse Generator** block equals $1/(8e3)$, **Pulse Width (% of Period)** of the **Pulse Generator** block equals 10, so that the sampling signal $p(t)$ is the same as shown in Fig. 7.5.

4.C-1   To implement a BPF by **Analog Filter Design**, set the parameter **Filter type =** **BPF**. The two new parameters **Lower pass band edge frequency** and **Upper pass band edge frequency**, which are defined as shown in Fig. 7.7, must be set as well. Set these two parameters to implement the BPF that has a center frequency of 8 kHz (which equals the sampling frequency) and a bandwidth W (which equals $2B = 8$ kHz). Capture your parameter setting window. Note that the unit is rad/s.

4.C-2   Modify the mdl/slx file as shown in Fig. 7.8. Set the parameters of the **Sine Wave** block properly so that the output of **Product 1** is equal to the right-hand side of equation (7.2), with the parameters A, ω, and θ derived in 4.B. Capture your parameter

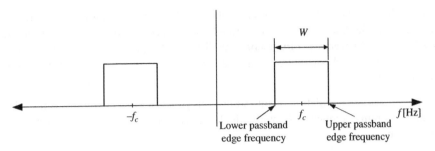

FIGURE 7.7   Definitions of the parameters of the **Analog Filter Design** block for the BPF design.

setting window of the **Sine Wave** block. Attention should be paid to the phase setting of the **Sine Wave** block so that it outputs $A\cos(\omega t + \theta)$, not $A\sin(\omega t + \theta)$.

4.C-3   Run the simulation for 0.02 seconds. Upon completing the simulation, **Autoscale** all the waveforms in the **Scope** display window and capture the window.

4.C-4   In order to check whether the answer to 4.B is correct, we can compare the waveforms in the captured window. (a) Which waveforms should we compare?

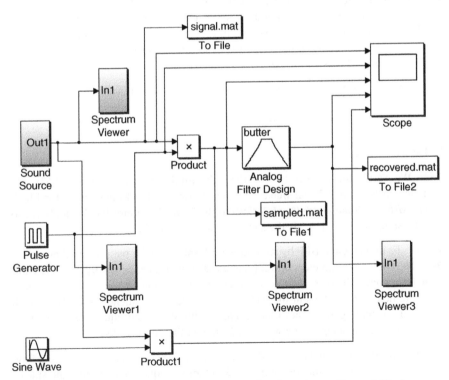

FIGURE 7.8   Design of frequency up-conversion through sampling and filtering.

(b) Is the answer to 4.B correct from the comparison? Note that the BPF introduces a certain amount of delay.

4.C-5   Run the simulation for 0.1 seconds. Capture the display window of the **Spectrum Viewer3/Spectrum Analyzer**.

4.C-6   Does the result in 4.C-5 match the answer to the problem in 4.A?

4.C-7   The signal can be up-converted to another center frequency above 8 kHz without increasing the sampling frequency F_S. Develop the method and determine the possible center frequencies.

4.D   From the problems completed so far, generalize the process to up-convert any arbitrary signal to a desired center frequency ω_0 without using a mixer with a sinusoid.

REFERENCES

[1] A. V. Oppenheim, *Applications of Digital Signal Processing*, Englewood Cliffs, NJ: Prentice-Hall, 1978.

[2] E. J. Baghdady, *Lectures on Communication Systems Theory*, New York: McGraw-Hill, 1960.

8

CORRELATION AND SPECTRAL DENSITY

- Calculate the correlation function of two time functions using numerical integration.
- Locate a pulse in severe noise using correlation.
- Estimate the shape and parameters of unknown periodic signals in severe noise using correlation.
- Investigate the relationship between the correlation function and the spectral density.

8.1 GENERATION OF PULSE SIGNALS

1.A [WWW]The following m-file generates a vector **psint**, the sampled version of a truncated 50 Hz sine waveform with a length of 0.05 seconds. The reference time vector **t** equals −5 to 5 seconds with a step size of 0.001 seconds. The truncation vector **tmp** is generated by Boolean operation introduced in Section 5.1 of Chapter 5.

```
clear
t_step=0.001;
t=-5:t_step:5;
tmp=(0<t) & (t<0.05);
figure(1)
plot(t,tmp); title('tmp');axis([-5 5 -2 2]);
```

Problem-Based Learning in Communication Systems Using MATLAB and Simulink, First Edition.
Kwonhue Choi and Huaping Liu.
© 2016 The Institute of Electrical and Electronics Engineers, Inc. Published 2016 by John Wiley & Sons, Inc.
Companion website: www.wiley.com/go/choi_problembasedlearning

```
tone=sin(2*pi*50*t);
psint=tone.*tmp;
figure(2)
plot(t,psint); title('sine pulse');axis([-5 5 -2 2]);grid on;
```

1.A-1 Add a comment to each line to explain what it does.

1.A-2 Execute the m-file above. Use *x* axis zoom-in button in the menu bar (or use **axis()** properly) to enlarge the [−0.1 0.1] portion of the two resulting figures. Capture the figures and comment on whether or not the waveform in the second figure is a truncated 50 Hz sine waveform with a length of 0.05 seconds.

1.B Add the following lines at the end of the m-file in 1.A to create the sampled vector **pt** of the time-limited square-wave pulse signal $p(t)$ in range of $0 \le t \le 0.05$ with a frequency of 50 Hz.

```
% Add the following to the m-file in 1.A.
pt=sign(psint);
figure(3)
plot(t,pt); title('pt');axis([-5 5 -2 2]);grid on;
```

1.B-1 Execute the m-file and capture the waveform of $p(t)$.

1.B-2 From the graph captured in 1.B-1, calculate the energy of $p(t)$ whose sampled vector is **pt**.

8.2 CORRELATION FUNCTION

2.A The process to shift the elements of a vector using **circshift()** was introduced in Section 1.F of Chapter 5. Execute the following lines of code in the command window. Record the results and justify the results for each case.

```
>> temp=rand(1,12)
>> temp2=circshift(temp',1)'
>> temp2=circshift(temp',4)'
>> temp2=circshift(temp',-5)'
```

2.B [WWW]The vector **pt** created in the m-file in 1.B is a sampling vector of the 50-Hz pulse signal $p(t)$ with a length of 0.05 seconds and a sampling interval of 0.001 seconds.

Suppose that the pulse signal $p(t)$ is transmitted and its delayed version $p(t - t_d)$ is received in additive white Gaussian noise $n(t)$. The delay t_d is due to the distance between the transmitter and the receiver. Thus the received signal $b(t)$ is expressed as $b(t) = p(t - t_d) + n(t)$. Let us denote the sampled vector of the received signal $b(t)$ by **bt**. In the following problems, we generate the sampled vector of the received signal $b(t)$.

2.B-1   [WWW]Add the following lines to the end of m-file created in 1.B. Add a comment to each of the lines in bold to explain what the variable on the left-hand side represents and justify how the right-hand side expression is properly formulated accordingly.

```
% Add the following to the m-file in 1.B.
randn(1,XXX); %XXX=the last three digits of your student ID number. This is irrele-
vant to the contents but be sure to add.
td=2+2*rand;
delayed_pt = circshift( pt', round(td/t_step) )' ;
nt=randn(1,length(delayed_pt));
bt= delayed_pt+nt;
figure(4)
plot(t, bt);axis([-5 5 -5 5]);
save mydelay.mat td
clear td delayed_pt
```

2.B-2   Note that **rand()** randomly generates a real valued number that is uniformly distributed between 0 and 1. From the line in the m-file in 2.B-1 that generates the delay time **td**, calculate the possible range of **td**. Do not use **load mydelay** to answer this question.

2.B-3   Execute the m-file and capture the waveform of $b(t)$ $(= p(t - t_d) + n(t))$. Estimate the location of the pulse, that is, estimate **td** from the captured waveform. You might need to zoom into any portion of the waveform for a close examination.

2.C   The autocorrelation of a signal $f(t)$ denoted by $r_f(\tau)$ is given as [1–3]

$$r_f(\tau) = \int_{-\infty}^{\infty} f^*(t)f(t + \tau)\,dt. \tag{8.1}$$

In the following problem, we calculate the autocorrelation function using numerical integration. Numerical integration was discussed in Section 2.1 of Chapter 2.

2.C-1   The following code fragment determines the autocorrelation function $r_n(\tau)$ of the noise signal $n(t)$ for a given $\tau = 1.5$.

(a) Determine the two quantities marked by '?'.
(b) Explain what the variable on the left-hand side represents and justify how the right-hand side expression is properly formulated accordingly.

```
% Execute the m-file in 2.B-1 first and execute the following code fragment in the com-
mand window.
>>tau=1.5;
>>shift_samples=round(tau/t_step);
>>nt_tau=circshift(nt', -shift_samples)';
>>rn_tau=sum(conj(?).*?)*t_step   % Refer to the numerical integration dis-
cussed in Section 1 of Chapter 2.
```

2.C-2   Execute the commands above in the command window and show the execution result of r_n (1.5).

2.C-3   Repeat the process in 2.C-1 to find r_n (τ) for $\tau = 0$, -1.5, and 2.72.

2.D   [WWW]In the code fragment below, the delay τ increases from -4 to 4 with a step size of 0.001. The autocorrelation r_n (τ) is calculated for each value of τ. Finally, r_n (τ) is plotted as a function of τ.

```
% Add the following to the m-file in 2.B-1.
tau_vector=[];
rn_vector=[];
for tau=-4:0.001:4
    tau_vector=[tau_vector tau];
    shift_samples=round(?/t_step);
    nt_tau=circshift(nt', -shift_samples)';
    rn_tau=sum(conj(?).*?)*t_step;
    rn_vector=[rn_vector rn_tau];
end
figure(5)
plot(tau_vector, rn_vector)
```

2.D-1   Complete the three places marked by '?' in the code fragment and add it to the end of the m-file in 2.B-1. Capture the completed m-file.

2.D-2   Execute the m-file and capture the noise autocorrelation function r_n (τ) displayed in **Figure 5.**

2.D-3   Summarize the characteristics of the white Gaussian noise on the basis of its autocorrelation function graph in 2.D-2. Discuss whether or not the graphical result is what you expected.

2.D-4   (a) [T]From equation (8.1), prove that the autocorrelation at $\tau = 0$ is equal to the signal energy. (b) From (a) and the autocorrelation function captured in 2.D-2, determine the energy of $n(t)$.

NOTE: With the ideal model (unrealistic), the background noise has infinite power and thus infinite energy. However, $n(t)$ in the m-file of 2.D corresponds to the sampled noise after a filter of finite bandwidth. Thus, within the time interval where the noise is truncated, it has a finite energy.

2.E   The m-file completed in 2.D can be modified to plot the autocorrelation function of any pulse signal $p(t)$ whose sampled vector is represented by **pt**.

2.E-1   [T]Write the expression for $r_p(\tau)$. You may modify equation (8.1).

2.E-2   [WWW]From the equation obtained in 2.E-1, properly modify the related lines of the m-file in 2.D to plot $r_p(\tau)$, instead of $r_n(\tau)$, and identify the modified lines. Let **rp_vector** denote the vector for the sampled $r_p(\tau)$.

2.E-3   Execute the modified m-file in 2.E-2 and capture the graph of $r_p(\tau)$. Prior to capturing the graph, zoom into the range of $[-0.2\ 0.2]$ along the x axis using **axis()** or the magnifying button in the menu bar in order to clearly see the shape of $r_p(\tau)$.

2.E-4   Find the energy of $p(t)$ from its autocorrelation function captured in 2.E-3. Is it consistent with the answer to 1.B-2?

2.E-5   [T]Prove that $r_f(\tau)$, the autocorrelation function of $f(t)$, given in equation (8.1) satisfies the Hermitian symmetry property, that is,

$$r_f(-\tau) = r_f^*(\tau). \tag{8.2}$$

2.E-6   Check whether the graphs of $r_n(\tau)$ and $r_p(\tau)$ captured in 2.D-2 and 2.E-3, respectively, match the results given by equation (8.2).

2.F   The cross-correlation function of two signals $f(t)$ and $g(t)$ denoted by $r_{fg}(\tau)$ is calculated as [1–3]

$$r_{fg}(\tau) = \int_{-\infty}^{\infty} f^*(t)\, g(t+\tau)\, dt. \tag{8.3}$$

2.F-1   [WWW]The code fragment below calculates $r_{pn}(\tau)$, the cross-correlation of $p(t)$ and $n(t)$, by using numerical integration. Recall that **pt** and **nt** are the sampled vectors of $p(t)$ and $n(t)$, respectively. The variable **rpn_tau** represents $r_{pn}(\tau)$ for a given value of $\tau(=\textbf{tau})$.

Determine what should be placed at '**?**' in the line '**rpn_tau=sum(conj (?).*?)*t_step;**' to implement the cross-correlation equation expressed in equation (8.3). After completing this line, add the code fragment to the end of m-file in 2.B-1 and capture the complete m-file.

```
% Add the following code fragment to the m-file in 2.B-1.
tau_vector=[];
rpn_vector=[];
for tau=-4:0.001:4
    tau_vector=[tau_vector tau];
    shift_samples=round(tau/t_step);
```

```
    nt_tau=circshift(nt', -shift_samples)';
    rpn_tau=sum(conj(?).*?)*t_step;
    rpn_vector=[rpn_vector rpn_tau];
end
figure(6)
plot(tau_vector, rpn_vector); axis([-4 4 -0.05 0.05]);
```

2.F-2   Execute the completed m-file above and capture the waveform of $r_{pn}(\tau)$ displayed in MATLAB **Figure 6**.

2.F-3   From the plots generated in 2.D-2, 2.E-3, and 2.F-2, determine the maximum values of $r_n(\tau)$, $r_p(\tau)$, and $r_{pn}(\tau)$.

2.F-4   Explain the results in 2.F-3. Why is the peak value of $r_{pn}(\tau)$ smaller than the peak values of $r_n(\tau)$ and $r_p(\tau)$?

2.G   The cross-correlation $r_{pb}(\tau)$ between the pulse $p(t)$ and the received signal $b(t)$ $(= p(t - t_d) + n(t))$ can be written as

$$r_{pb}(\tau) = r_p(\tau - t_d) + r_{pn}(\tau). \tag{8.4}$$

2.G-1   [T]Derive this equation.

2.G-2   From equation (8.4) and the shapes of $r_p(\tau)$ and $r_{pn}(\tau)$ captured in 2.E-3 and 2.F-2, we can more or less predict the approximate shape of $r_{pb}(\tau)$. Since the delay t_d is an unknown variable at this point, give t_d an arbitrary value to determine the shape of $r_{pb}(\tau)$.
   (a) Develop your process to determine the shape of $r_{pb}(\tau)$.
   (b) Where along the τ axis does the peak of $r_{pb}(\tau)$ occur?

2.G-3   From the answer in 2.G-2, develop a process to use $r_{pb}(\tau)$ to locate the pulse in the received signal, which consists of the pulse and a noise signal that is independent of the pulse. In other words, find t_d from $r_{pb}(\tau)$.

2.G-4   [WWW]The following code fragment generates **rpb_vector**, the sampled vector of $r_{pb}(\tau)$, using the numerical integration technique and generates its graph. The variable **bt_tau** represents the sampled vector of $b(t + \tau)$ for a given value of $\tau(=$**tau**). Using a similar code structure as the m-files in 2.F-1, complete the places marked by '?' and then add this code fragment to the end of the m-file completed in 2.B-1. Next, execute this m-file and capture the waveform of $r_{pb}(\tau)$ displayed in MATLAB **Figure 7**.

```
% Add the following code fragment to the m-file in 2.B-1
tau_vector=[];
rpb_vector=[];
for tau=-4:0.001:4
    tau_vector=[tau_vector tau];
    shift_samples=round(tau/t_step);
```

```
    bt_tau= circshift(bt', -shift_samples)';
    rpb_tau=sum(conj(?).*?)*t_step;
    rpb_vector=[rpb_vector rpb_tau];
end
figure(7)
plot(tau_vector, rpb_vector); axis([-4 4 -0.07 0.07]);grid on;
```

2.G-5   From the graph of $r_{pb}(\tau)$ captured in 2.G-4, obtain an estimate of the pulse delay t_d. Zoom into the desired portion of the figure will help increase the estimation accuracy.

2.G-6   Execute the following in the command window to obtain the actual value of t_d, that is, **td** in the m-file. Check whether the estimate in 2.G-5 is correct. There might be a slight mismatch between these two values because of approximation in numerical calculation or variation of the instantaneous noise.

```
>>load mydelay.mat
>>td
```

2.G-7   Each time the m-file is executed, **td** and the sampled noise vector are updated. Run the simulation multiple times to generate multiple **td** values. In each simulation, compare the estimated delay with the actual value to verify that the estimation method is correct. Capture your comparison results.

2.G-8   [A]Change the line '**pt=sign(psint);**' in the m-file into '**pt=psint;**' to replace the rectangular pulse by a sinusoidal signal for $p(t)$. Execute the m-file and estimate td. Does the estimate depend on the pulse type?

2.H   The pulse signal $p(t)$ is deterministic and is, in general, independent of the background Gaussian noise $n(t)$. Under this condition, $r_b(\tau)$, the autocorrelation of the received signal $b(t)$ $(= p(t - t_d) + n(t))$ can be expressed as

$$r_b(\tau) = r_p(\tau) + r_n(\tau). \tag{8.5}$$

2.H-1   [T]Derive equation (8.5).

2.H-2   Revisit the m-file in 2.B-1 and change the line '**nt=randn(1,length (delayed_pt));**' into '**nt=0.075*randn(1,length(delayed_pt));**'. Also execute the following two lines of code in the command window to calculate the energy of the revised noise sample vector **nt**.

```
>> nt=0.075*randn(1,length( delayed_pt));
>> sum(nt.^2)*t_step
```

2.H-3 [WWW]If you have finished 2.G-8, change the line **'pt=psint;'** in the revised m-file in 2.H-2 back to **'pt=sign(psint);'**. Then add the following code fragment to the end of the revised m-file. Execute the m-file and capture the graph of $r_b(\tau)$.

```
tau_vector=[];
rb_vector=[];
for tau=-4:0.001:4
    tau_vector=[tau_vector tau];
    shift_samples=round(tau/t_step);
    bt_tau=circshift(bt', -shift_samples)';
    rb_tau=sum(conj(bt).*bt_tau)*t_step;
    rb_vector=[rb_vector rb_tau];
end
figure(5)
plot(tau_vector,rb_vector);
grid on;
axis([-0.2 0.2 1.5*min(rb_vector) 1.5*max(rb_vector)]);
```

2.H-4 Assess whether the plot captured in 2.H-3 is consistent with equation (8.5). Identify first which problems and their answers are relevant for this assessment.

2.I [T]The correlation function of any power signal $f(t)$ is defined as [1–3]

$$R_f(\tau) = \lim_{T \to \infty} \frac{1}{T} \int_{-\frac{T}{2}}^{\frac{T}{2}} f^*(t) f(t + \tau)\, dt. \tag{8.6}$$

A periodic signal is a power signal. For periodic power signals with period P, $R_f(\tau)$ can be expressed as

$$R_f(\tau) = \frac{1}{P} \int_{-\frac{P}{2}}^{\frac{P}{2}} f^*(t) f(t + \tau)\, dt. \tag{8.7}$$

2.I-1 [T]Show that equation (8.6) is equivalent to equation (8.7) if $f(t)$ is periodic.

2.I-2 [T]Fig. 1.1 in the Chapter 1 illustrates a special case of a rectangular periodic function. Using equation (8.7), derive the correlation function of a periodic rectangular function $f(t)$ with period P, pulse width W ($W < P/2$), and height A.

2.I-3 [T]Sketch by hand $R_f(\tau)$ of the function given in 2.I-2 versus τ.

2.J Suppose that an unknown periodic signal $f(t)$ is received together with an additive noise signal $n(t)$, that is, $r(t) = f(t) + n(t)$. In this problem, we identify the periodic signal $f(t)$ and determine its parameters from the received signal $r(t)$.

2.J-1 [WWW]Download **rt_sampled.mat** from the companion website and execute the following lines of code. Record the name and the length of a vector contained in **rt_sampled.mat**.

```
>> clear
>> load rt_sampled.mat
>> whos
```

2.J-2   The vector **rt** contained in the data file **rt_sampled.mat** is the sampled vector of the received signal $r(t)$ with a sampling interval of 0.001 seconds and its length is 20 seconds. Execute the following to draw the waveform of $r(t)$ and capture it.

```
>> plot(0:0.001:20, rt)
```

2.J-3   Closely observe the waveform of $r(t)$ by zooming into the various portions of the figure. Based on your observation, describe the shape of the period signal $f(t)$ distorted by noise and estimate its parameters such as period or amplitude.

2.J-4   [T]The autocorrelation function of $r(t)$ is the sum of the autocorrelation functions of $f(t)$ and $n(t)$ expressed as

$$R_r(\tau) = R_f(\tau) + R_n(\tau). \tag{8.8}$$

Prove equation (8.8).

2.J-5   [WWW]The following m-file calculates $R_r(\tau)$, the autocorrelation function of the received signal $r(t)$, via numerical integration and plots it. Since the signal period is unknown and the received signal is given as a vector that represents only 20 seconds of the continuous time signal in the m-file, we use equation (8.6) and set T at 20 seconds, rather than infinity, in the numerical integration.

Add a comment to each of the lines in bold to explain what the variable on the left-hand side represents and justify how the right-hand side expression is properly formulated accordingly.

```
clear
load rt_sampled.mat

t_step=0.001;
Rr_vector=[];
tau_vector=[];
for tau=-4:0.01:4
    tau_vector=[tau_vector tau];
    shift_samples=round(tau/t_step);
```

```
rt_tau=circshift(rt', -shift_samples)';
Rr_tau=1/20*sum(conj(rt).*rt_tau)*t_step;
Rr_vector=[Rr_vector Rr_tau];
end
figure
plot(tau_vector, Rr_vector);grid on;
```

2.J-6   Execute the m-file above and capture the graph of $R_r(\tau)$.

2.J-7   Using the figure of $R_r(\tau)$ captured in 2.J-6 and equation (8.8), sketch $R_f(\tau)$ and justify your sketch.

2.J-8   Note that the sketch of $R_f(\tau)$ in 2.J-7 should be similar to the sketch in 2.I-3. Compare these two and estimate the parameters P, W, and A. With the estimated parameters, sketch the estimated periodic signal $f(t)$.

2.J-9   Develop a method to systematically determine the shape and estimate the parameters of an unknown periodic signal distorted by an additive noise.

8.3   ENERGY SPECTRAL DENSITY

3.A   In this subsection we calculate the energy spectral density (ESD) [4] using numerical integration.

3.A-1   [T]The ESD of an energy signal $f(t)$ is defined as $|F(\omega)|^2$, where $F(\omega)$ is the Fourier transform of $f(t)$. (a) Establish relationship between $r_f(\tau)$, the autocorrelation function of $f(t)$, and its ESD $|F(\omega)|^2$; that is, calculate $|F(\omega)|^2$ from $r_f(\tau)$ or vice versa. (b) Prove this relationship.

3.A-2   [WWW]Revisit the m-file in 2.E-2. If you have completed 2.G-8, change the line '**pt=psint;**' back to '**pt=sign(psint);**'. Recall that **pt** is the sampled vector of a pulse signal $p(t)$, whose graph has been captured in 1.B-1, and **rp_vector** is the numerically calculated vector of the autocorrelation function $r_p(\tau)$.

The code fragment below calculates and plots $|P(\omega)|^2$, the ESD of the pulse signal $p(t)$ using the relationship established in A-1. The variable **ESD_vector** denotes the numerically calculated vector of the ESD of $p(t)$. The frequency f increases from 0 Hz to 500 Hz with a 3-Hz step size, and the ESD at each frequency is via numerical integration and concatenated to the vector **ESD_vector**. Add a comment to the line in bold to explain what the variable on the left-hand side represents and justify how the right-hand side expression is properly formulated accordingly.

```
% Add the following code fragment to the m-file in 2.E-2.
tau=-4:0.001:4;
f_vector=[]; ESD_vector=[];
for f=0:3:500
```

```
    f_vector=[f_vector f];
    ESD_f=sum(rp_vector.*exp(-j*2*pi*f*tau))*0.001;
    ESD_vector=[ESD_vector ESD_f];
end
figure(8)
plot(f_vector,ESD_vector) grid on
xlabel('frequency [Hz]')
```

3.A-3   Add the code fragment above to the end of the m-file in 2.E-2 and execute the completed m-file. Capture the ESD graph of $p(t)$.

3.B   From the captured ESD graph, (a) find the frequency where the ESD reaches the maximum. (b) Is the answer to (a) consistent with the waveform of $p(t)$ captured in 1.B-1? Justify your answer.

3.C   [WWW]The following code fragment calculates $P(\omega)$, the Fourier transform of $p(t)$, via numerical integration and then calculates the ESD $|P(\omega)|^2$ directly from $P(\omega)$, rather than using the autocorrelation.

```
% Add the following code fragment to the m-file in 2.E-2.
f_vector=[]; P_vector=[];
for f=0:3:500
    f_vector=[f_vector f];
    P_f=sum(pt.*exp(-j*2*pi*f*t))*0.001;
    P_vector=[P_vector P_f];
end
figure(9)
plot(f_vector,abs(P_vector).^2)
grid on
xlabel('frequency [Hz]')
```

3.C-1   Add a comment to the line in bold to explain what the variable on the left-hand side represents and justify how the right-hand side expression is properly formulated accordingly.

3.C-2   Add the code fragment above to the end of the m-file in 2.E-2 and execute the m-file. Capture the resulting **Figure 9** window.

3.C-3   (1) Are the graphs in 3.C-2 and 3.A-3 the same? (2) This result validates the answer to which problem in this chapter?

3.C-4   Develop two different methods to calculate the ESD of an energy signal, one uses the autocorrelation function and one does not.

REFERENCES

[1] A. C. Aitken, *Statistical Mathematics*, 8th ed., Edinburgh: Oliver & Boyd, 1957.

[2] S. Dowdy and S. Wearden, *Statistics for Research*, New York: Wiley, 1983.

[3] G. U. Yule and M. G. Kendall, *An Introduction to the Theory of Statistics*, 14th ed., Glasgow: Charles Griffin, 1950.

[4] J. Y. Stein, *Digital Signal Processing: A Computer Science Perspective*, Hoboken, NJ: Wiley, 2000.

9

AMPLITUDE MODULATION

- Modulate and demodulate the double sideband-suppressed carrier (DSB-SC) amplitude modulation (AM) signals.
- Investigate the effects of frequency and phase errors on the demodulation performance.
- Modulate and demodulate AM signals by using the sampling and band pass filter (BPF) technique.

9.1 MODULATION AND DEMODULATION OF DOUBLE SIDEBAND-SUPPRESSED CARRIER SIGNALS

Over an additive noise channel, the received DSB-SC AM signal can be expressed as [1]

$$\phi(t) = f(t) \times A \cos(\omega_c t) + n(t), \tag{9.1}$$

where $f(t)$ is the information signal, $A \cos(\omega_c t)$ is the carrier wave, and $n(t)$ is the background noise.

In the receiver, the demodulated signal $y(t)$ is obtained through two major steps. First, a sinusoid $\cos(\omega_c t)$, called "local carrier," is generated with its frequency and

Problem-Based Learning in Communication Systems Using MATLAB and Simulink, First Edition.
Kwonhue Choi and Huaping Liu.
© 2016 The Institute of Electrical and Electronics Engineers, Inc. Published 2016 by John Wiley & Sons, Inc.
Companion website: www.wiley.com/go/choi_problembasedlearning

phase synchronized to those of the carrier wave of the received signal. The received signal is multiplied by the local carrier to generate the following signal:

$$g(t) = \phi(t)\cos(\omega_c t). \tag{9.2}$$

Second, $g(t)$ is passed through a low pass filter (LPF) to generate the demodulated signal $y(t)$ as

$$y(t) = LPF_{B_f}[g(t)], \tag{9.3}$$

where B_f is the bandwidth of $f(t)$. The notation $LPF_D[x(t)]$ denotes the output of an ideal (distortionless with zero-delay) LPF with a bandwidth D. Note that the notation $LPF_D[x(t)]$ will be used frequently in the remainder of this book.

1.A [T]Suppose that $n(t)$ is zero in equation (9.1). By substituting equation (9.1) into (9.2), and then into (9.3), show that $y(t) = (A/2) \times f(t)$. Assume that the carrier wave frequency ω_c is greater than B_f, the bandwidth of the signal.

1.B [WWW]The mdl/slx design shown in Fig. 9.1 is designed to simulate the DSB-SC AM system. In the transmitter, an audio signal from the **Sound source** block is modulated into a DSB-SC AM signal. In the receiver, the transmitted signal is received with an additive Gaussian noise and then demodulated to obtain the original audio signal.

Download **sound_CH9.mat** from the companion website to your MATLAB work folder. Design the mdl/slx file as shown in Fig. 9.1. Set the parameters of the blocks as follows. Leave the blocks not mentioned as they are; they will be set later.

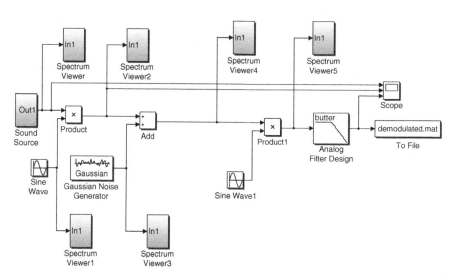

FIGURE 9.1   Simulink design of a DSB-SC AM system.

1. **Sound Source** : Use the subsystem designed in Section 1.6.A of Chapter 1. If you do not have it, design it again.
   - **From File/File name = sound_CH9.mat**
2. **Sine Wave**
   - **Frequency = 2*pi*(20+**The last digit of your student ID number**)*1e3**
   - **Phase = pi/2**
3. **Gaussian Noise Generator** : If you are using a Simulink version that does not have this block in the library, then use the design file on the companion website.
   - **Variance = 1e-3**
   - **Sample time = 1/16e4**
4. **To File**
   - **Save format = Array** (not required for old versions)
   - **File name = demodulated.mat**
   - **Sample time = 1/8192**
5. **Spectrum Viewer, Spectrum Viewer1, Spectrum Viewer2, ...** blocks: Use the subsystem created in Section 1.6.B of Chapter 1. If you do not have it, design it again. Make sure that the numbers in the names of these blocks occur in numerical order as shown in Fig. 9.1.

Save the completed mdl/slx file. Also make another copy for use in the remainder of this chapter. Preserve (do not revise) the original copy, which will be used in Chapter 12.

1.B-1   Capture the completed mdl/slx design window.

1.B-2   The output of the **Sound Source** block corresponds to the information signal $f(t)$ in equation (9.1). We want to modulate $f(t)$ into a DSB-SC AM signal and then demodulate the received signal $\phi(t)$ to recover $f(t)$.

Using equations (9.1)–(9.3), complete the quantities marked by '**?**' for the following eight blocks. The first two blocks are completed for your reference.

- **Sound Source** output $= f(t)$
- **Sine Wave** output $= A \cos(\omega_c t)$
- **Product** output $= ?$
- **Gaussian Noise Generator** output $= ?$
- **Add** output $= ?$
- **Sine Wave1** output $= ?$
- **Product1** output $= ?$
- **Analog Filter Design** output $= ?$

1.B-3   Determine the frequency (in Hz) of the carrier wave signal $A \cos(\omega_c t)$ according to the parameter setting in 1.B-1.

1.B-4   Explain why the parameter **Phase** of the **Sine Wave** block in B-1 is set to **pi/2**. You may use the identity $\cos(x) = \sin(x + \pi/2)$.

1.C   In this problem, we analyze the output waveform and spectrum of each block.

1.C-1   Run the simulation for 0.01 seconds by setting **Simulation stop time** to 0.01 seconds. First, **Autoscale** the **Scope** display window and then zoom into the range of [0.008 0.01] seconds along the x axis. (1) Capture the **Scope** display window. (2) Based on the captured window, explain whether or not the DSB-SC AM signal, that is, $f(t) \times A \cos(\omega_c t)$, the second waveform from the top in the **Scope**, is correctly generated.

1.C-2   Change the **Simulation stop time** to 3 seconds and run the simulation. Capture the **Spectrum Analyzer** display windows of **Spectrum Viewer, Spectrum Viewer1,** and **Spectrum Viewer2** in this order. Minimize the remaining three **Spectrum Viewer** windows at this point.

Before capturing the **Spectrum Analyzer** display windows, decrease the height of the window to get width:height ratio of about 7:1 for the graph portion as shown in Fig. 4.4. Do not **Autoscale**.

1.C-3   Discuss whether the spectrum shapes in the three captured windows in 1.C-2 are consistent with their corresponding equations in 1.B-2.

1.D   Analysis of the spectrum of the background noise.

1.D-1   [T]In Section 3 of Chapter 8, we have explored the relationship between the autocorrelation function $r_f(\tau)$ and the energy spectral density of an energy signal. This relationship holds for power signals as well, that is, relationship between the autocorrelation function $R_f(\tau)$ and power spectral density (PSD) $S_f(\omega)$ of a power signal. Write down this relationship.

1.D-2   [T]Sketch the autocorrelation function of the white Gaussian noise [1], which is often used to model the idealized additive background noise in a radio receiver (you may refer to Section 8.2.D of Chapter 8).

1.D-3   [T]Based on the answers to 1.D-1 and 1.D-2, sketch the PSD of the background noise.

1.D-4   Capture the display window of **Spectrum Viewer3/Spectrum Analyzer**.

1.D-5   Is the spectrum captured in 1.D-4 consistent with the sketch in 1.D-3?

1.D-6   Analyze the spectrum captured in 1.D-4 and determine whether the output of **Gaussian Noise Generator** is white.

1.E   Spectrum of received signals in a noisy environment.

1.E-1   Capture the display window of **Spectrum Viewer4/Spectrum Analyzer**.

1.E-2   Is the captured spectrum in 1.E-1 what you expected to see? Why?

1.E-3   Measure and record the carrier-to-noise ratio (CNR) of the received signal in dB. Refer to the following note for the definition of CNR.
NOTE: CNR is the difference in dB between the peak value of the signal PSD and the background noise PSD level.

1.F   In this problem, we complete the demodulation steps so that the output of the **Analog Filter Design** block generates the demodulated signal, that is, $y(t)$ in equation (9.3).

1.F-1   We want to make the output of the **Product1** block on the right-hand side of the design equal $g(t)$ given in equation (9.2). Determine the proper setting of the parameters **Frequency** and **Phase** of the **Sine Wave1** block of the design.

1.F-2   Set the parameters obtained in 1.F-1 and run the simulation for 3 seconds. Capture the display window of **Spectrum Viewer5/Spectrum Analyzer**. If the setting in 1.F-1 is correct, the spectrum plot should show three clusters of nonzero spectral components. Determine the center frequencies of these components from the **Spectrum Viewer5/Spectrum Analyzer** display window and justify these values on the basis of the mathematical expression of the signal at the input of **Spectrum Viewer5**.

1.F-3   Set the parameter **Passband edge frequency (rad/s)** of the **Analog Filter Design** block to **2*pi*500**. Run the simulation for at least 5 seconds (up to 20 seconds if your PC's computing power allows it). Upon completing the simulation, complete the place marked by '**??**' in the command below and execute it in the command window to play the demodulated signal.

```
>> load ??;soundsc(ans(2,:))
```

1.F-4   Repeat 1.F-3 for each of the following two cases: **Passband edge frequency (rad/s) = 2*pi*4e3** and **2*pi*40e3**. (a) Describe the sound difference among the cases with **Passband edge frequency (rad/s) = 2*pi*500**, **2*pi*4e3**, and **2*pi*40e3**. (b) Which value for the parameter **Passband edge frequency** results in the best sound quality?

1.F-5   Based on the spectrum in 1.F-2, explain why the sound quality with **Passband edge frequency (rad/s) = 2*pi*500** is worse than that with **Passband edge frequency (rad/s) = 2*pi*4e3**.

1.F-6   Based on the spectrum in 1.F-2, explain why the sound quality with **Passband edge frequency (rad/s) = 2*pi*40e3** is worse than that with **Passband edge frequency (rad/s) = 2*pi*4e3**.

1.F-7   Restore the parameter **Passband edge frequency (rad/s)** of the **Analog Filter Design** block back to **2*pi*4e3**. In the **Logging** tab (it may differ for different versions of MATLAB/Simulink) of the parameter window of the **Scope** block, uncheck **Limit data points to last**. Run the simulation for 0.5 seconds by setting **Simulation stop time** to 0.5 seconds. **Autoscale** the **Scope** display window and capture it.

1.F-8   Properly zoom into the **Scope** display window in 1.F-7 to check whether the demodulated signal $y(t)$ is, or at least approximates well $Af(t)/2$. If not, then the parameter setting is incorrect. In this case, correct the parameter setting and run the simulation again. If yes, capture the zoomed-in **Scope** display window.

9.2   EFFECTS OF THE LOCAL CARRIER PHASE AND FREQUENCY ERRORS ON DEMODULATION PERFORMANCE

2.A   The effect of the local carrier phase error.

2.A-1   [T]Consider the case when the phases the received signal and the local carrier differ by θ (phase error); that is, the local carrier of the receiver is $\cos(\omega_c t + \theta)$, instead of $\cos(\omega_c t)$, in equation (9.2). In this case, equation (9.2) should be modified as $g(t) = \phi(t)\cos(\omega_c t + \theta)$.

Substitute this modified equation into equation (9.3) and show that $y(t) = (A\cos\theta/2)f(t)$. Ignore the noise $n(t)$ as done in 1.A.

2.A-2   We will verify the result in A-1 with simulation. Set the parameters of the relevant blocks as follows.

1. **Gaussian Noise Generator**
   - **Variance** $= 0$ (only for verifying the answer to 2.A-1)
2. **Sine Wave1**
   - **Frequency** $= $ **2*pi*(20+**The last digit of your student ID number**)e3**
   - **Phase(rad)** $= $ **pi/2+pi/4**. (Maintain **Phase(rad)** of **Sine Wave** block in the transmitter $= $ **pi/2**.)
3. **Analog Filter Design**
   - **Passband edge frequency (rad/s)** $= $ **2*pi*4e3**

With the parameters above, what is the phase error θ between the received signal and the local carrier?

2.A-3   Set **Simulation stop time** to 0.25 seconds and run the simulation. **Autoscale** the **Scope** display window and capture it.

2.A-4   Does the result in A-3 verify that the demodulated signal is $y(t) = (A\cos(\theta)/2)f(t)$? That is, is the demodulated signal $y(t)$ equal to $f(t)$ with a factor of $A\cos(\theta)/2$. Zoom into $f(t)$ and $y(t)$ at the same time instant and measure the ratio of their magnitudes to estimate the scaling factor.

2.A-5   [A]Run the simulation for each of the following values of θ: **-pi/4, pi/3, pi/2, -pi/2**, and **pi**. (a) Capture the demodulated signal $y(t)$ for each case. (b) Do the simulation results match the theoretical results?

2.B   The effect of phase error in an environment with background noise.

In order to add a noise signal $n(t)$ to the received signal, restore the parameter **Variance** of the **Gaussian Generator** back to 1e-3.

2.B-1   Restore the parameter **phase** of the **Sine Wave1**(local carrier) block to **pi/2**. Run the simulation for 3 seconds. (a) Capture the display window of **Spectrum**

Viewer5/Spectrum Analyzer; (b) Measure the peak value of the baseband signal spectrum, the background noise level (noise PSD), and the CNR of the baseband signal in dB.

- Peak value of baseband signal spectrum = ? dB
- Background noise level (noise PSD) =? dB
- Baseband signal CNR = ? dB

2.B-2 Set the parameter **Phase** of the **Sine Wave1** (local carrier) block to **pi/2+7*pi/16** and repeat the steps in 2.B-1.

- Peak value of baseband signal spectrum = ? dB
- Background noise level (noise PSD) =? dB
- Baseband signal CNR = ? dB

2.B-3 From the results in 2.B-1 and 2.B-2, after the background noise is multiplied by the local carrier, how does the phase error affect the background noise PSD?

2.B-4 (a) Prove that the phase error θ results in a reduction of the baseband signal power after the local carrier multiplication and the reduction factor is $\cos^2 \theta$. (b) Verify the reduction factor using the results in 2.B-1 and 2.B-2.

2.B-5 (a) From the results of 2.B-3 and 2.B-4, explain that the CNR loss of the baseband signal after the local carrier multiplication is equal to $10 \log_{10} \cos^2 \theta [dB]$. (b) Measure the CNR loss in dB in the spectrum of 2.B-2 when compared with the spectrum of 2.B-1. (c) Does the measured CNR loss match the theoretical value, that is, $10 \log_{10} \cos^2 \theta \, [dB]$?

2.B-6 Restore the parameter **Phase** of the **Sine Wave1** (local carrier) block to **pi/2**. Run the simulation for 3 seconds and then execute the following line to play the demodulated signal. Pay attention to the sound quality.

```
>> load demodulated;soundsc(ans(2,:))
```

Set the parameter **Phase** of the **Sine Wave1** (local carrier) block to **pi/2+7*pi/16** and run the simulation again. Play the demodulated signal.

(a) How does the sound quality change according to the phase error?

(b) Explain why the sound quality changes?

2.B-7 (a) Discuss your expected sound of the demodulated signal if the parameter **Phase** of the **Sine Wave1** (local carrier) block is set to 0. (b) Mathematically justify your answer.

2.B-8 Set the parameter **Phase** of the **Sine Wave1** (local carrier) block to 0. Run the simulation and play the demodulated signal. Does it sound like what you expected?

2.B-9 Capture the display window of **Spectrum Viewer5/Spectrum Analyzer**. Explain the spectrum shape by deriving the mathematical expression of $g(t)$ in equation (9.2) for the case of phase setting in 2.B-8.

2.B-10 [A,T]In this problem, we mathematically justify the answer to 2.B-3. Suppose that the PSD of the background noise $n(t)$ is $N_0/2$. Denote the noise multiplied by the local carrier with phase error θ as $n_c(t)$, that is, $n_c(t) \overset{\Delta}{=} n(t)\cos(\omega_c t + \theta)$. We can show that regardless of the value of θ, the PSD of $n_c(t)$ equals $N_0/4$ through the following two steps: (a) show that the autocorrelation function of $n_c(t)$, i.e., $R_{n_c}(\tau) \left(= \lim\limits_{T \to \infty} 1/T \int_{-T/2}^{T/2} n_c^*(t)n_c(t - \tau)dt \right)$ equals $\left(N_0/4 \right)\delta(\tau)$; (b) apply the relationship between the autocorrelation function and the PSD.

2.C The effect of local carrier frequency error.

2.C-1 [T] Consider the case when the frequencies of the received signal and the local carrier differ by $\Delta\omega$ (frequency error); that is, in equation (9.2), the local carrier is $\cos((\omega_c + \Delta\omega)t)$, instead of $\cos(\omega_c t)$. In this case, equation (9.2) should be modified as $g(t) = \phi(t)\cos((\omega_c + \Delta\omega)t)$.

Substitute this modified equation into equation (9.3) and show that $y(t) = (A\cos(\Delta\omega t)/2)f(t)$. Ignore the noise $n(t)$ as done in 1.A.

2.C-2 (a) If the parameter **Frequency** of the **Sine Wave** block is set to **2*pi*25e3** and **Frequency** of the **Sine Wave1** block to **2*pi*(25e3+0.2)**, what should the demodulated signal sound like? (b) Mathematically justify your answer.

2.C-3 Set the parameter **Frequency** of **Sine Wave** to **2*pi*25e3** and **Frequency** of **Sine Wave1** to **2*pi*(25e3+0.2)**. Run the simulation for 10 seconds and play the demodulated signal.

Describe how the demodulated signal sounds differently from the case when **Frequency(rad/s)** of **Sine Wave1** is set to **2*pi*25e3**.

2.C-4 **Autoscale** the **Scope** display window and then capture it.

2.C-5 Does the demodulated signal captured in 2.C-4 match the theoretical result derived in 2.C-1?

2.C-6 Set **Frequency(rad/s)** of the **Sine Wave1** block to **2*pi*24.5e3**. Run the simulation again and play the demodulated signal. Describe the sound you hear.

2.C-7 Based on the simulation results so far, summarize the effects of frequency error on the demodulated signal.

9.3 [A]DESIGN OF AN AM TRANSMITTER AND RECEIVER WITHOUT USING AN OSCILLATOR TO GENERATE THE SINUSOIDAL SIGNAL

In Section 7.3 of Chapter 7, we investigated the spectral characteristics of the sampled signal. Sampling was accomplished by multiplying the information signal by a periodic pulse train. In Section 7.4 of Chapter 7, we introduced frequency up-conversion through sampling and BPF, rather than by multiplication with a sinusoid. In the following problems, we revisit this method and design an AM transmitter and receiver without using an oscillator to generate a sinusoid. In other words, we will implement AM and demodulation without using the **Sine Wave** block in Simulink.

3.A Design an mdl/slx file as shown in Fig. 9.2.

Set the parameters of the blocks as follows. Parameters for blocks not mentioned here will be set later.

1. **Sound Source/From File**
   - **File name = sound_CH9.mat**
2. **Pulse Generator**
   - **Period = 1/(2e4)**
   - **Pulse Width (% of period) = 5**

3.A-1 Set **Simulation stop time** to 5 seconds and run the simulation. Capture the **Spectrum Analyzer** display windows of **Spectrum Viewer, Spectrum Viewer1,** and **Spectrum Viewer2** (except **Spectrum Viewer3**) in the order listed.

3.A-2 You will see the line spectrum with equal spacing in the display window of **Spectrum Viewer1/Spectrum Analyzer**. Explain why the output of the **Pulse Generator** exhibits a line spectrum.

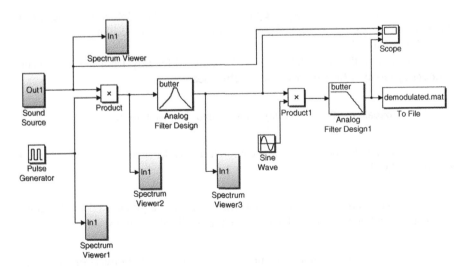

FIGURE 9.2 AM without using the sinusoidal signal.

3.A-3   (a) Measure the frequency spacing of the line spectrum. (b) Mathematically explain the frequency spacing.

3.A-4   By using the Fourier transform property of multiplication of two signals in the time domain, or by expanding the product of the information signal and the pulse train into multiple terms, each of which is the product of the information signal and a complex sinusoid, mathematically justify the spectrum shape shown in the display window of **Spectrum Viewer2/Spectrum Analyzer**.

3.B   Passing the sampled signal through a properly designed BPF will result in a signal whose spectrum shape is the same as that of the DSB-SC AM signal.

3.B-1   If we set the parameters of the **Analog Filter Design** block as shown below, the output of this block will be the DSB-SC AM signal with a carrier frequency at 20 kHz. The definitions of the parameters are shown in Fig. 7.7.
   Justify the settings of the following three parameters.

- **Filter type = Bandpass**
- **Lower passband edge frequency (rad/s) = 2*pi*16e3**
- **Upper passband edge frequency (rad/s) = 2*pi*24e3**

3.B-2   After setting the parameters given in 3.B-1, run the simulation. After the simulation is completed, **Autoscale** the display window of the **Scope** block and then zoom into the range of [4.995 5] seconds. (a) Capture the display window. (b) Does the output of the BPF look like the waveform of a DSB-SC AM signal? (c) Measure the carrier frequency from the waveform generated.

3.B-3   (a) Capture the display window of **Spectrum Viewer3/Spectrum Analyzer**. (b) Does the spectrum represent a DSB-SC AM signal with a carrier frequency of 20 kHz?

3.B-4   Even if we set the frequency (the inverse of the parameter **Period**) of the **Pulse Generator** block to a value lower than the current frequency of 20 kHz, the output of the **Analog Filter Design** block could still be a DSB-SC AM signal with a carrier frequency of 20 kHz. (a) Find such value(s) of the **Period** that are greater than the current setting of 1/2e4. (b) Justify your solution.

3.B-5   Apply the **Period** value(s) obtained in 3.B-4 to the **Pulse Generator** block and then run the simulation. (a) Capture the display windows of the four **Spectrum viewers**. (b) Is your answer to 3.B-4 correct?

3.B-6   A DSB-SC AM signal with carrier frequency of 40 kHz can be generated by changing only the parameters of the **Analog Filter Design** block. Properly set the parameters and run the simulation. Capture the display window of **Spectrum Viewer3/Spectrum Analyzer**.

3.C   Demodulation of the DSB-SC AM signal through sampling and BPF. First set the parameter **Period** of the **Pulse generator** block to **1/(2e4)** and then set the parameters of the **Analog Filter Design** block to the values given in 3.B-1.

3.C-1   The DSB-SC AM signal, that is, the output of the **Analog Filter Design** block, can be demodulated by properly setting the parameters of the **Sine Wave** and **Analog Filter Design1** blocks in the mdl/slx file. Set the parameters of these two blocks and capture the parameter setting windows.

3.C-2   Set the parameter **Simulation stop time** to 0.5 seconds and run the simulation. (a) **Autoscale** the display window of **Scope** and capture it. (b) Is the DSB-SC AM signal demodulated correctly?

3.C-3   In the **To file** block, set **File name=demodulated.mat, sample time**=1/8192, and run the simulation for 5 seconds. Then execute the following in the command window to play the demodulated signal. Judged by the sound, is the signal demodulated correctly?

```
>>clear;load ??;soundsc(ans(2,:))
```

3.C-4   We can demodulate a DSB-SC AM signal without using the **Sine Wave** block. Delete the **Sine Wave** block and revise the mdl/slx file so that the received DSB-SC AM signal can still be successfully demodulated. (a) Capture the revised mdl/slx. (b) Justify the parameter setting for the newly added blocks, if there are any.

3.C-5   Run the simulation using the mdl/slx revised in 3.C-4 for 0.5 seconds. (a) **Autoscale** all the waveforms in the display window of the **Scope** block and capture it. (b) Is the signal demodulated correctly?

REFERENCE

[1] S. Haykin and M. Moher, *An Introduction to Analog and Digital Communications*, 2nd ed., Hoboken, NJ: Wiley, 2006.

10

QUADRATURE MULTIPLEXING AND FREQUENCY DIVISION MULTIPLEXING

- Design a modulation and demodulation system for quadrature multiplexing (QM) amplitude modulation (AM).
- Design a frequency division multiplexing (FDM) system.
- Analyze the effects of phase and frequency errors in QM systems.
- Create stereo sound effects using intentional frequency error in QM methods.

10.1 QUADRATURE MULTIPLEXING AND FREQUENCY DIVISION MULTIPLEXING SIGNALS AND THEIR SPECTRA

Consider three audio signals $x(t)$, $y(t)$, and $z(t)$ all having the same bandwidth of 4 kHz. Each of them is DSB-SC amplitude modulated. Then the three AM signals are transmitted at the same time over a wireless channel. The receiver will observe a superposition of the three transmitted signals. As an example, let the received signal $\omega(t)$ be expressed as

$$\omega(t) = x(t)\cos(2\pi f_1 t + \theta_1) + y(t)\sin(2\pi f_1 t + \theta_1) + z(t)\cos(2\pi f_2 t + \theta_2) + n(t),$$

(10.1)

where $f_1 = 35000$, $\theta_1 = \pi/6$, $f_2 = 16000$, $\theta_2 = -\pi/1000$, and $n(t)$ are the background noise.

Problem-Based Learning in Communication Systems Using MATLAB and Simulink, First Edition.
Kwonhue Choi and Huaping Liu.
© 2016 The Institute of Electrical and Electronics Engineers, Inc. Published 2016 by John Wiley & Sons, Inc.
Companion website: www.wiley.com/go/choi_problembasedlearning

Signal $z(t)$ can be easily separated from $x(t)$ and $y(t)$ because they occupy different frequency bands that do not overlap. Such a scheme where different signals are transmitted at different nonoverlapping frequency bands is called "frequency division multiplexing" (FDM) [1].

1.A   Here, we analyze the passband bandwidth required for transmitting $x(t)$, $y(t)$, and $z(t)$ in equation (10.1).

1.A-1   [T] The bandwidth of the baseband signal $z(t)$ is 4 kHz. Calculate the bandwidth of $z(t)\cos(2\pi f_2 t + \theta_2)$, the DSC-SC AM signal of $z(t)$ in equation (10.1).

1.A-2   [T] The bandwidths of $x(t)$ and $y(t)$ are also 4 kHz. Calculate the bandwidth of $x(t)\cos(2\pi f_1 t + \theta_1) + y(t)\sin(2\pi f_1 t + \theta_1)$, of the sum of the two DSB-SC AM signals of $x(t)$ and $y(t)$ in equation (10.1).

1.A-3   [T] The information signals $x(t)$ and $y(t)$ of the same bandwidth are modulated with the same carrier frequency; thus their spectra overlap with each other and the total bandwidth required for transmitting both signals is the same as that for a single signal. This method is called "quadrature multiplexing" (QM).

Mathematically explain how we can demodulate the two quadrature multiplexed information signals free of interference even though they occupy the same frequency band.

1.B   In the following problems, we assume that $\omega(t)$ in equation (10.1) is generated by using the example of the three audio signals $x(t)$, $y(t)$, and $z(t)$ given at the beginning of Section 10.1. The sampled version of $\omega(t)$ is saved as a matrix and stored in a data file **wt.mat**, which is available for download from the companion website.

1.B-1   Download **wt.mat** from the companion website to your MATLAB work folder. Execute the following command in the command window. Determine the name and dimension (size) of the matrix saved in **wt.mat**.

```
>> clear;load wt.mat;whos
```

1.B-2   The first row of the matrix contains the sampling time instants with a sampling rate of 160 kHz; the second row contains the sampled values of $\omega(t)$ at the corresponding sampling instants.

From the number of samples of $\omega(t)$ and the sampling frequency (=160 kHz): (a) Calculate the total time duration of $\omega(t)$ in seconds; (b) Check whether or not it equals the last element of the first row. If so, explain why.

1.C   In this subsection, we analyze the spectra of QM and FDM signals. In Section 10.2, we will demodulate the audio signals $x(t)$, $y(t)$, and $z(t)$ from the sampled data **wt.mat**. Design an mdl/slx file as shown in Fig. 10.1.

Set the parameters of the blocks as follows. The main parameters for the demodulator will be determined in Section 10.2.

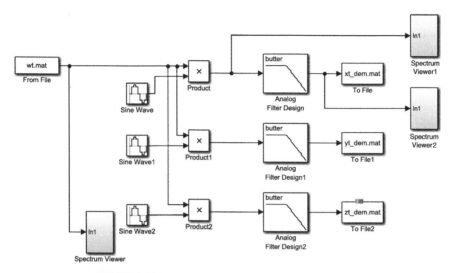

FIGURE 10.1   Demodulation of QM and FDM signals.

1. **From file**
   - **File name = wt.mat**
   - **Sample time = 1/(16e4)**
2. **Sine Wave, Sine Wave1, Sine Wave2**
   - **Sample time = 1/(16e4)**
3. **Product**: Keep the default setting
   - **Sample time = -1**
4. **To file, To file1, To file2**: Make sure that the number in the names of these blocks occurs in numerical order as shown in Fig. 10.1.
   - **Save format = Array** (not required for old versions)
   - **File name = xt_dem.mat** for **To file**, **yt_dem.mat** for **To file1**, and **zt_dem.mat** for **To file2** as shown in Fig. 10.1.
   - **Sample time =** 1/8192
5. **Spectrum Viewer, Spectrum Viewer1, Spectrum Viewer2**: Use the subsystem created in Section 1.6.B of Chapter 1. If you do not have it, design it before continuing. Make sure that the number in the names of these blocks occurs in numerical order as shown in Fig. 10.1.

1.C-1   Run the simulation for 5 seconds (set **Simulation stop time**=5). Capture the display window of **Spectrum Viewer/Spectrum Analyzer**, the one that is connected to the **From File** block whose output is the sampled version of $\omega(t)$. Before capturing the **Spectrum Analyzer** display windows, decrease the height of the window to get a width:height ratio of about 7:1 for the graph portion. Do not **Autoscale**.

1.C-2   The captured plot in 1.C-1 should show that the DSB-SC AM signal $z(t)$ $(= z(t)\cos(2\pi f_2 t + \theta_2))$ is centered at 16 kHz.

   (a) Measure the bandwidth of the DSB-SC AM signal $z(t)$. Ignore the mirror image in the negative frequency range.

   (b) Establish the relationship between the bandwidth of the baseband signal $z(t)$ and the bandwidth of its DSB-SC AM signal $z(t)\cos(2\pi f_2 t + \theta_2)$; that is, compare the bandwidths before and after DSB-SC AM.

1.C-3   Equation (10.1) shows that $\omega(t)$ consists of three terms associated with the audio information signals $x(t)$, $y(t)$, and $z(t)$. Therefore the output of the **From File** block, which is the sampled version of $\omega(t)$, contains three modulated terms of the audio signals. However, the captured plot in 1.C-1 shows only two blocks of spectral components. Explain why.

1.C-4   (a) Measure the total bandwidth occupied by the DSB-SC AM signals of $x(t)$ and $y(t)$ in the captured plot in 1.C-1. (b) What is the advantage of the DSB-SC AM signals of $x(t)$ and $y(t)$ compared with the DSB-SC AM signal of $z(t)$?

10.2   DEMODULATOR DESIGN

2.A   In this problem we finish designing the mdl/slx file from Section 10.1. First, we demodulate the audio information signals $x(t)$, $y(t)$, and $z(t)$ from **wt.mat**. Then, we save the demodulated data in **xt_dem.mat**, **yt_dem.mat**, and **zt_dem.mat** for $x(t)$, $y(t)$, and $z(t)$, respectively.

2.A-1   Set the parameters of the **Sine Wave, Sine Wave1, Sine Wave2** blocks and the **Analog Filter design** block as below: (a) Complete all places marked by '**?**' below. Equation (10.1) is a good starting point for determining most of these parameters. (b) Comment on your parameter settings.

   1. **Sine Wave**
- **Frequency (rad/s)** = ? (make sure you are using the proper unit)
- **Phase (rad)** = pi/2+pi/6 (explain the setting by using the identity $\cos(x) = \sin(x + \pi/2)$).

   2. **Sine Wave1**
- **Frequency (rad/s)** = ?
- **Phase (rad)** = pi/6 (explain the setting)

   3. **Sine Wave2**
- **Frequency (rad/s)** = ?
- **Phase (rad)** = ?

   4. **Analog Filter Design, Analog Filter Design1, Analog Filter Design2**
- **Passband edge frequency (rad/s)** = 2*pi*4e3 (explain the setting.)

TABLE 10.1   Blocks of Spectrum in Spectrum Viewer1.

Spectrum	Expression	Center frequency
Center spectral block	$x(t)/2$	0
1st symmetric pair (closest to the center)	$\dfrac{z(t)}{2}\cos[2\pi(f_1 - f_2)t + \theta_1 - \theta_2]$	$19\,(= f_1 - f_2)$ kHz
2nd symmetric pair		
3rd symmetric pair (remotest from the center)		

2.A-2   After the parameters in 2.A-1 are determined, capture the completed mdl/slx design window.

2.B   Run the simulation for 10 seconds. Perform the following steps after the simulation is complete.

2.B-1   Run the following in the command window to play the demodulated signal of $x(t)$. Describe the sound you hear.

```
>> load ??;soundsc(ans(2,:))
```

2.B-2   Play the demodulated signal of $y(t)$. Describe the sound you hear.

2.B-3   Play the demodulated signal of $z(t)$. Describe the sound you hear.

2.B-4   Capture the display window of **Spectrum Viewer1/Spectrum Analyzer**.

2.B-5   If the setting of the parameters of the **Sine Wave** block is done properly, we should see seven blocks of spectral components including the mirror images in the negative frequency range in the captured plot in 2.B-4.

Complete Table 10.1 (a portion of the table is completed already). To this end, substitute $\omega(t)$ in equation (10.1) into $\omega(t) \times \cos(2\pi f_1 t + \theta_1)$ and expand it by using the trigonometric identity. Then properly map each term into the corresponding block of spectrum in the captured plot in B-4.

2.B-6   Capture the display window of **Spectrum Viewer2/Spectrum Analyzer**.

10.3   EFFECTS OF PHASE AND FREQUENCY ERRORS IN QM SYSTEMS

3.A   [A]In a stereo audio system, the sound from the left speaker is not identical to the sound from the right speaker. Hence, in order to transmit a stereo audio signal, two audio signals should be transmitted simultaneously. Suppose that $x(t)$ and $y(t)$ in equation (10.1) form a stereo audio signal and are to be quadrature multiplexed and then transmitted. The transmission requires the same bandwidth as required for transmitting only one of them (refer to 1.A-3).

If we execute **soundsc(x)**, where **x** is a two-column matrix, then the sampled sound signals in the first and the second columns are played from the left and right speakers, respectively.

Execute the following lines of code to play the stereo sound pair $x(t)$, $y(t)$. Describe the sounds you hear from the left speaker and the right speaker.

```
>>clear;
>>load xt_dem.mat;st_data(1,:)=ans(2,:);
>>load yt_dem.mat;st_data(2,:)=ans(2,:);
>>soundsc(st_data')
```

3.B [A]As done in 3.A, in order to play a stereo sound from an audio receiver, the left-hand side and right-hand-side signal pair $x(t)$, $y(t)$ must be demodulated separately. This requires two sets of sine waves, multipliers, and LPFs in the mdl/slx design in Section 10.2. In other words, the stereo radio receiver with the QM method requires two separate AM demodulators for the left-hand side and right-hand side audio signals. If our goal is to check out both audio signals, rather than listening to the stereo signal, then we can use a mono AM receiver, that is, only one set of local oscillator, mixer, and LPF.

Revise the mdl/slx file designed in 2.A by adding eight additional blocks and deleting the three **spectrum viewers** as shown in Fig. 10.2. Note that another DSB-SC AM demodulator is added at the bottom of the design.

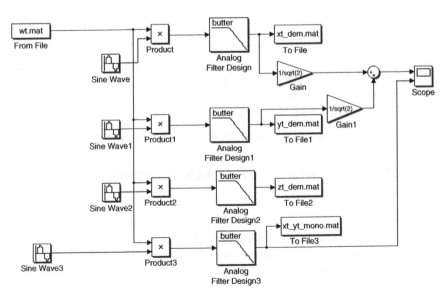

FIGURE 10.2   Mono receiver for QM modulated stereo sounds.

Set the parameters of the newly added blocks as follows.

1. **Sine Wave3**
    - **Frequency (rad/s) = 2*pi*35e3**
    - **Phase (rad) = pi/2+pi/6 - pi/4**
    - **Sample time = 1/(16e4)**

2. **Analog Filter Design3**
    - **Passband edge frequency (rad/s) = 2*pi*4e3**

3. **To File3**
    - **Save format = Array** (not required for old versions)
    - **File name = xt_yt_mono.mat**
    - **Sample time = 1/8192**

4. **Gain, Gain1**
    - **Gain = 1/sqrt(2)**

3.B-1 [A,T]The setting above results in the sum of $x(t)$ and $y(t)$, more precisely, $\frac{1}{2\sqrt{2}}(x(t) + y(t))$, be demodulated and saved in the file **xt_yt_mono.mat**. Mathematically justify this.

3.B-2 [A]Execute the completed mdl/slx file for 10 seconds. If the simulation speed is too slow, then remove the **Scope** block temporarily. Upon completing the simulation, execute the following in the command window to play the sound data in **xt_yt_mono.mat**. Describe the sound you hear.

```
>>load xt_yt_mono.mat;soundsc(ans(2,:))
```

3.B-3 [A]If you have removed the **Scope** block, add it back and connect it again. Run the simulation for 0.02 seconds (set **Simulation stop time** to 0.02). (a) **Autoscale** the **Scope** display window and capture it. It should show that the two signals in the scope are the same. (b) What does this result verify and why? Note that $x(t)/2$, not $x(t)$, is saved in **xt_dem.mat**.

3.B-4 [A]Let us summarize the results so far: if a pair of audio signals in stereo format is QM modulated, summarize the process to listen to the pair of audio signals together by using a mono radio receiver, that is, a receiver with only one set of sine wave generator, multiplier, and LPF.

3.B-5 [A,T]Change the value of **Phase (rad)** of **Sine Wave3** to **pi/2+pi/6-pi/3**. If we still want the two waveforms in **Scope** to be identical with this revised setting, then the gains of the two **Gain** blocks (**Gain** and **Gain1**) must be changed accordingly. Derive the proper gain values for the two **Gain** blocks.

3.B-6 [A]Apply the gains obtained in 3.B-5 to the **Gain** and **Gain1** blocks and run the simulation. (a) **Autoscale** and then capture the **Scope** display window. (b) Are the two signals identical? Since the two signals might still look very similar even with incorrect settings, the two waveforms must be closely examined in making this comparison.

3.C   [A]In this subsection, we create the surrounding sound effect by introducing an intentional frequency error in the stereo signal system with the QM method.

3.C-1   [A]Set the parameter **Frequency (rad/s)** of both **Sine Wave** and **Sine Wave1** to **2*pi*(35e3+0.05)**. Increase **Simulation stop time** to 10 seconds and run the simulation. Upon completing the simulation, execute the following commands again to listen the stereo sound. Be sure to use a headphone/earphone because the PC speaker does not produce well the surrounding effect.

    (a) Describe the sound you hear.

    (b) Compare the sound with the one you heard in 3.A.

```
>>clear;
>>load xt_dem.mat;st_data(1,:)=ans(2,:);
>>load yt_dem.mat;st_data(2,:)=ans(2,:);
>>soundsc(st_data')
```

3.C-2   [A,T]If the design is correct, the sound volumes of the two audio signals, a voice signal and a music signal, will not change but one of the audio signals will sound like moving from left to right and the other one moving right to left.

    (a) Mathematically explain why it sounds like that the sound sources are moving.

    (b) Mathematically explain why the total sound volume of the voice signals, that is, the sum of the voice signal power distributed in the left and right headphone/earphone, does not change. Also why the total sound volume of the music signal does not change?

    (c) From **Frequency (rad/s)** of **Sine Wave** and **Sine Wave1**, which are set to **2*pi*(35e3+0.05)**, calculate the moving cycle (in seconds) of the two sound sources.

    (d) Execute **soundsc(st_data')** again to play the sound signal generated in 3.C-1. Measure the moving cycle in seconds of the two sound sources. Is it consistent with the calculated results in (c)?

3.C-3   [A]How should the setting of **Frequency (rad/s)** of **Sine Wave** and **Sine Wave1** be changed in order to double the sound source moving speed of the signal in 3.C-1?

3.C-4   Verify the answer to 3.C-3 in simulation and play the sound. Does the played sound move twice faster?

REFERENCE

[1] H. P. E. Stern and S. A. Mahmoud, *Communication Systems: Analysis and Design*, Upper Saddle River, NJ: Prentice Hall, 2006.

11

HILBERT TRANSFORM, ANALYTIC SIGNAL, AND SSB MODULATION

- Understand and implement the Hilbert transform.
- Generate analytic signals using the Hilbert transform.
- Generate single-side band (SSB) modulation signals using the Hilbert transform.
- Generate SSB modulation signal using a band pass filter.
- Design an SSB demodulation system.

11.1 HILBERT TRANSFORM, ANALYTIC SIGNAL, AND SINGLE-SIDE BAND MODULATION

The Hilbert transform is a linear operator. The system that performs the Hilbert transform (called a "Hilbert transformer") is a linear system with a frequency transfer function defined as [1, 2]

$$H(\omega) \equiv -\mathrm{j} \times \mathrm{sgn}(\omega) = \begin{cases} -\mathrm{j}, & \omega > 0, \\ \mathrm{j}, & \omega < 0. \end{cases} \tag{11.1}$$

Let $F(\omega)$ and $\hat{f}(t)$ denote, respectively, the Fourier transform and Hilbert transform of an arbitrary signal $f(t)$. From equation (11.1), the Fourier transform of $\hat{f}(t)$ can be written as

$$\hat{F}(\omega) = H(\omega)F(\omega) = \begin{cases} -\mathrm{j}F(\omega), & \omega > 0, \\ \mathrm{j}F(\omega), & \omega < 0. \end{cases} \tag{11.2}$$

Problem-Based Learning in Communication Systems Using MATLAB and Simulink, First Edition.
Kwonhue Choi and Huaping Liu.
© 2016 The Institute of Electrical and Electronics Engineers, Inc. Published 2016 by John Wiley & Sons, Inc.
Companion website: www.wiley.com/go/choi_problembasedlearning

If $f(t)$ is a real signal, then the analytic signal $z(t)$ of $f(t)$ is defined as [1, 2]

$$z(t) = \frac{1}{2}[f(t) + j \times \hat{f}(t)].$$ (11.3)

The analytic signal is a conceptual complex signal, but it is widely used in signal analysis. We will examine the spectral characteristics of the analytic signal in 1.A and 1.B.

1.A [T]Prove that the Fourier transform of $z(t)$ can be calculated as

$$Z(\omega) = \begin{cases} F(\omega), & \omega > 0, \\ 0, & \omega \le 0. \end{cases}$$ (11.4)

1.B [T]Using equation (11.4), summarize the unique features of the spectrum of analytic signal with respect to the spectrum of the original real signal.

1.C [T]The impulse response of the Hilbert transformer can be calculated by taking the inverse Fourier transform of $H(\omega)$, and it is expressed as

$$h(t) = \frac{1}{\pi t}.$$ (11.5)

Equation (11.2) can be rewritten in the time domain as

$$\hat{f}(t) = h(t)?f(t).$$ (11.6)

Complete the operation in equation (11.6) left as a question mark.

1.D [T]The analytic signal of the AM signal $f(t) \cos \omega_c t$ (assuming that ω_c is sufficiently large) is $\frac{1}{2}f(t)e^{j\omega_c t}$. This can be verified easily by consulting your answer to 1.B.

1.D-1 Compare the spectrum shapes of $f(t) \cos \omega_c t$ and $\frac{1}{2}f(t)e^{j\omega_c t}$ and verify this conclusion.

1.D-2 However, the analytic signal of $f(t) \cos \omega_c t$ is not always $\frac{1}{2}f(t)e^{j\omega_c t}$. In order to guarantee that the analytic signal of $f(t) \cos \omega_c t$ is $\frac{1}{2}f(t)e^{j\omega_c t}$, the carrier frequency ω_c should be higher than the bandwidth of $f(t)$, as illustrated in Fig. 11.1. Explain why.

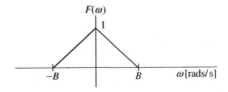

FIGURE 11.1   Fourier transform of $f(t)$.

1.E [T]Suppose that the spectrum of $f(t)$ is a triangle as shown in Fig. 11.1.

1.E-1   The Fourier transform of $f(t) \cos \omega_c t (\omega_c \geq B)$ is given as $\frac{1}{2}(F(\omega - \omega_c) + F(\omega + \omega_c))$. For the $F(\omega)$ shown in Fig. 11.1, sketch the Fourier transform of $f(t) \cos \omega_c t$.

1.E-2   Express the Fourier transform of $\hat{f}(t) \sin \omega_c t$ in terms of $\hat{F}(\omega)$.

1.E-3   The Fourier spectra of $\hat{f}(t) \sin \omega_c t$ and $f(t) \cos \omega_c t$ are identical for $|\omega| \leq \omega_c$; for $|\omega| \geq \omega_c$, they have opposite polarities. Prove this using the relationship between $\hat{F}(\omega)$ and $F(\omega)$. Also sketch their Fourier spectra and check your answer.

1.E-4   For the $F(\omega)$ shown in Fig. 11.1, sketch the Fourier spectrum of $\frac{1}{2}(f(t) \cos \omega_c t + \hat{f}(t) \sin \omega_c t)$. You may use your sketch in 1.E-3.

1.E-5   For the $F(\omega)$ shown in Fig. 11.1, sketch the Fourier spectrum of $\frac{1}{2}(f(t) \cos \omega_c t - \hat{f}(t) \sin \omega_c t)$. You may use your sketch in 1.E-3.

11.2   GENERATION OF ANALYTIC SIGNALS USING THE HILBERT TRANSFORM

The analytic signal $z(t)$ of a real signal $f(t)$ is given as

$$z(t) = \frac{1}{2}(f(t) + j\hat{f}(t)). \qquad (11.7)$$

In the following problems, the sampled vector of the Hilbert transform $\hat{f}(t)$ is generated from the sampled vector of an audio signal $f(t)$. The sampled vector of an analytic signal $z(t)$ is then generated by the relationship given in equation (11.7).

2.A   The sampled vector of an audio signal $f(t)$ and the corresponding sampling time vector are given in .mat format. The goal of this problem is to analyze this signal.

2.A-1   [WWW]Download **sound_CH11.mat** from the companion website to your MATLAB work folder. Then execute the following commands in the command window. Is **data(2,:)** an audio signal?

```
>>load sound_CH11.mat; soundsc(data(2,:))
```

2.A-2   Write and then execute the following m-file, which extracts from the file **sound_CH11.mat** the sampled vector of the audio signal $f(t)$, **ft_vector**, and the corresponding sampling time vector, **t_vector**.

```
clear ;
load sound_CH11.mat
t_vector=data(1,1:4000);
ft_vector=data(2,1:4000);
```

2.A-3   The variable **ft_vector** generated in 2.A-2 is the sampled vector of $f(t)$ with a sampling rate of 8000 Hz. The length of the vector is 4000 samples. The variable **t_vector** is a vector whose elements correspond to the sampling instants of each element of **ft_vector**. Therefore **t_vector**=[0:1/8000:0.5]. Execute the following command to plot $f(t)$ and capture the result.

```
>>figure; plot(t_vector,ft_vector)
```

2.B   The Hilbert transform $\hat{f}(t)$ (let **Hilbert_ft_vector** be its sampled vector) of $f(t)$ can be obtained by the following steps:

- Step 1. Use numerical integration (refer to Section 2.1 of Chapter 2) to obtain the Fourier transform of **ft_vector**. This generates a vector **Fw_vector** that is the sampled version of $F(\omega)$.
- Step 2. From **Fw_vector**, use equation (11.2) to generate a vector **Hilbert_Fw_vector** that is the sampled version of $\hat{F}(\omega)$.
- Step 3. Use numerical integration to obtain the inverse Fourier transform of **Hilbert_Fw_vector**. This generates a vector **Hilbert_ft_vector** that is the sampled version of $\hat{f}(t)$.

2.B-1   [WWW]The m-file below generates **Hilbert_ft_vector** via the three steps above. The time interval for numerical integration in Step 1 is set to 1/8000, the same as the sampling interval of **ft_vector**. The frequency interval for generating the discrete version of $F(\omega)$ is set to 2π rad/s. Thus the frequency interval for the numerical integration in Step 3 must also be set to 2π rad/s.

Identify the line numbers of the m-file that correspond to each of the three steps above.

Step 1: ? ~ ? line
Step 2: ? ~ ? line
Step 3: ? ~ ? line

```
clear ;
load sound_CH11.mat
t_vector=data(1,1:4000);
ft_vector=data(2,1:4000);

t_step=t_vector(1,2)-t_vector(1,1);
w_vector=[ ]; w_step=2*pi;
```

```
Fw_vector=[ ]; Hilbert_Fw_vector=[ ];
for w=(-2*pi*4000):w_step:(2*pi*4000)
    w_vector=[w_vector w];
    Fw=sum(?.*exp(-j*w*t_vector))*t_step;
    Fw_vector=[Fw_vector Fw];
    if ? >0
        Hilbert_Fw=-j*Fw;
    else
        Hilbert_Fw=j*Fw;
    end
    Hilbert_Fw_vector=[Hilbert_Fw_vector Hilbert_Fw];
end

w_step=2*pi;
Hilbert_ft_vector=[ ];
for t=t_vector
    Hilbert_ft=1/(2*pi)*sum(?.*exp(j*t*w_vector))*w_step;
    Hilbert_ft_vector=[Hilbert_ft_vector Hilbert_ft];
end

Data(1,:)=t_vector;
Data(2,:)=ft_vector;
save ft.mat Data

Data(1,:)=t_vector;
Data(2,:)=real(Hilbert_ft_vector);
save Hilbert_ft.mat Data
```

2.B-2   Create the m-file above. The variables indicated by the question mark '?' in the m-file are to be completed. Determine these variables and complete the m-file.

2.B-3   (a) For each of lines in bold, add a comment to explain what it does. Especially for the lines with '=', explain what the variable on the left-hand side represents and justify how the right-hand side expression is formulated accordingly. (b) Capture the commented m-file.

2.B-4   Execute the m-file above. Then execute the following lines in the command window. Capture the result.

```
>>figure
>>plot(t_vector,ft_vector)
>>hold on
>>plot(t_vector, Hilbert_ft_vector,'r')
```

2.B-5 [A]The plot obtained in 2.B-4 should show that if the amplitude of $f(t)$ is large, then the amplitude of $\hat{f}(t)$ is also large. This comes from the linearity property of the Hilbert transform.

(a) Explain why this observation supports the linearity of the Hilbert transform.
(b) From equation (11.6), mathematically explain why Hilbert transform is linear.

2.B-6 Execute the following command for the plot generated in 2.B-4 and then capture it.

```
>>axis([0 0.04 -0.5 0.5]); grid on
```

2.B-7 [A]The plot generated in 2.B-6 should show that if $f(t)$ crosses zero with a negative slope at $t = t_0$, then the positive local peak values of $\hat{f}(t)$ are located at $t = t_0$. Let us clarify this by graphically analyzing $\hat{f}(t) = f(t) * \frac{1}{\pi t}$.

Note that the convolution $a(t)*b(t)$ is equal to the integration of $a(\tau)b(t-\tau)$ over the entire range of the τ axis, where $b(t-\tau)$ is the t-second shifted version of $b(-\tau)$, the left-right reflected signal of $b(\tau)$. From this graphical representation of convolution, first, overlay $1/\pi(t-\tau)$ onto $f(\tau)$ in the plot of 2.B-6, with τ being the horizontal axis. Then, roughly sketch their product $f(\tau) \times 1/\pi(t-\tau)$ and observe the changes as a function of the shift t. Note that $1/\pi(t-\tau)$ has an odd symmetry at t. As this symmetry center moves to and then away from the zero-crossing points of $f(\tau)$, observe the resulting product.

Explain why $\hat{f}(t)$ has a positive local peak at the zero-crossing points of $f(t)$ with a negative slope.

2.B-8. [A]The observations made in 2.B-7 can be extended to assess how the shape of $\hat{f}(t)$ changes around the four typical points of particular interest of $f(t)$: local minima, local maxima, zero-crossing points with a positive slope, and zero-crossing points with a negative slope. Estimate the shape of $\hat{f}(t)$ around these points and confirm your answer in the plot captured in 2.B-6.

2.B-9. [A]Based on the results in 2.B-6 to 2.B-8: (a) What would you expect $\hat{f}(t)$ to be if $f(t)$ is $\cos(\omega_c t)$? (b) Verify your answer using equation (11.2).

2.B-10 Based on the answer to 2.B-9, explain why the Hilbert transformer is called as "$\pi/2$ phase shifter"?

11.3   GENERATION AND SPECTRA OF ANALYTIC AND SINGLE-SIDE BAND MODULATED SIGNALS

3.A Let us verify 1.A in simulation using the sampled waveforms of $f(t)$ and $\hat{f}(t)$ generated in Section 11.2. To this end, import the sampled vectors of these two signals into Simulink to generate the analytic signal and observe its spectrum. First, design an mld/slx file as shown in Fig. 11.2.

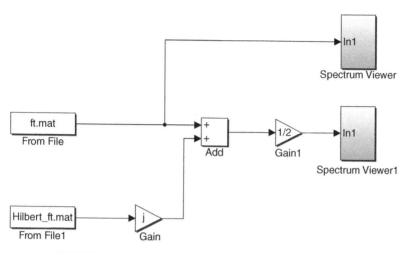

FIGURE 11.2 Simulink design to generate the analytic signal.

Set the parameters of the blocks as follows.

1. **From File, From File1**
   - **File name**: As shown in Fig. 11.2.
2. **Gain, Gain1**: As shown in Fig. 11.2.
3. **Spectrum Viewer, Spectrum Viewer1**: Use the subsystem created in Section 6.B of Chapter 1. It is recommended to add the **Spectrum Viewer** block first and set its parameter as
   - **Signal Specification(sub-block)/Sample time** = 1/4e4.

Then make a copy of **Spectrum Viewer** of the same parameter setting, which will be named **Spectrum Viewer1**.

3.A-1   Complete the expressions of the output of each block in Table 11.1.

3.A-2   Run the simulation for 0.5 seconds (**Simulation stop time** = 0.5) and capture the display windows of **Spectrum Viewer/Spectrum Analyzer** and **Spectrum Viewer1/Spectrum Analyzer**. Before capturing the **Spectrum Analyzer** display windows, be sure to decrease the height of the window to get a width:height ratio of about 7:1 for the graph portion as shown in Fig. 4.4. Also do not **autoscale**.

TABLE 11.1   Expressions of the Output of Each Block in Fig. 11.2.

Block name	Expression
From File	$f(t)$
From File1	
Gain	
Add	$f(t) + j \times ?$
Gain 1	

3.A-3 Does the result in 3.A-2 verify the answer to 1.A? Why?

3.A-4 Determine how the spectrum shape of **Spectrum Viewer 1** changes if the value of **Gain** is set to –j. Properly revise equation (11.3) and reformulate equation (11.4) to answer this question.

3.A-5 Set the value of **Gain** to –j and run the simulation again. Capture the display window of **Spectrum Viewer1/Spectrum Analyzer**.

3.B Now, let us verify 1.E-4 in simulation using the sampled waveforms of $f(t)$ and $\hat{f}(t)$ generated in Section 11.2. We will import the sampled vectors of these two signals into Simulink to generate the single-side band (SSB) modulation signal and observe its spectrum. First, design the following mdl/slx file.

Set the parameters of the blocks as follows.

1. **From File, From File1, Gain** (as shown in Fig. 11.3)
2. **Sine Wave**
   - **Frequency (rad/s)** = 2*pi*(7+The last digit of your student ID number)*1e3
   - **Phase (rad)** = pi/2
3. **Sine Wave1**
   - **Frequency (rad/s)** = 2*pi*(7+The last digit of your student ID number)*1e3
   - **Phase (rad)** = 0

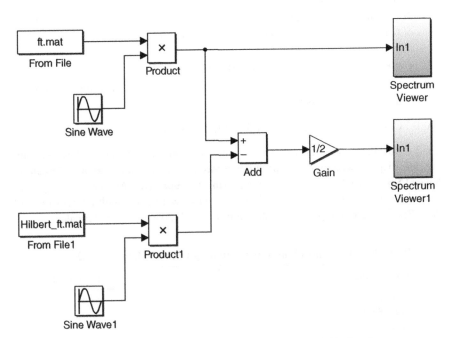

FIGURE 11.3 Simulink design for generating an SSB signal.

4. **Add**
   - **List of signs** $= +-$ (with no space)
5. **Spectrum Viewer, Spectrum Viewer1** (Copy and paste the ones used in 3.A.)
   - **Signal Specification (sub-block)/Sample time** $= 1/4e4$.

3.B-1   Complete the expressions that correspond to the output of each of the eight blocks in Table 11.2.

TABLE 11.2   Expressions of the Output of Each Block in Fig. 11.3.

Block name	Expression
From File	$f(t)$
From File1	
Sine Wave	$\cos(2\pi \times (7 + x) \times 1000t)$ where $x =$ the last digit of your student ID number.
Sine Wave1	
Product	
Product1	
Add	
Gain	

3.B-2   Run the simulation for 0.5 seconds (**Simulation stop time** $= 0.5$) and capture the display windows of **Spectrum Viewer/Spectrum Analyzer** and **Spectrum Viewer1/Spectrum Analyzer**.

3.B-3   Verify the answer to 1.E-5 using the simulation results in 3.B-2. Are the two results consistent with each other?

3.B-4   Predict how the spectrum shape of **Spectrum Viewer1** will change if **List of signs** of the **Add** block is set to **++** (with no space in-between the two + signs). Sketch your predicted spectrum.

3.B-5   Set the **List of signs** of the Add block to **++** and run the simulation. (a) Capture the display window of **Spectrum Viewer1/Spectrum Analyzer**. (b) Does the result verify the answer to 1.E-4?

11.4   IMPLEMENTATION OF AN SSB MODULATION AND DEMODULATION SYSTEM USING A BAND PASS FILTER

Fig. 11.4(a) shows the block diagram of an SSB modulation system using a band pass filter (BPF). The signal $\phi_{DSB-SC}(t)$ is the passband DSB-SC signal obtained by multiplying $f(t)$ with $\cos(\omega_c t)$, and $\phi_{USSB}(t)$ is the upper single-side band (USSB) signal generated by passing $\phi_{DSB-SC}(t)$ through a BPF. The operation is best explained in the

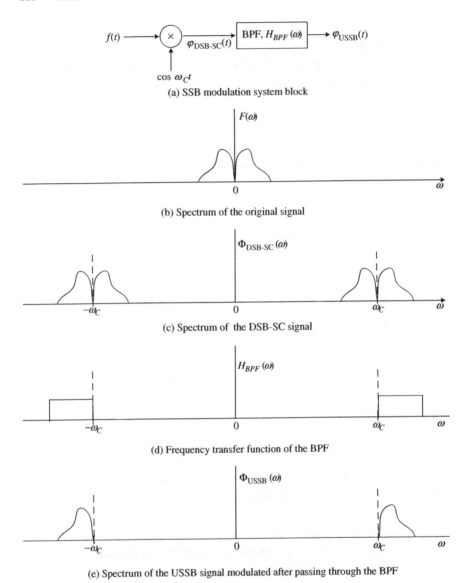

(a) SSB modulation system block

(b) Spectrum of the original signal

(c) Spectrum of the DSB-SC signal

(d) Frequency transfer function of the BPF

(e) Spectrum of the USSB signal modulated after passing through the BPF

FIGURE 11.4   USSB modulator using a BPF and the spectra at each stage.

frequency domain. An example of the spectrum of $f(t)$ is shown in Fig. 11.4(b). The spectrum of the corresponding DSB-SC signal $\phi_{DSB-SC}(t)$ is shown in Fig. 11.4(c). Let $H_{BPF}(\omega)$ denote the frequency response of the BPF. With the ideal BPF $H_{BPF}(\omega)$ as shown in Fig. 11.4(d), the passband lower edge, which corresponds to the **Lower passband edge frequency** of the **Analog Filter Design** block, equals the carrier frequency, and the bandwidth of $H_{BPF}(\omega)$ is set to be greater than the bandwidth of $f(t)$.

The USSB modulation signal as shown in Fig. 11.4(e) is generated by passing $\phi_{\mathrm{DSB-SC}}(t)$ through $H_{\mathrm{BPF}}(\omega)$.

Next, we design a system that generates SSB signals at the output of the **From File** block using a BPF.

4.A   Design an mdl/slx as shown in Fig. 11.5.

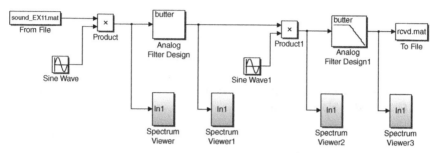

FIGURE 11.5   Simulink design for SSB signal generation with a BPF.

Set the parameters of the blocks as follows. The parameters of any blocks not mentioned below do not need to be set at this stage; they will be set later.

1. **From File**
   - **File name = sound_CH11.mat**
2. **To File** :
   - **Save format =Array** (not required for old versions)
   - **File name = rcvd.mat**
   - **Sample time** = 1/8192
3. **Sine Wave**
   - **Frequency (rad/s)** = 2*pi*7e3
   - **Phase (rad)** = pi/2
4. **Analog Filter Design**
   - **Filter type = Bandpass**
   - **Filter order** = 32
   - **Lower passband edge frequency (rad/s)** = 2*pi*7e3
   - **Upper passband edge frequency (rad/s)** = 2*pi*11e3
5. **Spectrum Viewer, Spectrum Viewer1, Spectrum Viewer2, Spectrum Viewer3**:
   Copy and paste the ones used in 3.A.
   - Be sure to set **Signal Specification(sub-block)/Sample time** = 1/4e4.

4.A-1   The signal $f(t)$ in Fig. 11.4(a) corresponds to the output of the **From** file block in the mdl/slx file. The outputs of which blocks correspond to the DSB-SC signal $\phi_{\mathrm{DSB-SC}}(t)$, the carrier wave $\cos(\omega_c t)$, and the USSB signal $\phi_{\mathrm{USSB}}(t)$, respectively, in Fig. 11.4(a)? Complete Table 11.3.

TABLE 11.3   Corresponding Block's Output to the Signal in Fig. 11.4(a).

Signal in Fig. 11.4(a)	Corresponding block's output
Information signal $f(t)$	**From File**
DSB-SC signal $\phi_{DSB-SC}(t)$,	
Carrier wave $\cos(\omega_c t)$,	
USSB signal $\phi_{USSB}(t)$	

4.A-2   Answer the following questions:

1. Determine the carrier frequency on the basis of the parameter setting in the mdl/slx file.
2. What is the reason to set **Lower passband edge frequency (rad/s)** of the **Analog Filter Design** block to 2*pi*7e3?

4.A-3   Run the simulation for 10 seconds (**Simulation stop time** = 10). Upon completing the simulation, capture the two display windows of **Spectrum Viewer/Spectrum Analyzer** and **Spectrum Viewer1/Spectrum Analyzer**.

4.A-4   (a) From the result in 4.A-3, measure the bandwidths of $\phi_{DSB-SC}(t)$ and $\phi_{USSB}(t)$. For consistency, here let us define bandwidth as the distance in frequency between the two points where the spectrum crosses the x axis from the positive frequency range. (b) Is the USSB modulation done correctly?

4.B   In this problem, we will demodulate $\phi_{USSB}(t)$.

4.B-1   [T]From 1.E-5, the USSB modulation signal $\phi_{USSB}(t)$ of $f(t)$ is expressed as $\frac{1}{2}(f(t)\cos\omega_c t - \hat{f}(t)\sin\omega_c t)$. Substitute it into the left-hand side of the following equation:

$$LPF_{B_f}[\phi_{USSB}(t)\cos(\omega_c t)] = \frac{1}{4}f(t), \qquad (11.8)$$

where B_f is the bandwidth of $f(t)$.

NOTE: The notation $LPF_D[x(t)]$ denotes the output of an ideal (distortionless with zero-delay) LPF with bandwidth D for input $x(t)$. This notation will be used at many other places in this book.

4.B-2   Generating a USSB demodulation is equivalent to implementing $LPF_{B_f}[\phi_{USSB}(t)\cos(\omega_c t)]$. The output of the **Analog Filter Design** block on the left-hand side of the mdl/slx file in 4.A corresponds to $\phi_{USSB}(t)$, and the three blocks on the right-hand side, **Sine Wave1, Product1,** and **Analog Filter Design1,** are required to create $LPF_{B_f}[\phi_{USSB}(t)\cos(\omega_c t)]$. Assuming that the parameters of these three blocks are set correctly (will set these in the next problem), determine the expressions of each block's output in Table 11.4.

TABLE 11.4   Expressions of the Output of Each Block in Fig. 11.5.

Block in Fig. 11.5	Expression of the output
Analog Filter Design	$\phi_{\text{USSB}}(t)$
Sine Wave1	
Product1	
Analog Filter Design	

4.B-3 Properly set the parameters of **Sine Wave1** and **Analog Filter Design1** to generate $\text{LPF}_{B_f}[\phi_{\text{USSB}}(t)\cos(\omega_c t)]$. Capture the parameter setting windows of each block.

4.B-4 Run the simulation for 10 seconds again and capture the two display windows of **Spectrum Viewer2/Spectrum Analyzer** and **Spectrum Viewer3/Spectrum Analyzer**.

4.B-5 Spectrum Viewer2 should show three blocks of spectra. Mathematically explain this.

4.B-6 Explain whether or not the captured spectrum of **Spectrum Viewer3** is what you expect to see.

4.B-7 Execute the following in the command window to play the demodulated signal. Does it sound the same as the original signal?

```
>>clear; load rcvd.mat; soundsc(ans(2,:))
```

4.C USSB modulation and demodulation are implemented in 4.A and 4.B. The parameter settings of this implementation can be changed to perform L(Lower)SSB modulation and demodulation. For this, set the parameters of **Analog Filter Design** (the one on the left) as

- **Lower passband edge frequency (rad/s)** = 2*pi*3e3
- **Upper passband edge frequency (rad/s)** = 2*pi*7e3

4.C-1 Justify the settings above for LSSB systems.

4.C-2 (a) Determine the parameter(s) of the block(s) on the right-hand side of the mdl/slx file that need to be changed for the LSSB demodulator, and (b) justify your answers.

4.C-3 After setting the parameters for the **Analog Filter Design block**, run the simulation for 10 seconds. Capture the display windows of **Spectrum Viewer1/Spectrum Analyzer**, **Spectrum Viewer2/Spectrum Analyzer**, and **Spectrum Viewer3/Spectrum Analyzer**.

4.C-4 (a) Judged from the figures generated in 4.C-3, are the LSSB modulation and demodulation results what you expected? (b) Justify your answer.

4.D [A]Examine how the frequency error, that is, the difference between the carrier frequency generated in the modulator and the local carrier frequency generated in the demodulator, influences the demodulated signal in the SSB system.

4.D-1 [WWW]Download **sound_billy.mat** from the companion website to your MATLAB work folder. Then execute the following in the command window. Does **data(2,:)** saved in **sound_billy.mat** sound like a voice signal?

```
>>clear; load sound_billy.mat; soundsc(data(2,:),11025)
```

4.D-2 Let us perform SSB modulation and demodulation on a voice signal. Change the variable **File name** of the **From file** block in the mdl/slx file in 4.B (or 4.C) to **sound_billy.mat**. Run the simulation for 10 seconds. After completing the simulation, execute the following command and assess whether the demodulated signal sounds the same as the original voice signal.

```
>>clear; load rcvd.mat; soundsc(ans(2,:))
```

4.D-3 Now we consider the case with frequency errors. Change the parameter **Frequency (rad/s)** of the **Sine Wave1** block to **2*pi*(7e3+150)** to introduce a frequency error of 150 Hz. Run the simulation for 10 seconds. After the simulation is completed, play the demodulated signal and describe the difference between the sounds of the demodulated signal and the original. Does the demodulated signal have a higher or lower tone than the original signal?

4.D-4 Change the variable **Frequency (rad/s)** of the **Sine Wave1** block to 2*pi*(7e3-150) to change the frequency error to −150 Hz. Run the simulation for 10 seconds and then play the demodulated signal. Describe the difference between the sounds of the demodulated signal and the original signal. Does the demodulated signal have a higher or lower tone than the original signal?

4.D-5 There should be a noticeable difference between the sounds of the demodulated signals in 4.D-3 and 4.D-4. (a) Assume that the original voice signal has the spectrum shape in 1.D-2 for simplicity. Sketch the spectra of the demodulated signal with a positive as well as a negative frequency offset. (b) Explain how and why the sound of the demodulated signal changes if the frequency error is positive or negative.

REFERENCES

[1] R. Bracewell, *The Fourier Transform and Its Applications*, 2nd ed., New York: McGraw-Hill, 1986.

[2] J. Duoandikoetxea, *Fourier Analysis*, Providence, RI: American Mathematical Society, 2000.

12

VOLTAGE-CONTROLLED OSCILLATOR AND FREQUENCY MODULATION

- Analyze the impact of signal clipping in amplitude modulation systems.
- Investigate the operation of voltage-controlled oscillator.
- Design a frequency modulation transmitter and demodulator.

12.1   [A]IMPACT OF SIGNAL CLIPPING IN AMPLITUDE MODULATION SYSTEMS

Like some components in a communications system transmitter, the amplifiers and analog-to-digital converters in the receiver also have their linear operation regions. When the instantaneous input signal exceeds the dynamic operation region of these components, for example, due to slow automatic gain control responses, the input signal will be clipped. In this problem, we study the impact of signal clipping (due to components/processes in the transmitter or in the receiver) in amplitude modulation (AM) systems.

1.A   [WWW]Open the DSB-SC amplitude modulation and demodulation system mdl/slx file designed in Section 1.B of Chapter 9; if you do not have this file, complete Section 1.B of Chapter 9 first. Modify the mld/slx file as shown in Fig. 12.1:

Problem-Based Learning in Communication Systems Using MATLAB and Simulink, First Edition.
Kwonhue Choi and Huaping Liu.
© 2016 The Institute of Electrical and Electronics Engineers, Inc. Published 2016 by John Wiley & Sons, Inc.
Companion website: www.wiley.com/go/choi_problembasedlearning

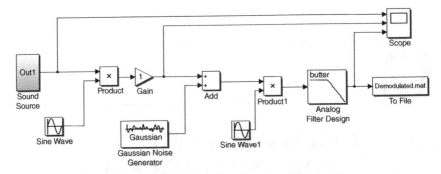

FIGURE 12.1   Simulink design for AM in an additive white Gaussian noise channel.

First, complete the following parameter settings:

1. Set **Frequency** and **Phase** of **Sine Wave** the same as those of **Sine Wave1**.
2. Set **passband edge frequency(rad/s)** of **Analog Filter Design** to **2*pi*4e3**.
3. Set the parameters of **To File** as follows.
    - **File name = demodulated.mat**
    - **Sample time = 1/8192**

Then, make the following changes:

1. Remove all **Spectrum Viewer** blocks to speed up the simulation.
2. Download **sound_CH12.mat** from the companion website to your MATLAB work.
3. Set the parameter **File name** of **Sound Source/From File**(sub_block) = **sound_CH12.mat**.
4. Add the **Gain** block to the output of the transmitter to scale the DSB-SC signal amplitude.
5. Set the parameter **Variance** of the **Gaussian Noise Generator** block to 1e-6 (it was set to 1e-3 in Chapter 9).

1.A-1   Execute the mdl/slx file for 5 seconds (set **simulation stop time=5**). Then execute the following to play the demodulated signal. Can you clearly hear the audio sound?

```
>>clear;load demodulated.mat;soundsc(ans(2,:))
```

1.A-2   Increase the parameter **Variance** of the **Gaussian Noise Generator** block to 1e2. Then execute the mdl/slx file for 5 seconds and play the demodulated signal. Describe the sound quality.

1.A-3   In order to achieve the same sound quality as the scenario where the signal is not amplified and **Variance** of the **Gaussian Noise Generator** block is set to 1e-6, the parameter **Gain** of the **Gain** block should be set to 1e4 (=10000). Justify this setting.

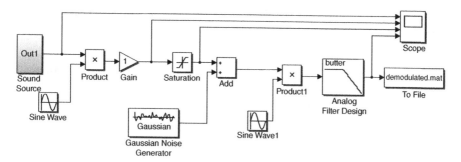

FIGURE 12.2 Simulink design for an AM system in the presence of amplitude clipping.

1.A-4 Set the parameter of the **Gain** block to `1e4`. Run the simulation and play the demodulated signal. Is the original sound quality restored?

1.B From the results in 1.A, the quality of the demodulated signal is not determined by the signal power or noise power alone but by the signal-to-noise power ratio. To maintain the same quality, if the noise power increases, the signal power must be increased proportionally.

For practical applications, however, maintaining a strong received signal to meet a high received signal-to-noise ratio (SNR) might not always be possible. For example, component nonlinearity might cause signal clipping if the input signal is too strong. Also, for wireless systems, a strong received signal is often not possible. This section focuses on studying the impact of signal clipping in AM systems.

Modify the mdl/slx file completed in 1.A as follows.

1. Add the **Saturation** block as shown in Fig. 12.2. Set the two parameters, **Upper limit** and **Lower limit** of the **Saturation** block to 10 and −10, respectively.
2. Increase the input port number of the **Scope** block to 4 and connect the output of the **Saturation** block to the third input as shown in Fig. 12.2.

The **Saturation** block in the mdl/slx file implements the clipping effect in the transmitter, for example, due to power amplifier nonlinearity. If the input is larger than the parameter **Upper limit** (currently set at 10), the output will be equal to **Upper limit**; if the input is smaller than the parameter **Lower limit** (currently set at −10), the output will be equal to **Lower limit**. Otherwise, the output equals the input.

In 1.B-1 to 1.B-5 below, the processes in 1.A-1 to 1.A-4 will be repeated with the revised mdl/slx file, where a **Saturation** block is added.

1.B-1 As in 1.A-1, change the parameter **Gain** back to 1 and the parameter **Variance** of the **Gaussian Noise Generator** block back to 1e-6. Execute the mdl/slx file for 5 seconds. Autoscale all the waveforms in the **Scope** display window and capture the **Scope** display window.

1.B-2 Explain why the input and output of the **Saturation** block are identical.

1.B-3 Play the demodulated signal. Is the audio clearly hearable?

1.B-4   As in 1.A-2, increase the **Variance** of the **Gaussian Noise Generator** block to 1e2 and execute the mdl/slx file again. Play the demodulated signal. Describe the sound quality.

1.B-5   In order to achieve an SNR identical to the case of 1.B-1, set the value of **Gain** to 1e4 as in 1.A-3 and 1.A-4. Run the simulation again. Play the demodulated signal. Is the original sound quality restored as in 1.A-4?

1.B-6   Autoscale all the waveforms in the **Scope** display window and then set the y axis range of the third waveform, that is, the output of the **Saturation** block, in the **Scope** window to [−20 20]. Capture the Scope display window.

1.B-7   The signals before and after the **Saturation** block, that is, the second and third waveforms in the captured windows in 1.B-6, should show that even though the signal is scaled in the transmitter, the transmitted signal (the third waveform) is still smaller than the desired scaled signal (the second waveform) due to clipping by the saturation block. This results in an SNR that is lower than what is desired in the receiver. Thus the received signal (the fourth waveform) is still severely distorted by noise.

   Next, we disconnect the **Gaussian noise generator** block in the mdl/slx file simulated in 1.B-6. The SNR is thus infinity since the noise power is zero. (a) If the demodulated audio signal is played in this case, do you expect it to sound like the original signal? (b) Justify your answer.

1.B-8   In the mdl/slx file simulated in 1.B-6, disconnect the **Gaussian noise generator** block from the **Add** block in order to emulate the ideal case of an infinite SNR. Run the simulation again and play the demodulated signal. (a) Is the sound what you expected to hear as discussed in B-7? (b) If not, what has caused it?

1.B-9   Summarize some of the drawbacks of AM, keeping in mind that many blocks along the communications chain could be nonlinear.

12.2   OPERATION OF THE VOLTAGE-CONTROLLED OSCILLATOR AND ITS USE IN AN FM TRANSMITTER

2.A   This section focuses on voltage-controlled oscillator (VCO) [1–3] operation by using a Simulink design. We apply various inputs to the VCO and determine how the VCO output changes for each of these input signals. Design an mdl/slx file as shown Fig. 12.3.

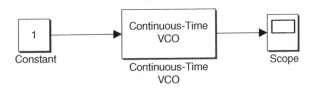

FIGURE 12.3   A VCO test system.

Set the parameters of the **Scope** block via the following steps before proceeding to the next problems. Note that the settings for the **Scope** block might be slightly different for different Simulink versions.

1. Double click the **Scope** block to open the **Scope** display window.
2. Click the icon **Parameters** in the menu bar for the **Scope** display window to open the **Scope** parameter window.
3. Click the **General** tab in the **Scope** parameter window. In the **Sampling** option at the bottom, select **Sample time** and set it to 1e-4.
4. Click the **Logging** (**History** or **Data History** in some old Simulink versions) tab in the **Scope** parameter window. Uncheck the option **Limit data points to last**.

Next, set the parameters of the **Continuous time VCO** block as follows. Hereafter, for simplicity, we will simply call the **Continuous time VCO** block **VCO**.

- **Quiescent frequency** = X + 1, X = Last digit of your student ID number
- **Input sensitivity** = 1

2.A-1   After completing the settings of the **Scope** and the **VCO** blocks, run the simulation for 1 second.

(a) Capture the **Scope** display window.
(b) Measure the frequency of the **VCO** output on the basis of the waveform. Counting the number of cycles per second is one of the ways to measure the frequency. Determine the measured frequency in [rad/s], not [Hz].

2.A-2   Denote x = **Constant value** of the **Constant** block, y = **Quiescent frequency**, and z = **Input sensitivity** of VCO, respectively. Run the simulation for each of the 12 combinations of the three parameters listed in Table 12.1. Count the number of cycles per second to determine the VCO output frequency. Complete the fourth and fifth columns of the table with the measured values. Note that the unit of frequency is rad/s.

TABLE 12.1   VCO Output Frequency According to the Parameters.

x (VCO input)	y (Quiescent frequency)	z (Input sensitivity)	Number of cycles per second of VCO output	ω_i (= VCO output frequency)[rad/s]
1	0	4		
2	0	2		
2	0	4		
4	0	2		
1	1	4		
2	1	2		
2	1	4		
4	1	2		
1	4	4		
2	4	2		
2	4	4		
4	4	2		

2.B   Based on the measured result in 2.A, show that the **VCO** output frequency ω_i in rad/s can be expressed as

$$\omega_i = (2\pi \times y) + (2\pi \times x \times z). \tag{12.1}$$

2.C   [T]Let us determine the relationship between the input information signal and the output (the modulated signal) of the frequency modulator.

2.C-1   The instantaneous frequency denoted by $\omega_i(t)$ is defined as the derivative of the instantaneous phase $\theta(t)$ w.r.t. time t as $\omega_i(t) = d\theta(t)/dt$. From this definition, derive the instantaneous phase $\theta(t)$ from $\omega_i(t)$:

$$\theta(t) = ? \tag{12.2}$$

2.C-2   In frequency modulation (FM), the relationship between the input information signal to be modulated, $f(t)$, and the instantaneous frequency of the output modulated signal, $\omega_i(t)$, is given as

$$\omega_i(t) = \omega_c + k_f f(t), \tag{12.3}$$

where ω_c is the carrier frequency and k_f is FM index.
    First, substitute equation (12.3) into the expression of $\theta(t)$ as a function of $\omega_i(t)$ as derived in 2.C-1. Then, simplify this expression to obtain $\theta(t)$, instantaneous phase of the frequency modulated signal, as a function of the information signal $f(t)$. Finally, substitute $\theta(t)$ into $\phi_{FM}(t) = A\cos(\theta(t))$ to express the FM signal $\phi_{FM}(t)$ as a function of ω_c, k_f, and $f(t)$.

2.D   The VCO can be implemented as a frequency modulator. The fundamental relationships that enable this implementation are the VCO operation equation formulated in 2.B and equation (12.3).

2.D-1   Show that if x, y, and z given below are the **input**, **Quiescent frequency**, and **input sensitivity** of the **VCO** block,

$$x = f(t), \quad y = \frac{\omega_C}{2\pi}, \quad z = \frac{k_f}{2\pi}, \tag{12.4}$$

then the **VCO** output frequency in 2.B is equal to the instantaneous frequency of the FM signal given in equation (12.3).

2.D-2   The result in 2.D-1 shows that if an information signal $f(t)$ is applied as the input of the **VCO** block whose parameters **Quiescent frequency** and **input sensitivity** are set properly, then the **VCO** output frequency will be equal to the right-hand side of equation (12.3). This implies that the **VCO** operation is mathematically equivalent to FM, which allows us to employ the VCO as a frequency modulator.

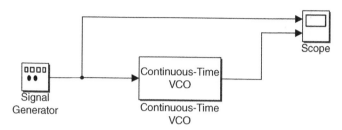

FIGURE 12.4   **VCO** test system II.

Consider an example with a goal to have the **VCO** generate an FM signal with a carrier frequency of $\omega_c = 3e4$ rad/s and a modulation index of $k_f = 50$ from its input $f(t)$. Determine the proper values for the parameters **Quiescent frequency** and **Input sensitivity** of the **VCO** block.

2.E   Replace the **Constant** block with the **Signal Generator** block as shown in Fig. 12.4.

Set the parameters of each block as follows.

1. **Signal Generator**
   - **Waveform = sawtooth**
   - **Amplitude** = 1
   - **Frequency** = 2
   - **Units = Hertz**

2. **VCO**
   - **Quiescent frequency** = (80 + Last digit of your student ID number)
   - **Input sensitivity** = 40

3. **Scope**: Copy the **Scope** block from the mdl/slx file completed in 2.A and inherit all parameters of the block except the number of input ports, which should be increased to 2 as shown Fig. 12.4. Connect the output of **Signal Generator** to the first input port.

2.E-1   Execute the mdl file for 2 seconds (**Simulation stop time** = 2). Capture the **Scope** display window.

2.E-2   At around $t = 0.4$ seconds, zoom into the range of a couple of periods of the **VCO** output along the x axis. (a) At $t = 0.4$ seconds, measure the **VCO** input voltage and the **VCO** output frequency. (b) In the VCO operation equation completed in 2.B, replace the variables x, y, and z with the measured **VCO** input voltage, the current **Quiescent frequency**, and the **Input sensitivity** of VCO, respectively. What is the theoretical **VCO** output frequency at $t = 0.4$? (c) Does the measured **VCO** output frequency match the theoretical value? Ignore small measurement errors.

2.E-3   Repeat the simulation for **Input sensitivity** = 20 and 80. (a) Capture the **Scope** display window for each case. (b) Describe how the VCO output waveform changes in response to the change of **Input sensitivity**. (c) In terms of how the parameter **Input**

sensitivity affects the VCO output, is the result you observed in simulation what you expected to see? Justify your answer.

12.3   IMPLEMENTATION OF NARROWBAND FM

3.A   [T]The FM signal is expressed as

$$\phi_{FM}(t) = A \cos\left(\omega_c t + k_f \int_0^t f(\tau)d\tau\right),$$ (12.5)

where $f(t)$ is the information signal, ω_c is the carrier frequency, and k_f is the modulation index.

3.A-1   We can rewrite equation (12.5) as $\phi_{FM}(t) = X \cos(\omega_c t) + Y \sin(\omega_c t)$ using the trigonometric identity. Find the expressions for the quantities X and Y.

3.A-2   According to Carson's rule [4], if the magnitude of $k_f \int_0^t f(\tau)d\tau$ (defined as β in the related textbooks) in equation (12.5) is kept in a small range, say much smaller than 1, then $\phi_{FM}(t)$ is called a "narrowband FM" (NBFM) signal.

Show that if $\phi_{FM}(t)$ in equation (12.5) is an NBFM signal, that is, the magnitude of $k_f \int_0^t f(\tau)d\tau$ is much smaller than 1, then equation (12.5) can be approximated as

$$\phi_{NBFM}(t) \approx A \cos(\omega_c t) + k_f \int_0^t f(\tau)d\tau \times (-A \sin(\omega_c t))$$

$$= A \cos(\omega_c t) - Ak_f \int_0^t f(\tau)d\tau \times \sin(\omega_c t).$$ (12.6)

3.A-3   Prove the following relationship:

$$\frac{d}{dt}LPF_{B_f}[\phi_{NBFM}(t) \times (-\sin(\omega_c t))] \approx \frac{Ak_f}{2}f(t),$$ (12.7)

where B_f denotes the bandwidth of $f(t)$. The notation $LPF_D[x(t)]$ denotes that the output of an ideal LPF (distortionless and zero-delay) with bandwidth is D for input $x(t)$. The carrier frequency ω_c is assumed to be greater than the signal bandwidth B_f as always.

3.A-4   The NBFM signal $\phi_{NBFM}(t)$ can be generated following equation (12.6) without using a VCO. The Simulink blocks that perform the differentiation and integration are **Difference** and **Integrator**, respectively.

Identify all the required Simulink blocks in order to generate $\phi_{NBFM}(t)$ on the basis of equation (12.7).

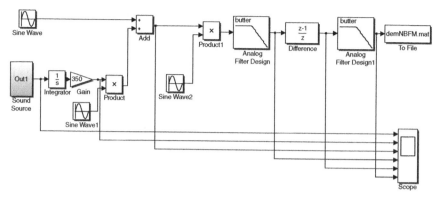

FIGURE 12.5 Simulink design for NBFM.

3.A-5 Identify the four required Simulink blocks to demodulate $\phi_{NBFM}(t)$ according to the left-hand side of equation (12.7).

3.B Design the NBFM signal generation and demodulation system as shown in Fig. 12.5, where the information signal $f(t)$ is the output of the **Sound Source** block in the mdl/slx designed in Section 12.1. The output of the **Add** block corresponds to $\phi_{NBFM}(t)$, and the output of the **Difference** (differentiator) block corresponds to the demodulated signal.

Set the parameters of each block as follows.

1. **Sound Source**: Copy the block from the mdl/slx file designed in Section 12.1. The parameter File name of the From file subblock is set as:
   - **From file**(subblock)/**File name = sound_CH12.mat**
2. **Gain**
   - **Gain** $= 350$
3. **Sine Wave**
   - **Frequency (rad/s)** $= 2{*}\text{pi}{*}10e3$
   - **Phase (rad)** $= \text{pi}/2$
4. **Sine Wave1, Sine Wave2**
   - **Amplitude** $= -1$
   - **Frequency (rad/s)** $= 2{*}\text{pi}{*}10e3$
   - **Phase (rad)** $=0$
5. **Analog Filter Design, Analog Filter Design1**
   - **Pass band edge frequency (rad/s)** $= 2{*}\text{pi}{*}4e3$
6. **To File**
   - **Save format = Array** (not required for old versions)
   - **File name = demNBFM.mat**
   - **Sample time** $= 1/8192$.

7. **Scope**: Add a new scope from the Simulink library with the default parameter setting of **Limit data points to last** = 5000, but increase the number of input ports to 6 (**Number of ports** = 6) as shown in Fig. 12.5.

3.B-1   The output of the **Add** block corresponds to $\phi_{\mathrm{NBFM}}(t)$ in equation (12.6). Complete the output expressions of each block of the transmitter listed in Table 12.2.

TABLE 12.2   Expressions of the Output of Each Block in the Transmitter Part of Fig. 12.5.

Block name	Output expression
Sound Source	$f(t)$
Integrator	$\int_0^t f(\tau)d\tau$
Gain	$350 \int_0^t f(\tau)d\tau$
Sine Wave	$\cos(2\pi \times 10000t)$
Sine Wave1	
Product	
Add	

3.B-2   The output of **Difference** (differentiator block) corresponds to the demodulated signal given in equation (12.4). Now complete the output expressions of each block of the receiver listed in Table 12.3. To determine the output of **Analog Filter Design**, rewrite the output of **Product 1** by using trigonometric identities and keep the term that remains after LPF.

3.C   Run the simulation for 0.2 seconds.

3.C-1   Autoscale all the waveforms in the **Scope** display window and then zoom into the range of [0.195~0.2] along the x axis and capture the window.

3.C-2   (a) In the captured window in 3.C-1, the first waveform is the information signal $f(t)$ and the third waveform is the NBFM signal $\phi_{\mathrm{NBFM}}(t)$. Summarize how the amplitude of the NBFM signal $\phi_{\mathrm{NBFM}}(t)$ varies in response to the change of $f(t)$?

(b) The term $-Ak_f \int_0^t f(\tau)d\tau \times \sin(\omega_c t)$ in the NBFM signal $\phi_{\mathrm{NBFM}}(t)$ given in equation (12.6) is not a constant amplitude sinusoid because the scaling factor

TABLE 12.3   Expressions of the Output of Each Block in the Receiver Part of Fig. 12.5.

Block	Output expression
Sine Wave2	$-\sin(2\pi \times 10000t)$
Product 1	$[\cos(2\pi \times 10000t) - 350 \int_0^t f(\tau)d\tau \times \sin(2\pi \times 10000t)] \times [-\sin(2\pi \times 10000t)]$
Analog Filter Design	
Difference	

$-Ak_f \int_0^t f(\tau)d\tau$ is time-varying. Explain why the amplitude of $\phi_{\mathrm{NBFM}}(t)$ is nearly constant despite the non–constant amplitude sinusoidal term $-Ak_f \int_0^t f(\tau)d\tau \times \sin(\omega_c t)$. The properties of NBFM signals might be a good starting point to explain this observation made.

3.C-3   The constant amplitude of the NBFM signal $\phi_{\mathrm{NBFM}}(t)$ as observed in 3.C-2 is consistent with the properties of FM signals; for FM, the signal's frequency, rather than its amplitude, changes in response to the information signal.

From the waveforms captured in 3.C-1, is the change in the frequency of the NBFM signal $\phi_{\mathrm{NBFM}}(t)$ according to $f(t)$ noticeable to the naked eyes?

3.C-4   Autoscale all the waveforms in the **Scope** display window. (a) Capture the display window. (b) Is the information signal $f(t)$ demodulated properly?

3.C-5   Because of numerical calculation employed by the **Difference** (differentiator) block, there are some distortions in the output of the **Difference** block. Thus an LPF (**Analog Filter Design1**) is added in the final stage of the demodulator to remove this distortion. Does the LPF output have the same shape as that of the information signal $f(t)$?

3.D   Run the simulation for 5 seconds. Complete the quantity marked by '**??**' in the following commands and execute them to play the demodulated signal. Describe the sound you hear.

```
>>clear;load   ??;soundsc(ans(2,100:40961)) % Note that it is not ans(2,:)
as usual !
```

3.E   In this subsection, we perform FM modulation and demodulation in the presence of amplitude clipping as done in Section 12.1. Insert a **Saturation** block in the mdl/slx file as shown in Fig. 12.6.

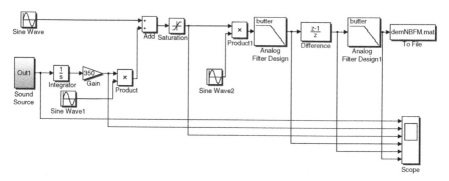

FIGURE 12.6   Simulink design for the FM system.

Set the parameters of the saturation block as follows.

- **Upper limit** $= 0.1$
- **Lower limit** $= -0.1$

3.E-1   Run the simulation for 5 seconds. **Autoscale** all the waveforms in the **Scope** display window and then zoom into the range of [4.995~5] seconds along the x axis. Right-click the **Saturation** block output (the third waveform) and then select the **Axes properties** option and adjust the y axis range to [−0.2 0.2]. (a) Capture the **Scope** display window. (b) Is the transmitted signal (the **Saturation** block output) clipped?

3.E-2   (a) With the **Saturation** block added, is the information signal $f(t)$ demodulated successfully? (b) If yes, explain why.

3.E-3   Play the demodulated signal. Describe the sound you hear.

3.E-4   The double side-band with a large carrier (DSB-LC) AM signal for the information signal $f(t)$ is written as [5]

$$\phi_{\text{DSB-LC}}(t) = A\cos(\omega_c t) + Amf(t)\cos(\omega_c t). \tag{12.8}$$

Equations (12.6) and (12.8) have a similar signal structure: they both consist of two terms. The first term $A\cos(\omega_c t)$ is a sinusoidal term that does not bear the information signal, which is called the "pilot." This term is common to both equations (12.6) and (12.8). The second terms, that is, $-Ak_f \int_0^t f(\tau)d\tau \times \sin(\omega_c t)$ in equation (12.6) and $Amf(t) \times \cos(\omega_c t)$ in equation (12.8), are similar in that they both carry the information signal.

In 3.E-2 and 3.E-3, we have observed that the information signal in equation (12.6) can be demodulated successfully even though the transmitted signal is clipped. (a) Noticing the similar structures of the signals in equations (12.8) and (12.6), discuss whether or not the signal in equation (12.8) (which is a DSB-LC AM signal) can also be demodulated successfully if the transmitted signal is clipped. (b) If not, explain why?

REFERENCES

[1] R. E. Best, *Phase-Locked Loops: Design, Simulation and Applications*, 6th ed., New York: McGraw-Hill, 2007.

[2] W. F. Egan, *Phase-Lock Basics*, Hoboken, NJ: Wiley, 1998.

[3] D. H. Wolaver, *Phase-Locked Loop Circuit Design*, Upper Saddle River, NJ: Prentice Hall, 1991.

[4] J. R. Carson, "Notes on the Theory of Modulation," *Proceedings IRE*, Vol. 10, No. 1, 1922, pp. 57–64.

[5] F. G. Stremler, *Introduction to Communication Systems*, Upper Saddle River, NJ: Prentice Hall, 1990.

13

PHASE-LOCKED LOOP AND SYNCHRONIZATION

- Understand the function and operation of the phase-locked loop (PLL).
- Design a PLL and use it to perform synchronization in the presence of phase and frequency errors.
- Design an FM demodulator using the PLL.
- Implement a near-ultrasonic wireless data transmission system from a mobile phone to a PC.

13.1   PHASE-LOCKED LOOP DESIGN

Design an mdl/slx file for the phase-locked loop (PLL) system [1–5] as shown in Fig. 13.1 and save it as **PLLtest.mdl/slx**.

Set the parameters of each block as follows.

1. **Sine Wave**
   - **Frequency (rad/s)** = 2*pi*93.XXe3 (XX denotes the last two digits of your student ID number.)

2. **Continuous Time VCO**
   - **Quiescent frequency (Hz)** = 93.XXe3 (XX denotes the last two digits of your student ID number.)
   - **Input sensitivity** = 1000

Problem-Based Learning in Communication Systems Using MATLAB and Simulink, First Edition.
Kwonhue Choi and Huaping Liu.
© 2016 The Institute of Electrical and Electronics Engineers, Inc. Published 2016 by John Wiley & Sons, Inc.
Companion website: www.wiley.com/go/choi_problembasedlearning

3. **Analog Filter Design**
   - **Design method = Chebyshev II**
   - **Filter type: Lowpass**
   - **Filter order = 32**
   - **Stopband edge frequency (rad/s) = 2*pi*50e3**
   - **Stopband attenuation in dB = 40**
4. **Gain = 8**
5. **Gain1 = 1e5**

1.A   In Fig. 13.1, the **Sine Wave** block is added to generate a test input to the PLL. All the remaining blocks consist of a PLL.

1.A-1   Denote the output of the Sine Wave block as $PLL_{IN}(t)$. Check the parameter settings of the **Sine Wave** block and complete the following expression:

$$PLL_{IN}(t) = \sin(?). \tag{13.1}$$

1.A-2   Right-click the **Continuous Time VCO** block (hereafter, we call it **VCO** for simplicity) and select Help, which shows how the **VCO** output is mathematically related to its input. Using this information and the current parameter setting as well as the current input of **VCO** ('0' since there is no input connected currently), complete the mathematical expression of the output of **VCO** denoted by $VCO_O(t)$.

$$VCO_O(t) = ?. \tag{13.2}$$

1.B   Phase detector and its operation.
The phase detector (PD) detects the input signal's phase. This is accomplished by multiplying the reference sinusoidal signal with the input and then extracting the low frequency component in the multiplier output. In Fig. 13.1, the **Product** block and **Analog Filter Design** block form a typical type of PD. In this subsection, we investigate the operation of the PD.

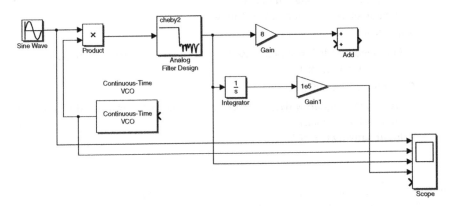

FIGURE 13.1   PLL system under construction.

1.B-1 Based on the answers to 1.A-1, 1.A-2, and the **Stopband edge frequency** setting of **Analog Filter Design**, derive the mathematical expression of the PD output (= output of **Analog Filter Design**). For all derivations in this chapter, assume that **Analog Filter Design** is ideal, that is, distortionless with zero delay.

1.B-2 Denote input phase (=the parameter **Phase** of the **Sine Wave** block) by θ_{IN} and the VCO initial phase (=the parameter **Initial Phase** of the **VCO** block) by θ_{VCO}. They are set to 0 in the current mdl/slx file.

Now, consider the case when θ_{IN} and θ_{VCO} are given some typical values.

(a) Show that the PD output can be expressed as a function of the phase difference as

$$PD_O(t) = 0.5 \sin(\theta_{\mathrm{IN}} - \theta_{\mathrm{VCO}}). \tag{13.3}$$

(b) Show that for a small phase difference, the PD output is approximately proportional to the phase difference.

(c) Is this result consistent with what a PD is supposed to function?

1.B-3 Set the **Simulation (stop) time** to 0.001 seconds. Run the simulation of **PLL-test.slx/mdl** for each of the values of θ_{ERROR} ($\triangleq \theta_{\mathrm{IN}} - \theta_{\mathrm{VCO}}$ = **Phase** of **Sine Wave**– **Initial Phase** of **VCO**) listed in Table 13.1. For each value of θ_{ERROR} listed in the first column of the table, set the pair of parameters $(\theta_{\mathrm{IN}}, \theta_{\mathrm{VCO}})$ to any values so that $\theta_{\mathrm{IN}} - \theta_{\mathrm{VCO}}$ is equal to θ_{ERROR}. Examples are provided for the cases of θ_{ERROR} = -pi and pi/6, where the numbers 3.7 and −4.1 can be replaced by any other arbitrary numbers.

After each simulation, **Autoscale** the output waveform of **Analog Filter Design** in the **Scope** display window and read the steady state output of the PD after

TABLE 13.1   PD Output Test Result.

	Arbitrary setting of $(\theta_{\mathrm{IN}}, \theta_{\mathrm{VCO}})$ such that $\theta_{\mathrm{IN}} - \theta_{\mathrm{VCO}}$ is equal to θ_{ERROR}.		PD output $PD_O(t)$	
θ_{ERROR}	θ_{IN} (= **Phase** of **Sine Wave**)	θ_{VCO} (= **Initial Phase** of VCO)	Experimental (**Analog Filter Design** output)	Theoretical $(PD_O(t) = 0.5 \sin(\theta))$
-pi	3.7 (example)	3.7 + pi (example)		
-2*pi/3				
-pi/2				
-pi/3				
0				
pi/6	−4.1 + pi/6 (example)	−4.1 (example)		
pi/4				
2*pi/3				
3*pi/4				

convergence. If the waveform fluctuates slightly around a constant, treat that constant as the converged value.

(a) Complete the following table with the simulated and theoretical values of the PD output.

(b) Are the simulated values consistent with the theoretical ones?

1.C   Closed-loop connection in PLL and operation with phase error.
In this subsection we make a closed feedback loop connection and investigate how PLL operates when there exists a constant phase error.

1.C-1   (a) We introduce a phase error θ_{ERROR} ($\triangleq \theta_{\text{IN}} - \theta_{\text{VCO}}$) that equals $\pi/4$ as an example. First, in the design **PLLtest.slx/mdl**, set θ_{IN} (= **Phase (rad)** of **Sine Wave**) and θ_{VCO} (= **Initial Phase** of **VCO**) to the values you set for $\theta_{\text{ERROR}} = \pi/4$ in Table 13.1.

(b) Connect the output of the **Add** block to both of the input of **VCO** and the open input (the fifth input) of the **Scope** block as shown in Fig. 13.2. Does the revised system now form a closed loop?

(c) Run the simulation with the revised **PLLtest.slx/mdl**. **Autoscale** the PD output waveform in the **Scope** display window and capture it.

1.C-2   In 1.C-1 (c), the PD output should converge to 0 irrespective of the phase settings of the PLL input and VCO output. The PD output approaching 0 means that the phase error between the PLL input and the VCO output phase approaches 0.

The closed-loop structure of the PLL exploits the feedback mechanism. The subblocks are cascade connected in a loop and work jointly so that the PD output, that is, the phase error, converges to 0.

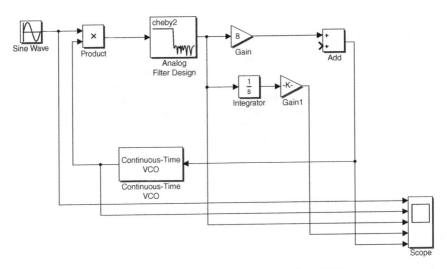

FIGURE 13.2   Closed-loop connection in the PLL.

Next we investigate how the output of each block influences the output of its next block in the loop. This allows us to clearly see how the PD output converges to 0, that is, how the VCO output phase tracks the phase of the PLL input.

(a) Substitute the values of θ_{IN} and θ_{VCO} set in 1.C-1(a) into equation (13.3) and calculate $PD_O(t = 0)$. Also determine the polarity of the PD output.

(b) Note that the PD output is scaled by **Gain** (= 8) and then fed into the **VCO** input in the design. Given the polarity of the PD output obtained in (a), is VCO input positive or negative?

(c) Based on the **VCO** properties, should the instantaneous frequency of the **VCO** output increase or decrease given the polarity of the **VCO** input obtained in (b)?

(d) From the relationship between the instantaneous frequency and phase, will the **VCO** output phase increase or decrease given the answer in (c)? Note that here we are not referring to the initial setting of θ_{VCO} but the internally updated θ_{VCO} via feedback.

(e) Assume that the increment or decrement of the updated θ_{VCO} is smaller than the initial phase gap, which equals the difference between θ_{IN} and initial setting of θ_{VCO}. From the answer in (d), does the PD output get closer to 0 than its initial value, which equals PDo ($t = 0$) as calculated in (a)? Justify your answer by substituting the updated θ_{VCO} into equation (13.3), even though the exact value of the updated θ_{VCO} is unknown.

(f) As the feedback process described in (b)–(e) continues, will the PD output successively get closer to 0?

(g) Imagine a special case that the increment or decrement of the updated θ_{VCO} is greater than the current phase gap. This implies that VCO phase will be overcompensated, causing the phase error polarity to be opposite from the initial phase error polarity. Go through the feedback processes in (b)–(f) for this case and show that eventually the phase error will also converge to 0.

1.C-3   Note that for convenience of explaining the PLL operation, the VCO output is expressed in the form of a cosine function $\cos(\omega t + \theta_{VCO})$, with θ_{VCO} being the VCO output phase, whereas the PLL input is expressed in the form of a sine function $\sin(\omega t + \theta_{IN})$ with θ_{IN} being the PLL input phase. Thus the inputs of the VCO and PLL have a phase difference of $\pi/2$. This choice allows the PD output to be described by equation (13.3). This also leads to the following relationship, which is assumed for other problems in this section:

The phase error of the PD output is expressed as

$$\theta_{IN} - \theta_{VCO} = \text{Actual phase difference between the PLL input and}$$

$$\text{VCO output waveforms} + \pi/2 \tag{13.4}$$

$$= \omega \times (t_{VCO\_out\_peak} - t_{PLL\_in\_peak}) + \pi/2, \tag{13.5}$$

where ω is the PLL input frequency, which equals 2*pi*93.XXe3 with XX being the last two digits of your student ID number, and is currently equal to the VCO quiescent frequency, $t_{VCO\_out\_peak}$ is the time instant of a positive peak of the VCO output, and $t_{PLL\_in\_peak}$ is the time instant of a positive peak of the PLL input near $t_{VCO\_out\_peak}$.

(a) Provide a more detailed explanation of equation (13.4).

(b) Provide a more detailed explanation of equation (13.5).

(c) A zero phase error in the PD output means that the positive peak of the VCO output waveform is a quarter wavelength ahead of the positive peak of the PLL input waveform. Explain this in more detail.

1.C-4   When the PD output converges to 0, which implies no phase error, we call this status "PLL is locked." In this problem we observe how the VCO output waveform changes until PLL has been completely locked.

First in the **Scope** display window, **Autoscale** the **Sine Wave** output waveform and the **VCO** output waveform.

(a) Zoom into the beginning part of the waveforms, for example, approximately the time range of [0 2e-5] seconds. Capture the **Scope** display window.

(b) From the waveforms in the **Scope** display window, measure the phase error using equation (13.5). To improve reading accuracy, sufficiently zoom into the target peak. Denote the measured phase error by ph_err. If **ph_err** is out of the range $[-\pi \ \pi]$, then perform **mod(ph_err+pi, 2*pi)- pi** in the command line. Record the measured phase error.

(c) Is the measured value in (b) approximately equal to $\pi/4$, which is the initial phase error θ_{ERROR} ($\triangleq \theta_{IN} - \theta_{VCO}$) set in 1.C-1(a)?

(d) Again, zoom into the last part of the waveforms, say the time interval of [9.8e-4~1e-3] seconds. Capture the Scope display window.

(e) From the waveforms in the **Scope** display window, measure the phase error using equation (13.5). Sufficiently zoom into the target peak to obtain accurate reading. Record the measured phase error.

(f) Is the measured value in (e) approximately equal to 0?

(g) Is the phase correctly locked?

(h) As the system locks into the input signal phase, has the VCO output phase ($= \phi$) or the PLL input phase ($= \varphi$) changed from its initial value?

1.C-5   Execute **PLLtest.slx/mdl** for each of the following cases: $\theta_{ERROR} = -2\text{*pi/3}$, -pi/2, 0, pi/6 and pi.

(a) For each case, capture the last portion of the **VCO** output and the PLL input waveforms (approximately in the range of [9.8e-4 1e-3]) and check whether or not the PLL is locked.

(b) For each case, determine whether PLL is locked using a method described in 1.C-3(c).

1.C-6   Set the parameter **phase** of the **Sine Wave** block to 3*pi/4 and **Initial phase** of VCO to 0. Execute **PLLtest.slx/mdl** for each of the following values of the **Gain** block (not the **Gain1** block): 2, 8, 16, 32, and 64. After each simulation, **Autoscale** the PD output waveform in the **Scope** display window and observe the waveform. No need to capture the waveforms.

(a) Based on the simulation result for the case of **Gain** = 2, summarize the problems if the gain is set too small? Explain the reason that causes this problem.

(b) Based on the simulation result for the case of **Gain** = 64, summarize the problems if the gain is set too large? Explain the reason that causes this problem.

1.D PLL operation with frequency error.

We introduce a 1-kHz frequency error in the PLL input as an example. Change the parameter **Frequency** of the **Sine Wave** block to (93e3+1e3) Hz (= ? rad/s) and change **Quiescent frequency** of **VCO** to 93e3 Hz.

1.D-1 Execute **PLLtest.slx/mdl** for each of the following values of the **Gain** block (not the **Gain1** block): 2, 8, 16, 32, and 64. After each simulation, **Autoscale** the PD output waveform in the **Scope** display window and observe the waveform. Measure and record the value that the PD output converges to for each value of **Gain**. Sufficiently zoom into the last portion of the PD output to accurately read the converged values.

1.D-2 Try various gain values of the **Gain** block to find a proper value with which the PLL will lock, that is, the PD output converges to 0. Record this value if it is found; if such value does not exist, then provide an explanation.

1.D-3 The current PLL structure cannot lock on to the phase if there exists a frequency error between the PLL input and the **VCO** output. Next we investigate why.

(a) If the PLL successfully locks on to its input phase, then the **VCO** output frequency should be automatically synchronized with the PLL input frequency because the frequency is the derivative of the phase. Consider the design **PLLtest.slx/mdl** completed in 1.D-2 and its current setting of the frequency error between the PLL input frequency and the VCO quiescent frequency. To ensure successful phase lock, how much should the VCO output frequency be increased or decreased?

(b) Determine the VCO input that results in the VCO output frequency change obtained in (a). Note that input sensitivity of VCO is set to 1000 Hz/V. To understand how VCO operates, alternatively, you may use help in the parameter setting window of VCO, or refer to Section 12.2 of Chapter 12.

(c) A successful phase lock means that the phase of the input signal is being tracked closely. Thus the PD output should converge to zero. In the current PLL structure, the PD output is scaled by **Gain** and then connected to the **VCO** input. With this structure, is it possible that the PD output converges to 0 while the **VCO** input is equal to the value obtained in (b)?

(d) From the discussion in (a)–(c), explain why the current PLL structure cannot lock on to the phase if there is a frequency error.

1.D-4 Set the value of **Gain** to 8 and connect the output of the **Gain1** block to the open input of the **Add** block as shown in Fig. 13.3. Then execute **PLLtest.slx/mdl**. Does the PD output converge to 0?

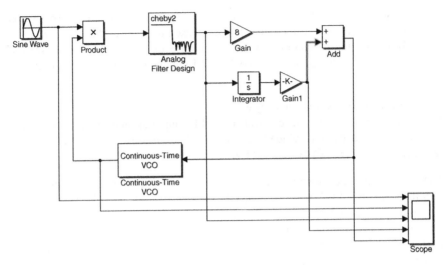

FIGURE 13.3   PLL structure for the case with frequency error.

1.D-5   With the revised structure in 1.D-4, it is now possible that the PD output converges to 0 while the **VCO** input is equal to the value obtained in 1.D-3(b) because of the newly added blocks. Justify this.

1.D-6   Capture the **VCO** input waveform in the **Scope** display window. Does the **VCO** input converge to the value obtained in 1.D-3(b)?

13.2   FM RECEIVER DESIGN USING THE PLL

2.A   Make sure that in the **PLLtest.slx/mdl** file revised in 1.D-4, **Quiescent frequency** of **VCO** is set to 93e3 Hz and then save the file.

In **PLLtest.slx/mdl**, right-click all blocks one by one except **Sine Wave** and **Scope** while holding down the Shift key. This selects all blocks except **Sine Wave** and **Scope**. Create a subsystem with the selected blocks and name it **PLL**. Make sure that the **VCO** input is connected to **Out4** among the output ports of **PLL** as shown in Fig. 13.4, which shows the inside of subsystem **PLL**.

2.A-1   Suppose that the frequency of the PLL input is $(93 + x)$ kHz. Show that if PLL is locked, then the **VCO** input of the **PLL** subsystem will converge to x.

2.A-2   Execute **PLLtest.slx/mdl** for each of the following PLL input frequencies (**Frequency** of the **Sine Wave** block): 89.8, 91.7, 93.5, 95.3, and 97 kHz. (a) After each simulation, accurately measure and record the converged value of the **VCO** input. (b) Are the measured values consistent with the result in 2.A-1?

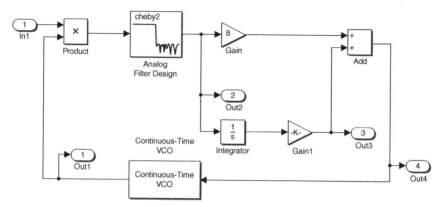

FIGURE 13.4   Making a subsystem PLL.

2.A-3   Extend 2.A-1 to the case of slowly time-varying frequency errors.

(a) Prove that if the PLL input frequency is equal to $\omega_i(t) = \omega_c + k_f f(t)$, where $\omega_c = 2\pi \times 93e3$ rad/s and $f(t)$ is a slowly time-varying signal, then the **VCO** input will be $k_f f(t)/(2\pi \times k_v)$, where k_v is **Input sensitivity [Hz/V]** of **VCO**.

(b) From (a), if the PLL input is an FM signal with a carrier frequency of 93 kHz, then the PLL is equivalent to an FM demodulator. Justify this conclusion.

2.A-4   If **PLL** is used as an FM demodulator, which subblock input inside the subsystem **PLL** is equal to the FM demodulated signal? Ignore the scaling factor in the demodulated signal.

2.B   Execute the following in the command window to create **sound_file.mat**, which will be used as the information signal to be modulated in the next problem. Show the execution result of the last command **ls *.mat**.

```
>> clear
>> load handel
>> tmp=(1:length(y))/Fs;
>> snd(1,:)=tmp;
>> snd(2,:)=y;
>> save sound_file.mat snd
>> ls *.mat
```

2.C   Modify **PLLtest.slx/mdl** as shown in Fig. 13.5 and save it as **PLL_FMdemod .slx/mdl**.

Set the parameters of the newly added blocks as follows.

1. **From File**

   • **File name** = sound_file.mat

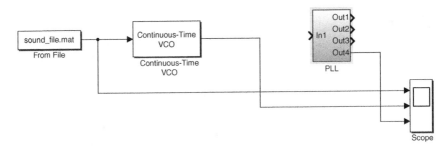

FIGURE 13.5   FM signal generation using a PLL.

2. **Continuous Time VCO**
   - **Quiescent frequency** = 93e3 kHz
   - **Input sensitivity** = 1000

2.C-1   Open the **Scope** display window and click the **parameter** icon in the menu bar to open the **Scope** parameters window. In the **Main** (**General** in some old versions) tab, set **Sample time** to **1/93e4**. Then, in the **Logging** (**History** or **Data History** in some old versions) tab, be sure to change **Limit data points to last** to 50000, not 5000.

Capture the parameter setting windows.

2.C-2   The output of the newly added **VCO** block corresponds to the frequency-modulated signal of the information from the **From File** block. For details, refer to 2.D in Chapter 12.

Recall that the FM signal is expressed as

$$\phi_{FM}(t) = A \cos\left(\omega_c t + k_f \int_0^t f(\tau)d\tau + \theta_0\right). \tag{13.6}$$

In 2.D-2 of Chapter 12, we studied the relationship between the modulation index k_f and the **VCO** input sensitivity k_v.

Based on the parameter setting of the new **VCO** block, determine the carrier frequency ω_c and the modulation index k_f.

2.C-3   Set the **Simulation stop time** to 0.02 seconds and execute **PLL_FMdemod.slx/mdl**. **Autoscale** all three waveforms in the **Scope** display window and then horizontally zoom into the time period of [1e-3 1.8e-3] seconds by using the horizontal zoom-in button in the menu bar. The second (middle) waveform is the FM signal.

(a) Capture the **Scope** display window.
(b) From the FM signal waveform, is there a noticeable variation in the instantaneous frequency? In other words, can the information imbedded in the FM signal be estimated by examining the waveform?

2.C-4   Connect the output of the new **VCO** block, that is, the FM signal, to the input of the **PLL** block and execute **PLL_FMdemod.slx/mdl** again.

(a) **Autoscale** all the waveforms in the **Scope** display window. If the waveform is not properly scaled even after autoscaling, manually zoom in by using the vertical zoom-in button in the menu bar. Repeat this step for the remaining problems if any waveform is not properly autoscaled.

Capture the **Scope** display window.

(b) Are the shapes of the demodulated signal by the PLL (the third waveform) and the information signal (first waveform) similar? Note that the demodulated signal displayed in **Scope** might not look like a clean line since **Gain** and **Gain1** are not optimized. Ignore this in making this comparison. Also ignore signal-scaling factor.

2.D   In this subsection, a frequency error will be introduced in the FM signal.

2.D-1   Change the **Quiescent frequency** of **VCO** of the FM modulator to 94.5 kHz. Recall that the **Quiescent frequency** of **VCO** in the FM demodulator is set to 93 kHz. Calculate the frequency error of the received FM signal?

2.D-2   How is the frequency error introduced in 2.D-1 expected to change the demodulated waveform as compared with the demodulated waveform captured in 2.C-4?

2.D-3   Run the simulation with **PLL_FMdemod.slx/mdl** revised in 2.D-1. **Autoscale** all the waveforms in **Scope** display window and capture it.

2.D-4   Is the captured waveform in 2.D-3 what you expected as discussed in 2.D-2?

2.D-5   Let us denote the information signal from the **From file** block by $f(t)$.

(a) Express the instantaneous frequency of the PLL input (= output of **VCO** in the FM modulator). Note that the parameter **Quiescent frequency** of **VCO** in the FM modulator has been changed to 94.5 kHz.

(b) To derive the demodulated signal (= **VCO** input in **PLL**), repeat the derivation done in 2.A-1 after replacing the PLL input frequency by the answer obtained in (a). Document the derived expression. Is it consistent with the captured waveform in 2.D-3?

2.D-6   Change the **Quiescent frequency** of **VCO** of the FM modulator to 91.8 kHz. Run the simulation with **PLL_FMdemod.slx/mdl**.

(a) **Autoscale** all the waveforms in the **Scope** display window and capture them.

(b) Is the demodulated signal what you expected to see? Justify it.

2.E   [A]In this subsection we investigate whether or not the PLL will still work for clipped input signals or for the nonsinusoidal input signals.

2.E-1   Insert the **saturation** block into the **PLL_FMdemod.slx/mdl** as shown in Fig. 13.6.

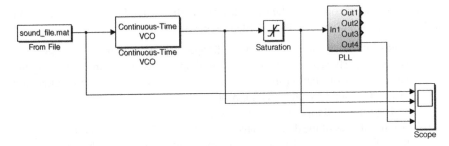

FIGURE 13.6   FM system in the presence of amplitude clipping.

Set the parameters of the blocks as follows:

1. **VCO** in the FM modulator
   - **Output amplitude(V)** = 10
   - **Quiescent frequency (Hz)** = 93e3
2. **Saturation**
   - **Upper limit** = 0.X, **Lower limit** = −0.X (X is the last digit of your student ID number.)

Run the simulation with **PLL_FMdemod.slx/mdl** for 1e-4 seconds (**Simulation stop time** = 1e-4).

  (a) **Autoscale** all the waveforms in **Scope** display window and capture them.

  (b) Compare the FM signal (= **VCO** output, 2nd waveform) with the clipped FM signal (= **Saturation** output, 3rd waveform). Explain why the clipped FM signal has the captured waveform shape.

2.E-2 Set the **simulation stop time** to 0.02 and run the simulation again.

  1. **Autoscale** all the waveforms and capture them.

  2. Do the shapes of the demodulated signal (the fourth waveform) and the information signal (the first waveform) look similar? As discussed in 2.C-4(b), the demodulated signal displayed in **Scope** might not look like a clean line since **Gain** and **Gain1** are not optimized. Ignore this in making this comparison. Also ignore signal scaling factor.

2.E-3 Can the PLL design completed, named **PLL_FMdemod.slx/mdl**, be used to demodulate an FM signal if the FM signal is modulated by a nonsinusoidal periodic signal? Justify your answer.

13.3 [A]DATA TRANSMISSION FROM A MOBILE PHONE TO A PC OVER THE NEAR-ULTRASONIC WIRELESS CHANNEL

In this section we implement near-ultrasonic wireless data transmission from a mobile phone to a PC. In this system, the mobile phone transmits an FM signal in a band near the ultrasonic frequencies wirelessly. A PC receives the FM signal and then samples and demodulates the signal.

3.A   In this subsection, we transmit an FM signal at a frequency near the ultrasonic band from a phone by simply playing it in the phone.

3.A-1   [WWW]Download file **FMmodWordX.wav** (**X** = Last digit of your student ID number) from the companion website to your PC. Then move the downloaded .wav file into the memory or disk of your mobile phone via email or USB cable. Do not accept any file format conversion or encoding if you are asked during the process of transferring the .wav file to your phone. For iPhones, Gmail (Google mail) attachment and outlook attachment are the two ways confirmed working.

   This .wav file is one of the sample FM signals with a carrier frequency in the near-ultrasonic band.

   (a) Identify the size (or playing time) of the .wav file downloaded into your phone.
   (b) Search the literature and determine the minimum frequency of the ultrasonic band.

3.A-2   Play the downloaded file **FMmodWordX.wav** using any media player on the phone. Describe the sound you hear.

3.A-3   [WWW]The m-file below samples the input signal to the internal microphone (mic) of a laptop or desktop PC and then saves the sampled signal into a .mat file. The last line plots the spectrum of the sampled signal.

   Go through the following experimental steps:

Step 1. Place the phone within 5 cm from the internal mic of the laptop or desktop PC where MATLAB is running. In the audio setting in your PC, select the internal mic as the default and disable any sound effects to avoid the distortion and unnecessary filtering during the recording process. Refer to the following substeps to disable the sound effect.

   Step 1-a. In your PC, go to "Control Panel/Sound" and select the "Recording" tab.
   Step 1-b. Select the internal mic (which you need to set as the default mic) and click the button "Properties" at the bottom.
   Step 1-c. Select the "Enhancements" tab. If "Enhancements" does not exist, no need to do the next substep (Step 1-d).
   Step 1-d. Check "Disable all sound effects."

Step 2. Play the downloaded file **FMmodWordX.wav** with maximum volume in the phone.

Step 3. After you start playing the file, immediately (preferably within 1 second after the play starts) execute the m-file below to sample the near-ultrasonic FM signal contained in the file **FMmodWordX.wav**. The m-file takes about 5 seconds to finish executing. Place the phone within 5 cm form the mic until the execution completely finishes.

```
clear
fs=40000;
T=5;

r = audiorecorder(fs, 16, 1);
record(r); pause(T);stop(r); y=getaudiodata(r)';
recorded(1,:)=1/fs*(1:length(y));
recorded(2,:)=y';

save FM_record.mat recorded
pwelch(y,[],[],[],fs);
```

Now let us continue the following steps:

(a) Capture the spectrum plot of the digitized received signal in the displayed **Figure 1** window.

(b) If the preceding steps have been completed successfully, then the spectrum plot should have a clear local spectrum peak located at a frequency in the near-ultrasonic band. As the m-file is being executed, if the environment is noisy, that is, other interfering audio signals are present at the same time, then the spectra of these interfering audio signals will also appear in the spectrum plot. If a local spectrum peak at near-ultrasonic frequencies cannot be clearly identified, then you cannot move forward to the next steps. In this case, repeat the steps above with a shorter distance between the phone and the laptop or desktop PC, a higher volume of the media player, and while the environment is less noisy. Also orient the speaker of the phone toward the mic of the laptop or the desktop PC. Sufficiently zoom into the spectrum peak at near-ultrasonic frequencies to estimate the FM carrier frequency. Record the carrier frequency in [rad/s].

3.B [WWW]Now design an FM demodulator in Simulink as shown in Fig. 13.7, which demodulates the information signal from the digitized received FM signal contained in the file **FM_record.mat**. For PLL, use the subsystem designed in Section 13.2.

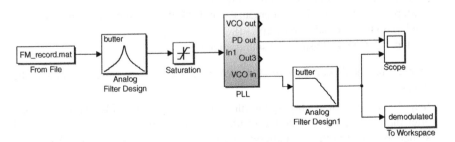

FIGURE 13.7 Demodulation of the information signal from the digitized received FM signal.

Set the parameters of the blocks as follows:

1. **From File**
   - **File name** = **FM_record.mat**
2. **Analog Filter Design** (connected to **From file**)
   - **Filter type** = **Bandpass**
   - **Lower passband edge frequency (rad/s)** = Answer to 3.A-3(b) − 2*pi*1e3
   - **Upper passband edge frequency (rad/s)** = Answer to 3.A-3(b) + 2*pi*1e3
3. **Analog Filter Design1** (connected to **PLL**)
   - **Filter type** = **Lowpass**
   - **Passband edge frequency (rad/s)** = 2*pi*4e3
4. **To workspace** (Be sure that it is not **To file** as before.)
   - **Variable name** = demodulated
   - **Sample time** = 1/(8192/3)
   - **Save format** = Array
5. **Scope**: Open the **Scope** display window and click the **parameter** icon in the menu bar to open the **Scope** parameters window. In **Logging** (**Data History** in some older versions) tab, unselect the option **Limit data points to last**.

Set the parameters of the subblocks inside PLL as follows.

1. **PLL/VCO**
   - **Quiescent frequency (Hz)** = Answer to 3.A-3(b)
   - **Input sensitivity (Hz/V)** = 1000
2. **PLL/Analog Filter Design**
   - **Stopband edge frequency (rad/s)** = 2*pi*4e3.
3. **PLL/Gain**
   - **Gain** = 0.25.
4. **PLL/Gain1**
   - **Gain** = 50.

3.B-1   Answer the following questions:

   (a) Based on the parameter setting of **Analog Filter Design** (connected to **From file**), explain the role of **Analog Filter Design**.
   (b) Explain why the **Sign** block needs to be inserted between **Analog Filter Design** and **PLL**.
   (c) Explain why **Quiescent frequency (Hz)** of **VCO** should be set to the value obtained in 3.A-3(b).

3.B-2   Set simulation stop time to 5 seconds and run the simulation.

   (a) After simulation is completed, **Autoscale** the two waveforms in the **Scope** display window. Capture the **Scope** display window.

(b) Using the DC-bias value in the demodulated signal (the second waveform) and the VCO's input sensitivity, calculate the difference between the visually measured carrier frequency (answer to 3.A-3(b)) and the exact carrier frequency. Record your calculated frequency error.

3.B-3   Execute the following command. If the demodulation is successful, you will hear a certain word. Record the word you hear.

```
>>soundsc(demodulated)
```

3.B-4   [WWW]Demodulate the other sample file, **FMmodWordX.wav** with different values of **X** from your ID, which are also made available for download from the companion website. Each sample file has a different carrier frequency. Hence be sure to properly change the related parameters in the mdl/slx file of Fig. 13.7 for each sample file.

(a) Record the words you hear from the demodulated FM wav files.
(b) Obtain the carrier frequency of each file and list them in a table along with the answer to (a).

3.B-5   [A]Identify the .wav file with which the demodulated signal gives the clearest sound of the word. With this file, repeat the demodulation experiment in a noisy environment. During the steps in 3.A-3, let the laptop or desktop PC independently play an audio sound loudly to act as an interfering signal to the desired FM signal.

(a) Before this experiment begins, do you still expect successful demodulation of the message in the presence of the audio signal interference?
(b) Perform the experiment in a noisy environment as mentioned above. Is the result consistent with what you expected and why?

3.B-6   [A]Repeat the demodulation experiment for different distances between the phone and the laptop or desktop PC. What is the maximum distance that allows you to successfully demodulate the signal and hear the word?

REFERENCES

[1] A. Blanchard, *Phase-Locked Loop*, New York: Wiley, 1976.

[2] R. E. Best, *Phase-Locked Loops: Design, Simulation and Applications*, 6th ed., New York: McGraw-Hill, 2007.

[3] W. F. Egan, *Phase-Lock Basics*, Hoboken, NJ: Wiley, 1998.

[4] D. H. Wolaver, *Phase-Locked Loop Circuit Design*, Upper Saddle River, NJ: Prentice Hall, 1991.

[5] J. R. Carson, "Notes on the Theory of Modulation," *Proceedings IRE*, Vol. 10, No. 1, 1922, pp. 57–64.

14

PROBABILITY AND RANDOM VARIABLES

- Generate uniform and Gaussian random variables (RVs).
- Empirically obtain the probability density function (PDF) of RVs and comparison with the theoretical PDF [1, 2].
- Compare the empirically obtained statistics with the theoretical values.

14.1 EMPIRICAL PROBABILITY DENSITY FUNCTION OF UNIFORM RANDOM VARIABLES

1.A [WWW]The following m-file plots the empirically obtained probability distribution of a fair die.

```
clear
Nsim=6;
count=zeros(1,6);
for n=1:Nsim
   x=ceil(rand*6);
   if x==1
        count(1)=count(1)+1;
```

Problem-Based Learning in Communication Systems Using MATLAB and Simulink, First Edition.
Kwonhue Choi and Huaping Liu.
© 2016 The Institute of Electrical and Electronics Engineers, Inc. Published 2016 by John Wiley & Sons, Inc.
Companion website: www.wiley.com/go/choi_problembasedlearning

```
   elseif x==2
       count(2)=count(2)+1;
    ...
   end
end
Px=count/?;
stem(1:6,Px)
```

1.A-1 Explain how the command **ceil(rand*6)** simulates the occurrence of each face of a die. Use **'help ceil'** or **'help rand'** in the command line if needed.

1.A-2 The m-file above executes **x=ceil(rand*6)** repeatedly for **Nsim** times and then stores the number of the event (occurrence) of **x** being equal to **n** in **count(n)**. Finally, the vector **Px** has the simulated (empirical) probability of each case (face of the die) as its elements. For example, **Px(3)** is the simulated probability of the die face 3. Fill in all the incomplete parts and show the completed m-file.

1.B (a) Execute the m-file above and capture the resulting figure. (b) Is the distribution shown in the graph consistent with the theoretical distribution of a fair die? If not, why?

1.C Let XXX denote the last three digits of your student ID. Run the simulation with the m-file above after the line **'Nsim=6'** is replaced by **'Nsim=(1000+XXX)'**. Repeat this after the line **'Nsim=6'** is replaced by **'Nsim=(10000+XXX)'**.

(a) Capture the empirical probability density function (PDF) plots for both cases.
(b) What are the differences among the three PDF plots? Give reasons that have caused such differences.

14.2 THEORETICAL PDF OF GAUSSIAN RANDOM VARIABLES

2.A [T]Let $f_X(x)$ denote the PDF of a random variable (RV) X. Write the PDF $f_X(x)$ of the Gaussian RV X with mean m_X and variance σ_X^2 [1,2].

2.B [WWW]The m-file below plots $f_X(x)$ of a Gaussian RV X with zero mean and variance 1. The vectors **x** and **Px** denote the sampled version of variable x and the sampled version of the PDF $f_X(x)$, respectively. The vector **x** is set from -5 to 5 with a step size of 0.01, that is, **'x=-5:x_step:5'**. Complete the **Px** generation line on the basis of the $f_X(x)$ in 2.A. Capture the completed line.

```
clear
x_step=0.01;
x=-5:x_step:5;
Px=1/sqrt(?)*exp( -((x-?).^2)/?);
plot(x,Px);
```

2.C   Execute the m-file for each of the following cases:

- mean = 0, variance = 3
- mean = 0, variance = 0.2
- mean = 2, variance = 1
- mean = −3, variance = 1

2.D   Do the shapes of the PDF plots in 2.C reflect the different means and variances of the cases simulated?

14.3   EMPIRICAL PDF OF GAUSSIAN RVs

Using MATLAB functions such as **random()**, **rand()**, **and randn()**, we can generate various kinds of RVs. By conducting a large number of independent experiments, we can obtain the empirical PDF of an RV. Although the built-in function **histogram()** is convenient for generating the empirical distribution, we will go through the detailed steps to obtain the distribution to gain an in-depth understating of the PDF concept.
   The overall procedure is as follows.

- Step 1. Partition the x axis of a PDF plot into a number of the small segments (also called "bins" in a histogram). Create the vector **x** that determines the boundary of the partitions, that is, the first partition is [**x(1) x(2)**], the second partition is [**x(2) (x3)**], and so on.
- Step 2. Tick the center of each bin.
- Step 3-1. Independently generate outcomes for the RV, say, **Nsim** times.
- Step 3-2. In parallel with Step 3-1, for each random generation of an outcome, check which partition the outcome falls into and increase the occurrence counter for the corresponding partition.
- Step 4. After finishing **Nsim** runs, normalize the number of occurrence for each bin by **Nsim**. The normalized count for each bin corresponds to the PDF of the RV sampled at the center of the bin.

3.A   [WWW]The MATLAB function **randn**, every time it is invoked, generates a sample (also called "an outcome") of the Gaussian RV with zero mean and unit variance. The following m-file obtains the empirical PDF of the zero mean, unit variance Gaussian RV by using randn in the procedure described above.

```
clear
rand(1XX); % XX= last two digits of your student ID. This is not relevant to the aim
of the code but mandatory.
Nsim=100;
xstep=0.01;
xmin=-5;
xmax=5;
```

```
x=xmin:xstep:xmax;
Number_of_partitions=(xmax-xmin)/xstep;
PartitionCenters=(xmin+xstep/2):xstep:(xmax-xstep/2);

CountAtEachPartition=zeros(1,Number_of_partitions);

for n=1:Nsim

  random_sample=randn;

  for k=1:Number_of_partitions

    kth_partition_left_end= x(k);
    kth_partition_right_end=x(k+1);

    if (kth_partition_left_end<=random_sample)&(random_sample<kth_partition_
      right_end)
      CountAtEachPartition(k)=CountAtEachPartition(k)+1;
    end
  end

end
Px=CountAtEachPartition/xstep/Nsim;
figure
plot(PartitionCenters,Px,'r')
grid on
```

3.A-1   Identify the parts or lines for each of the four steps and mark them in the m-file (using comments for each line, for example, % Step 2, etc.). Capture the marked m-file.

3.A-2   What does the condition checking line **'if (kth_partition_left_end<=random_sample) & (random_sample<kth_partition_right_end)'** do?

3.B   Execute the m-file in 3.A. Then execute **hold on** in the command window and execute the m-file in 2.B to plot the theoretical PDF together with the empirical PDF. Capture the resulting plot.

3.C   Repeat 3.B for **Nsim** = 1e4 and for **Nsim** = 1e6.

3.D.   The empirical PDF should not look very similar to the theoretical PDF when **Nsim** is small. Why?

14.4   GENERATING GAUSSIAN RVs WITH ANY MEAN AND VARIANCE

4.A   [T]Consider a Gaussian RV X. If RV Y is given as $Y = aX + b$, where a and b are some constants; then Y is another Gaussian RV. Let the mean and variance of X be m_X and σ_X^2, respectively. Show that the mean and variance of Y are $am_X + b$ and $a^2\sigma_X^2$, respectively.

4.B   Based on the derivation above, complete the line below to create **z**, a sample of a Gaussian RV with mean -3 and variance $= 0.01$, using the function **randn**.

```
z= ?*randn+? ;
```

4.C   Repeat 3.B for each of the following two cases. Be sure to set **Nsim = 1e6** and properly change the line '**random_sample=randn;**' in the m-file in 3.A and the mean and variance in the m-file in 2.B.

- mean $= 0$, variance $= 0.2$
- mean $= -3$, variance $= 1$

14.5   VERIFYING THE MEAN AND VARIANCE OF THE RV REPRESENTED BY MATLAB FUNCTION RANDN()

Using '**X=randn(1,N)**', we can create an N-element vector **X**, whose elements are independently drawn observations on the zero-mean, unit-variance Gaussian RV. In 5.A and 5.B, we calculate the empirical mean and variance of the RV using the elements of **X** to verify the theoretical mean and variance of the RV.

5.A   The empirical mean of an RV is defined as

$$m = \frac{\sum_{k=1}^{N} x(k)}{N}, \tag{14.1}$$

where $x(k)$ denotes the kth element of **X**; that is, the kth observation on the RV.

(a) Complete the line to calculate **m**, the sample mean of the RV, in the following lines of code.
(b) Capture the execution result.

```
>>N=1XXX; %XXX= the last three digits of you student ID number.
>>X=randn(1,N);
>>m = sum(?)/? ;
```

5.B   Denote the variance of the RV by v, calculated as

$$v = \frac{\sum_{k=1}^{N} x^2(k)}{N} - m^2. \tag{14.2}$$

Complete the line below to calculate v and capture the result.

```
>> v = ??/N - m^2 ; % m is the value obtained in 5.A
```

5.C   (a) Are the empirically obtained mean and variance of **randn()**, that is, m and v, almost identical to the theoretical values?
   (b) If not, increase the value of **N** to 10 times of its current value and recalculate m and v. Capture the results.

5.D   Empirically calculate the mean and variance of RV, **z**, in 4.B using the method completed in 5.A to 5.C. To this end, generate an **N**-element vector, whose elements are independently drawn observations on **z**. So change the line that creates **z** in 4.B to '**Z=?*randn(1,N)+?**' and then calculate the mean and variance of **Z** according to equations (14.1) and (14.2).

   (a) Capture the empirically calculated mean and variance of **Z**.
   (b) Are they almost identical to the theoretical values, that is, mean $= -3$ and variance $= 0.01$? Small numerical errors should be taken into consideration in making this judgment.

14.6   CALCULATION OF MEAN AND VARIANCE USING NUMERICAL INTEGRATION

In 2.B, we created a vector **Px** that is the sampled version of $f_X(x)$, the PDF of a zero-mean, unit-variance Gaussian RV X. In this section, we calculate the mean and the variance of X using numerical integration. Numerical integration was studied in Section 2.1 of Chapter 2.

6.A   The mean of an RV X is calculated from its PDF $f_X(x)$ as

$$m_X \triangleq E[X] = \int x f_X(x) dx. \tag{14.3}$$

Append the line below, which calculates the mean by numerical integration, to the end of the m-file completed in 2.B. (a) Execute the revised m-file and capture the calculated mean. (b) Is the result approximately equal to the theoretical value?

```
m_x=sum(x.*Px)*x_step
```

6.B   The variance of an RV X is calculated from its PDF $f_X(x)$ as

$$\sigma_X^2 \triangleq E\left[(X - m_X)^2\right] = \int (x - m_X)^2 f_X(x)dx. \qquad (14.4)$$

Complete the line below, which calculates the variance by numerical integration, and append it at the end of the m-file completed in 6.A. (a) Execute the revised m-file and capture the calculated variance. (b) Is the result approximately equal to the theoretical value?

```
v=sum(???.*Px)*x_step
```

6.C   (a) Provide the detailed steps for obtaining the second and third equalities in the following equation:

$$\begin{aligned}
\sigma_X^2 &= \int (x - m_X)^2 f_X(x)dx \\
&= \int x^2 f_X(x)dx - m_X^2 \qquad (14.5) \\
&= E[X^2] - m_X^2.
\end{aligned}$$

(b) Complete the following line, which calculates the second line of equation (14.5), by numerical integration and append it at the end of the m-file completed in 6.B. Execute the revised m-file and capture the calculated variance.

```
sum(???)*x_step-m_x^2;
```

(c) Is the resultequal to the value calculated in another form in 6.B?

6.D   (a) Execute the m-file in 6.C for each pair of the following mean and variance values:

- mean $= 0$, variance $= 3$
- mean $= 0$, variance $= 0.001$
- mean $= -2$, variance $= 1$

(b) Depending on the values of the mean and variance, the calculation errors differ. Justify this observation.

14.7 [A]RAYLEIGH DISTRIBUTION

One of the commonly encountered fading channel [3–5] models in wireless communications is the Rayleigh fading [5] model, which will be employed in Chapters 25, 27, 28, and 29. In the simplest form, the received signal over a Rayleigh fading channel is expressed as

$$r = \alpha s + n, \tag{14.6}$$

where s is the transmitted signal, n the Gaussian noise, and α the fading coefficient modeled as a Rayleigh RV whose PDF is expressed as [6]

$$f_\alpha(x) = \begin{cases} \dfrac{x}{\sigma^2} \exp\left(-\dfrac{x^2}{2\sigma^2}\right), & x \geq 0, \\ 0, & x < 0. \end{cases} \tag{14.7}$$

The mean square value α,$E[\alpha^2]$, is equal to $2\sigma^2$.

7.A Solve the following:

(a) Show that $\int_{-\infty}^{\infty} f_\alpha(x)dx = 1$.

(b) Calculate $E[\alpha^2]$ using $f_\alpha(x)$ and show that it is equal to $2\sigma^2$.

(c) Repeat (a) and (b) using symbolic math.

7.B The Rayleigh RV α is typically generated by using two independent Gaussian RVs as

$$\alpha = \sqrt{z^2 + y^2}, \tag{14.8}$$

where z and y are independent Gaussian RVs with mean 0 and variance σ^2.

(a) For the case of $\sigma^2 = 1$, complete the following command line, which creates **alpha_sample**, a sample of α according to equation (14.8). You may use MATLAB function **randn()**.

```
alpha_sample= ?;
```

(b) Repeat (a) for the case $\sigma^2 = 0.5$.

7.C We plot an empirical PDF of the Rayleigh RV α with $\sigma^2 = 1$ via the method used in Section 14.3. Here we set **xstep=0.1, xmin=0, xmax=5**, and **Nsim=10000**. Properly revise the m-file completed in 3.A to generate α according to equation (14.8) and check to which partition the generated value for α belongs. Also properly revise the part that plots the theoretical PDF given in equation (14.7).

  (a) Capture the revised m-file.
  (b) Capture the resulting figure. If the empirical PDF curve is not smooth, sufficiently increase the value of **Nsim**.

7.D Repeat 7.C for the cases of $\sigma^2 = 0.5$ and 3.

REFERENCES

[1] T. M. Cover and J. A. Thomas, *Elements of Information Theory*, Hoboken, NJ: Wiley-Interscience, 2006.

[2] D. M. Bertsekas and J. N. Tsitsiklis, *Introduction to Probability*, Belmont, MA: Athena Scientific, 2002.

[3] T. S. Rappaport, *Wireless Communications: Principles and practice*, 2nd ed., Upper Saddle River, NJ: Prentice Hall, 2002.

[4] D. Tse and P. Viswanath, *Fundamentals of Wireless Communication*, Cambridge, UK: Cambridge University Press, 2005.

[5] B. Sklar, "Rayleigh Fading Channels in Mobile Digital Communication Systems Part I: Characterization," *IEEE Communications Magazine*, Vol. 35, No. 7, 1997, pp. 90–100.

[6] A. Papoulis and S. Pillai, *Probability, Random Variables and Stochastic Processes*, New York: McGraw-Hill, 2002.

15

RANDOM SIGNALS

- Integration of the Gaussian probability density function and the Q-function.
- The weighted sum of random variables and properties of Gaussian variables.
- The central limit theorem.
- Ensemble average, autocorrelation functions of random processes.
- Statistical properties of additive white Gaussian noise (AWGN).

15.1 INTEGRATION OF GAUSSIAN DISTRIBUTION AND THE Q-FUNCTION

The distribution of a Gaussian random variable X with mean m and variance σ^2 is denoted by $X \sim N(m, \sigma^2)$. The Q-function $Q(k)$ is defined as the probability of the Gaussian random variable X being greater than $m + k\sigma$ [1]. It equals the integration of $f_X(x)$ from $m + k\sigma$ to infinity as

$$Q(k) \overset{\Delta}{=} \Pr\{X \geq (m + k\sigma)\}, \quad \text{where } X \sim N(m, \sigma^2)$$
$$= \int_{m+k\sigma}^{\infty} f_X(x)\mathrm{d}x. \tag{15.1}$$

1.A By replacing $m + k\sigma$ with t in equation (15.1), we can express the probability of Gaussian random variable X being greater than any real number t, that is,

Problem-Based Learning in Communication Systems Using MATLAB and Simulink, First Edition.
Kwonhue Choi and Huaping Liu.
© 2016 The Institute of Electrical and Electronics Engineers, Inc. Published 2016 by John Wiley & Sons, Inc.
Companion website: www.wiley.com/go/choi_problembasedlearning

$\Pr\{X \geq t\}$, by the Q-function as

$$\Pr\{X \geq t\} = Q(?), \quad \text{where } X \sim N(m, \sigma^2). \tag{15.2}$$

Determine the quantity marked by '**?**' in terms of t, m, and σ in equation (15.2).

1.B In recent versions of MATLAB, there is a built-in function **qfunc()** for calculating the Q-function $Q(k)$. The Q-function can also be calculated by using the complementary error function **erfc(x)** as [1,2]

$$Q(x) = \frac{1}{2}\text{erfc}\left(\frac{x}{\sqrt{2}}\right). \tag{15.3}$$

The relationship given by equation (15.3) will be used for problems in this chapter.

The m-file below calculates $Q(x)$ by using **erfc()** and equation (15.3) for each element of the vector **x=0:0.01:6**. It also plots the graph of $Q(x)$ versus x with the y axis shown in the logarithmic scale. Complete the m-file and execute it. Capture the completed m-file and the resulting figure.

```
clear;
x=0:0.01:6;
Qx=??;
semilogy(x, Qx);
```

1.C Use equations (15.2) and (15.3) to complete the following equation:

$$\Pr\{X \geq t\} = 0.5 \times \text{erfc}(?), \quad \text{where } X \sim N(m, \sigma^2). \tag{15.4}$$

1.D Calculate the following probabilities of the Gaussian random variables using equation (15.4) with the MATLAB built-in function **erfc()**:

1.D-1

$$\Pr\{X \geq 1.5\}, \quad \text{where } X \sim N(1, 0.5).$$

1.D-2

$$\Pr\{X \geq 2.5\}, \quad \text{where } X \sim N(2, 1).$$

1.D-3

$$\Pr\{X \geq 0\}, \quad \text{where } X \sim N(1.5, 0.25).$$

1.D-4

$$\Pr\{X \le -1.5\}, \quad \text{where } X \sim N(1, 0.5).$$

Note that the probability density function (PDF) of the Gaussian random variable is symmetric about the mean.

1.E Verify the probability obtained in 1.D-1 with simulation. Use the process discussed in Section 4.B of Chapter 14 to generate samples of the Gaussian random variable $X \sim N(1, 0.5)$.

1.E-1 [WWW]The m-file below simulates $\Pr\{X \ge 1.5\}$, which is represented by the variable **P** in the m-file. This code generates X repeatedly until the number of outcomes when X 1.5 or greater reaches a preset number **Nid**. Complete the m-file and capture it.

```
clear
Nid=1XXX;  %XXX= last three digits of your student ID.
cnt=0;
trials=0;
while cnt < Nid
  X=?*randn+?;
  if X >= 1.5
    cnt=cnt+1;
  end
  trials=trials+1;
end
P=?/?
```

1.E-2

    (a) Execute the m-file above and check the simulated value for $\Pr\{X \ge 1.5\}$ in the command window. (b) Is the result consistent with the one obtained by using the function **erfc()** in 1.D-1? Note that there always exists a simulation error because of the finite number of trials.

1.F Revise the m-file in 1.E-1 to simulate the probabilities in 1.D-2, 1.D-3, and 1.D-4, respectively. Execute the m-file for each case and verify that the simulation results are consistent with the ones obtained by using the function **erfc()**.

15.2 PROPERTIES OF INDEPENDENT RANDOM VARIABLES AND CHARACTERISTICS OF GAUSSIAN VARIABLES

This section investigates important properties of independent random variables and the unique properties of Gaussian variables [3, 4].

2.A [T]Let X_1 and X_2 be two independent RVs and $f_{X_1}(x)$ and $f_{X_2}(x)$ represent, respectively, their PDFs. Let $X_3 = X_1 + X_2$. Using the properties of independent RVs, we can calculate the PDF of $X_3, f_{X_3}(x)$, from $f_{X_1}(x)$ and $f_{X_2}(x)$. Establish this relationship.

2.B [T]Now we consider the more general case of the sum of K independent RVs as $S = \sum_{k=1}^{K} X_k$. We can extend the theorem in 2.A to calculate the PDF of $S, f_S(x)$, from $f_{X_1}(x), f_{X_2}(x), \dots, f_{X_K}(x)$. Establish this relationship.

2.C [T]Let Y be a linear combination (a weighted sum) of K independent RVs expressed as $Y = \sum_{i=1}^{K} (a_i X_i + b_i)$, where a_i and b_i are some constants. The mean of Y is derived as

$$m_Y = E[Y] = E\left[\sum_{i=1}^{K} (a_i X_i + b_i)\right] = \sum_{i=1}^{K} [a_i E[X_i] + b_i]$$

$$= \sum_{i=1}^{K} [a_i m_{X_i} + b_i]. \tag{15.5}$$

Prove that the variance of Y can be expressed as

$$\sigma_Y^2 = \sum_{i=1}^{K} a_i^2 \sigma_{X_i}^2. \tag{15.6}$$

2.D [WWW]Consider $X_1 \sim N(1, 2)$, $X_2 \sim N(-2, 9)$, $X_3 \sim N(0, 4)$ and $Y = X_1 + 3X_2 - 2X_3$. Numerically, X_1, X_2, X_3, and Y can be generated by using **randn** as

```
X1=sqrt(?)*randn+? ;
X2=sqrt(?)*randn+? ;
X3=sqrt(?)*randn+? ;
Y= ? ;
```

The following m-file empirically calculates the mean **mY** and variance **vY** of Y by repeatedly generating **Nid** samples of Y stored in the $1 \times$ **Nid** vector Y.

```
clear
Nid=1XXX; %XXX is the last three digits of your student ID.
for n=1:Nid
  X1=sqrt(?)*randn+? ;
  X2=sqrt(?)*randn+? ;
  X3=sqrt(?)*randn+? ;
  Y(n)= ? ;
end
mY=?; %Refer to 5.A of Chapter 14. Do not use the MATLAB built-in function mean().
vY=?; %Refer to 5.B of Chapter 14. Do not use the MATLAB built-in function var().
```

2.D-1 Determine all the quantities marked by '?' to complete the m-file. Execute the completed m-file and capture the simulation result.

2.D-2

(a) Calculate the mean and variance of Y using equations (15.5) and (15.6). (b) Are the simulated values in 2.D-1 consistent with the theoretical values?

2.E Suppose that X_i ($i = 1, 2, 3, ..., K$) are K independent Gaussian RVs and their linear combination is expressed as $Y = \sum_{i=1}^{K} a_i X_i$.

2.E-1 [T]According to the theorem for Gaussian RVs, what distribution should RV Y follow?

2.E-2 [T]For the PDF of Y obtained in 2.E-1, $f_Y(x)$, is completely determined by the mean and the variance of Y. Consider the example of Y given in 2.D, whose mean and variance have been derived in 2.D-2 (a). Determine its PDF $f_Y(x)$.

2.E-3 [T]The probability of the RV Y given in 2.D being 10 or greater, Pr $\{Y \geq 10\}$, can be calculated by using equation (15.4) as follows. Complete the following two lines of code and record the execution result.

```
>>m=? ; v= ?; t=?;
>>Pr=?*erfc(?)
```

2.E-4 [WWW]The m-file below simulates Pr $\{Y \geq 10\}$ for the RV Y given in 2.D.

(a) Document the complete m-file.
(b) Record the simulation result.
(c) Is it consistent with the theoretical result in 2.E-3?

```
clear
Nid=1XXX; % XXX= the last three digits of your student ID number.
cnt=0;
Trials=0;
while cnt < Nid
  X1=sqrt(?)*randn+? ;
  X2=sqrt(?)*randn+? ;
  X3=sqrt(?)*randn+? ;
  Y= ? ;
  if Y >= ?
    cnt=cnt+1;
  end
  Trials=Trials+1;
end
P=?/?
```

2.E-5   Repeat 2.E-3 and 2.E-4 for $\Pr\{Y \le -15\}$. You may use the property that the Gaussian distribution is symmetric about its mean. Show that the theoretical values and the simulation results are consistent with each other.

15.3   CENTRAL LIMIT THEORY

Suppose that X_i $(i = 1, 2, \ldots, M)$ are independent RVs, all having the same uniform distribution over the set $[1, 2, 3, \ldots, 6]$. Let Y be the sum of X_i $(i = 1, 2, \ldots, M)$ expressed as

$$Y = \sum_{i=1}^{M} X_i. \tag{15.7}$$

3.A   For the case of $M = 2$, Y can be numerically generated as

```
X(1)=ceil(rand*6);
X(2)=ceil(rand*6);
Y=sum(X)
```

3.A-1
  (a) Record all possible values that Y can take for $M = 2$. (b) Given any value of M, the minimum possible value of Y is M. What is the maximum possible value of Y as a function of M?

3.A-2   [WWW]The following m-file repeatedly generates the samples of Y with $M = 2$ for Nid*100 times and plots the distribution of Y based on the outcome. Complete the places marked by '?' and then execute the m-file. Capture the resulting plot.

```
clear
M=2;
Nid=1XXX; % XXX = last three digits of your student ID number.
rand(Nid); % Irrelevant to the main goal of this m-file, but be sure to insert this line.
Nsim=Nid*100;

Possible_Y=M:1:? ;   %Fill in ? with the maximum possible value of Y accord-
ing to the answer to 3.A-1(b).
count=zeros(1,length(Possible_Y));

for n=1:Nsim
  for ii=1:M
    X(ii)=ceil(rand*6);
  end
  Y=sum(?);
```

```
  for k=1:length(Possible_Y)
    if Y==Possible_Y(k)
      count(k)=count(k)+1;
    end
  end
end
P_Y=?/Nsim;
plot(Possible_Y,P_Y)
xlabel('Y')
ylabel('Pr[Y]')
```

3.B Execute the modified m-file for each of the following M values: 4, 8, 16, and 50. (a) Capture the distribution of Y for each value of M. (b) Describe how the distribution shape changes as M increases.

3.C We repeat 3.B for another distribution of X_i. If we replace the line '**X(ii)=ceil(rand*6)**' by '**X(ii)=ceil((rand^2)*6)**' in the m-file completed in 3.B, then the possible values of X_i (which equals **X(ii)** in the m-file) will still be from the set [1, 2, 3,...,6]. However, it is not uniformly distributed.

3.C-1 Replace the line '**X(ii)=ceil(rand*6)**'by '**X(ii)=ceil((rand^2)*6)**', set **M=1**, and determine the empirical distribution of X_i.

(a) Execute the revised m-file and capture the distribution of X_i.
(b) Describe the shape of the distribution of X_i, that is, how X_i is distributed over [1, 2, 3,...,6]), in comparison with the uniform distribution case.
(c) Intuitively explain why X_i is more likely to take on the smaller values from the set [1, 2, 3,...,6].

3.C-2

(a) If the distribution of X_i is nonuniform as observed in 3.C-1, then will $Y=\sum_{i=1}^{M} X_i$ change as M increases? Will the distribution of $Y=\sum_{i=1}^{M} X_i$ still converge to Gaussian as M increases like the case where $\{X_i\}$ are uniformly distributed?
(b) Execute the modified m-file for the following values of M: 4, 8, 16, and 50. Capture the distribution of Y for each value of M.
(c) Describe how the distribution shape changes as M increases.

3.D Do the results in 3.B and 3.C verify the central limit theorem (CLT) [3,4]?

3.E In 3.B and 3.C, in order to verify the CLT, we resorted to the empirically generated distribution. A problem with the empirical distribution method is that it needs a large number of random variable generations and it is rather time-consuming. In this section we verify the CLT with a different approach that does not require generating outcomes of RVs.

3.E-1 Execute the following two lines of code in the command window. In the first line, be sure to replace the elements of the vector **ID** by the digits of your student

ID number. The length of **pX** does not matter. In the rest of this section, we treat the vector **pX** as a discrete probability distribution function of random variable **X**. For example, **pX(1)**, **pX(2)**, ... denote the probability of random variable **X** being equal to 1, 2,..., respectively.

    (a) Capture the generated **pX**.

    (b) From the generated **pX**, calculate $\Pr(X = 3)$.

```
>> ID=[2,0,8,4,3,8,1,2] % Example assuming your student ID is 20843812.
>> pX=ID/sum(ID) % To normalize pX such that sum(pX) =1.
```

3.E-2   In order to check the overall shape of the probability distribution function of **X**, execute **plot(pX)** in the command line and capture the plot.

3.E-3   The MATLAB command **conv(a,b)** outputs the convolution result of the vectors **a** and **b**. In the command window,

    (a) Execute '**pX=conv(pX,pX);plot(pX)**' and capture the plot of the convolution result.

    (b) Repeat (a) for 5 times. Do not create an m-file to execute the commands at once; execute the commands one by one. After each execution, check the convolution plot and capture it.

```
>> pX=conv(pX,pX);plot(pX) %Execute it and capture the plot.
>> pX=conv(pX,pX);plot(pX) %Execute it and capture the plot.
... Repeat for 5 times
```

3.E-4   Set **pX** in 3.E-1 to a different vector of an arbitrary length with any nonnegative real-valued elements. Repeat 3.E-3 with this **pX** and capture the convolution plots.

3.E-5   We should see Gaussian distribution–shaped curves in 3.E-3 and 3.E-4, which are not obtained by generating samples to obtain a histogram but by convolution. Vector **pX** is a deterministic vector, and we plotted the final **pX** itself, rather than the histogram of the elements of **pX**. The results in 3.E-3 and 3.E-4 still properly verify the CLT, which can be done using the results in 2.B. Provide the details of this verification.

3.E-6   Create at least 6 arbitrary nonnegative real-valued vectors as shown below. The vectors do not have to be of the same length.

```
>> pX1=[1,  0.3,  4.3,  8,  3 ];
>> pX2=[2,  0,  5,  1,  8 ];
>> pX3=[9, 11.5,  6.7, 1.02, 8];
...
>> pX6=[0.1,  4,  3,  0,  2];
```

Recursively convolve these vectors as below:

```
>> pX=pX1;
>> pX=conv(pX, pX2); plot(pX)
>> pX=conv(pX, pX3); plot(pX)
...
>> pX=conv(pX, pX6); plot(pX)
```

    (a) Capture the plot after each convolution.
    (b) Using the result in (a), generalize the CLT.

3.E-7

    (a) Repeat 3.E-1~3.E-3 by setting **ID**=[10, 1, 1, 1, 1, 1, 1, 10].
    (b) Design a vector with nonnegative elements so that it will never converge to the typical shape of Gaussian PDF despite the recursive convolutions as done in 3.E-3.
    (c) Verify the answer(your design) in (b).
    (d) If you correctly designed a vector in (b), theoretically explain why your designed vector does not converge to the typical shape of Gaussian PDF.

15.4 GAUSSIAN RANDOM PROCESS AND AUTOCORRELATION FUNCTION

4.A [WWW]Consider a certain random process $x(t)$. At each iteration of the '**for**' loop in the m-file below, a realization (observation or outcome) of $x(t)$ is generated, and then sampled; the samples are stored in the vector xt. Assume that the sampling interval is 1 second. The following m-file plots the sampled waveform of a newly realized $x(t)$ every time you press any key.

```
clear
figure
ID=[2,5,3,8,1,5,7,1];% Replace the elements of the vector ID with the full dig-
its of your student ID number.
for trial=1:100
xt=conv(randn(1,100),ID)+4.35;
%The line above is for realization of x(t). Don't worry about the right-hand side of the
command at this point.
plot(xt)
xlabel('t [sec]')
  pause
end
```

4.A-1   Replace the elements of the vector **ID** with the full digits of your student ID number and execute the m-file.

(a) Whenever you press any key, a sampled waveform of a new realization of $x(t)$ will be plotted. Capture at least four sampled waveforms of $x(t)$. Press Ctrl-C key to terminate the execution of the m-file.

(b) Based on the captured plots in (a), determine whether or not $x(t)$ is a random process. Justify your answer.

4.A-2   Is it possible to determine whether or not $x(t)$ is a Gaussian random process [3,4] based only on the captured plots in 4.A-1(a)? Justify your answer.

4.A-3   [WWW]Let us collect the realizations of $x(t)$ at $t = t_0$. In the m-file below, we repeatedly generate observations of $x(t)$, expressed in the simulation code as **xt**. For each observation, the **t0**-th element of **xt**, that is, the observation evaluated at $t = t_0$, is saved in the vector **xt_at_t0** as a new element. In the m-file below, t_0 is set to 23 for illustration. Finally, the m-file returns the mean of $x(t)|_{t=t_0}$ and plots the distribution of $x(t)|_{t=t_0}$ using histogram.

(a) Identify the variable that corresponds to one observation of $x(t)|_{t=t_0}$ in the m-file below.

(b) If we plot **xt_at_t0** using the command **plot(xt_at_t0)**, what does the x axis of the resulting figure represent.

(c) The line **mean(xt_at_t0)** returns the average of the random process $x(t)$. Is this the time average or the ensemble average [3]? Justify your answer.

```
clear
figure
ID=[2,5,3,8,1,5,7,1]; % Replace the elements of the vector ID with the full dig-
its of your student ID number.
t0= 23 ;
for trial=1:50000
%You may set total trial number to be larger than 50000 if the PC comput-
ing power allows.
%You may set total trial number to be smaller than 50000 for speedy simula-
tion, but the simulation error increases.
  xt=conv(randn(1,100), ID)+4.35;
  xt_at_t0(trial)=xt(t0);
end
mean(xt_at_t0)
hist(xt_at_t0,100)
```

4.A-4   Execute the m-file above. Capture the result from executing the command **mean(xt_at_t0)** and the PDF graph generated by **hist(xt_at_t0,100)**.

4.A-5   Change **t0** to an arbitrary integer between 1 and 100 and execute the m-file. Try at least four different values of **t0** and capture the execution results

4.A-6 Is the process $x(t)$ realized through the m-file a Gaussian random process? Justify your answer on the basis of the execution results, not based on the code lines in the m-file.

4.B Autocorrelation function of random process.

The autocorrelation of a random process $x(t)$ denoted by $R_x(t_1, t_2)$ is defined as [3,4]

$$R_x(t_1, t_2) = E[x(t_1)x^*(t_2)]$$
$$= E[x(t_1)x(t_2)] \quad \text{when } x(t) \text{ is a real-valued random process.} \quad (15.8)$$

4.B-1 [WWW]Consider the same random process $x(t)$ that was considered in 4.A. The m-file below calculates the autocorrelation between the samples $x(t_1)$ and $x(t_2)$, $R_x(t_1, t_2)$. For each of lines, add a comment to explain what the variable on the left-hand side represents and justify how the right-hand side expression is properly formulated accordingly. Capture the commented m-file.

```
clear
ID=[2,5,3,8,1,5,7,1]; % Replace the elements of the vector 'ID' with the full dig-
its of your student ID number.
t1= 51 ;
t2= 87 ;
for trial=1:50000
%You may set total trial number to be larger than 50000 if the PC comput-
ing power allows.
%You may set total trial number to be smaller than 50000 for speedy simula-
tion, but the simulation error increases.
    xt=conv(randn(1,100), ID)+4.35;
    xt1_mult_xt2(trial)=xt(t1)*xt(t2);
end
mean(xt1_mult_xt2)
```

4.B-2 Does the command **mean()** in the last line calculate the time average or the ensemble average? Justify your answer.

4.B-3 Let us execute the m-file for various pairs of (**t1**, **t2**).

(a) Consider at least four pairs of the time instants (**t1**, **t2**) that satisfy the time difference constraint $t2 - t1 = 3$ and range constraints $8 < t1 < 100$ and $8 < t2 < 100$. For example, $t1 = 37$ and $t2 = 40$. Execute the m-file for each of the four different pairs. Record the simulation results of $R_x(t_1, t_2)$ for the four cases in a table.

(b) Assume that the time difference constraint is $t2 - t1 = 5$ and repeat the process discussed in (a). Record the simulated $R_x(t_1, t_2)$ values for the four cases in the same table created in (a).

(c) From the results in (a) and (b), summarize the main properties of the correlation function $R_x(t_1, t_2)$ of the random process $x(t)$ considered. If the simulation results are not what you expected to see, then increase the total number of trials in the m-file to reduce the simulation error.

4.B-4   [WWW]In the m-file below, the variable **tau** represents the time difference **t2−t1** for generating $R_x(t_1, t_2)$, that is, **t2** is set to **t1 + tau**. It calculates the autocorrelation by fixing **t1** $= 51$ and changing the value of **tau** and plots the autocorrelation as a function of **tau**.

For each of the lines, add a comment to explain what the variable on the left-hand side represents and justify how the right-hand side expression is properly formulated accordingly. Capture the commented m-file.

```
clear
figure(1)
ID=[2,5,3,8,1,5,7,1]; % Replace the elements of the vector 'ID' with the full dig-
its of your student ID number.
t1= 51 ;

R_vector=[];
tau_vector=[];
for tau= (-t1+1):(-t1+100)
    t2=t1+tau;
    for trial=1:10000
%You may set total trial number to be larger than 10000 if the PC comput-
ing power allows.
%You may set total trial number to be smaller than 10000 for speedy simula-
tion, but the simulation error increases.
        xt=conv(randn(1,100), ID)+4.35;
        xt1_mult_xt2(trial)=xt(t1)*xt(t2);
    end
    tau_vector=[tau_vector tau];
    R_at_tau= mean(xt1_mult_xt2);
    R_vector=[R_vector R_at_tau];
end
plot(tau_vector,R_vector);
axis([-50 50 min(R_vector) max(R_vector)]);
hold on
xlabel('tau')
grid on
```

4.B-5

(a) Execute the m-file above and capture the resulting graph. (b) After analyzing and understanding the m-file, for a point x_0 along the x axis of the graph, what does the corresponding value of that point along the y axis represent?

4.B-6   Set **t1** to a different integer arbitrarily chosen between 40 and 60 and execute the m-file again. Run the simulation for at least four different values of **t1** and capture the resulting graphs of $R_x(t_1, t_1 + x_0)$.

4.B-7   From the graphs of $R_x(t_1, t_1 + x_0)$ generated in 4.B-6, determine whether $R_x(t_1, t_1 + x_0)$ is a function of t_1 or x_0 or both.

4.B-8   [T]Summarize the definition of "wide-sense stationary random processes" [3,4].

4.B-9   Is the random process $x(t)$ generated by the m-file wide-sense stationary? Analyze the results in 4.B-7 to justify the answer.

4.B-10   Based only on the results obtained so far, determine the value of $E[x(77)x(82)]$ without further simulation.

4.B-11   [A]Let L denote the length of the vector **ID** in the m-file. From the autocorrelation figures generated in 4.B-6, the autocorrelation is found to be nearly 0 for **tau**<-L or L<**tau**. This implies that if $x(t)$ is evaluated at two time instants that are at least L seconds apart, then the RVs corresponding to the random process at these two instants are uncorrelated. This can be explained by examining the line '**xt=conv(randn(1,100), ID)+4.35**' in the m-file. Justify this conclusion in detail.

4.C   Statistical properties of additive white Gaussian noise.

4.C-1   The MATLAB command **randn(1,b)** generates a 1×**b** vector whose elements are realizations of independent and identically distributed Gaussian random variables with zero mean and unit variance.
   If the line '**xt=conv(randn(1,100), ID)+4.35**' in the m-file in 4.B-4 is changed into '**xt=randn(1,100);**', without simulating this case, discuss how the autocorrelation will change.

4.C-2   Change the line '**xt=conv(randn(1,100), ID)+4.35**' of the m-file in 4.B-4 into '**xt=randn(1,100);**' and execute the m-file. Is the resulting autocorrelation plot what you expected to see as discussed in 4.C-1?

4.C-3   Based on only the autocorrelation plot captured in 4.C-2, can we determine whether $x(t)$ is a Gaussian random process or not? Why?

4.C-4   Now we analyze the m-file and identify which line(s) shows that $x(t)$ is a Gaussian random process, although this conclusion cannot be made by observing only the autocorrelation function.

4.C-5   Change the line '**xt=conv(randn(1,100), ID)+4.35**' in the m-file in 4.A-3 into '**xt=randn(1,100)**' and execute the m-file for at least four different values of t0. Capture the result of each case. Based on these results, determine whether $x(t)$ is Gaussian.

4.C-6   The distribution graphs captured in 4.C-5 do not say anything about the whiteness of $x(t)$, but it is true that $x(t)$ is white. Review problem 4.C-2 and its associated autocorrelation results and then explain why $x(t)$ is a white process.

4.C-7   Also, from the line '**xt=randn(1,100)**' in the m-file, explain why $x(t)$ is white.

4.D   The output characteristics of a linear system with a Gaussian random process as its input.

4.D-1   Modify the line '**xt=conv(randn(1,100), ID) + 4.35**' in the m-file in 4.A-3 into '**xt=conv(randn(1,100), ID)**'.

Change **ID** to an arbitrary real-valued vector of any length and execute the m-file. Capture the distribution plots for the several different values of t0. Repeat this for at least four different values of **ID**.

4.D-2   From the plots obtained in 4.D-1, summarize the common characteristics of the distribution of the random process $x(t)$ at $t = t_0$.

4.D-3   Summarize the analytical relationship among the input, the output, and the impulse response of a linear system in the time domain.

4.D-4

(a) From the relationship summarized in 4.D-3 and the line '**xt=conv(randn(1,100), ID)**' in the m-file, the vector **xt** can be considered as the output of a linear system whose input is equal to the vector generated by **rand(1,100)**. Then, which quantity (variable) represents the impulse response of this linear system?

(b) From (a) and the characteristic of the distributions observed in 4.D-2, summarize an important property of Gaussian random process when it passes through a linear system.

REFERENCES

[1] M. Abramowitz and I. A. Stegun, *Handbook of Mathematical Functions with Formulas, Graphs, and Mathematical Tables*, New York: Dover, 1965.

[2] L. C. Andrews, *Special Functions of Mathematics for Engineers*, Bellingham, WA: SPIE Press, 1992.

[3] A. Papoulis, *Probability, Random Variables, and Stochastic Processes*, New York: McGraw-Hill, 1965.

[4] A. Leon-Garcia, *Probability and Random Processes for Electrical Engineering*, 2nd ed., Boston, Mass, North Reading, MA: Addison-Wesley, 1994.

16

MAXIMUM LIKELIHOOD DETECTION FOR BINARY TRANSMISSION

- Derivation of the likelihood function.
- Maximum likelihood detection.

16.1 LIKELIHOOD FUNCTION AND MAXIMUM LIKELIHOOD DETECTION OVER AN ADDITIVE WHITE GAUSSIAN NOISE CHANNEL

The additive white Gaussian noise (AWGN) is a random process that is widely used to model the background noise in a communications system receiver [1,2]. The name AWGN well describes the characteristics of the background noise: it is "additive" (to the received desired signal), "white" (because of its flat spectral density being over a very wide range of frequencies), and follows the "Gaussian" distribution.

If the channel impulse response $h_c(t)$ is a delta function, and the received signal is corrupted by only AWGN, then this channel is called an "AWGN channel." Consider a binary digital communications system where signals $s_1(t)$ and $s_2(t)$ represent data bits '1' and '0', respectively. The received signal $r(t)$ over an AWGN channel is expressed as

$$r(t) = \begin{cases} s_1(t) + n(t), & \text{if transmitted bit is } '1', \\ s_2(t) + n(t), & \text{if transmitted bit is } '0', \end{cases} \quad (16.1)$$

where $n(t)$ is the AWGN.

Problem-Based Learning in Communication Systems Using MATLAB and Simulink, First Edition.
Kwonhue Choi and Huaping Liu.
© 2016 The Institute of Electrical and Electronics Engineers, Inc. Published 2016 by John Wiley & Sons, Inc.
Companion website: www.wiley.com/go/choi_problembasedlearning

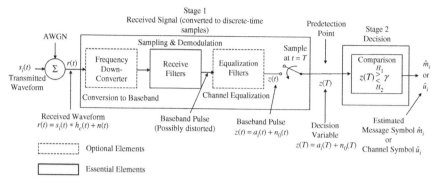

FIGURE 16.1   Basic demodulation/detection steps.

Fig. 16.1 shows the basic demodulation and detection steps in a typical digital communications system. In the first stage, the received signal is frequency down-converted into baseband, filtered, and then sampled to generate the decision variable $z(T)$. If $r(t)$ is baseband modulated, then frequency down conversion is not needed. Over an AWGN channel, the equalization filter is not needed. Thus the filtered baseband signal $z(t)$ is expressed as

$$z(t) = \begin{cases} a_1(t) + n_0(t), & \text{for bit } '1', \\ a_2(t) + n_0(t), & \text{for bit } '0', \end{cases} \tag{16.2}$$

where $a_i(t)$, $i = 1, 2$ is the filtered output of the signal term $s_i(t)$, $i = 1, 2$ and $n_0(t)$ is the filtered noise signal.

Consider the simplest case where the input is a single random variable. The filtered baseband signal $z(t)$ is sampled at the sampling instant $t = T$ to generate the decision variable $z(T)$, which is simply denoted by z since there is only one transmitted symbol, and is expressed as

$$z = \begin{cases} a_1 + n_0 & \text{if } s_1(t) \text{ (bit} = '1') \text{ is transmitted,} \\ a_2 + n_0 & \text{if } s_2(t) \text{ (bit} = '0') \text{ is transmitted,} \end{cases} \tag{16.3}$$

where a_i denotes $a_i(T)$ and n_0 denotes the Gaussian noise $n_0(T)$ with zero mean and variance σ_0^2, that is, $n_0 \sim N(0, \sigma_0^2)$.

1.A

(a) [T]From the study in 4.A of Chapter 14: if $X \sim N(0, \sigma^2)$ and $Y = Ax + b$, then $Y \sim N(?, ?)$. Complete the quantities marked by '?'.

(b) [T]Equation (16.3) shows that the decision variable z is equivalent to a Gaussian random variable whose mean is determined by the transmitted bit. Let $f_{z|s_1}(z)$ and $f_{z|s_2}(z)$ denote the conditional probability density functions (PDFs) of z, assuming that $s_1(t)$ (bit = '1') and $s_2(t)$ (bit = '0') are transmitted, respectively. Determine the expressions of $f_{z|s_1}(z)$ and $f_{z|s_2}(z)$.

1.B [T]In Stage 2 of Fig. 16.1, the symbol (bit in this problem) decision, that is, to determine whether the transmitted bit is '1' or '0' because binary signal is considered, is made on the basis of the decision variable z. In this problem we focus on maximum likelihood (ML) detection [3–5], which minimizes the bit error rate (BER) over an AWGN channel, assuming that '1' and '0' are equally probable. The ML detector chooses the bit from the alphabet to maximize the likelihood function [3–5]. For the binary case considered, the bit decision can be made by comparing the likelihood ratio denoted by $\Lambda(z)$ with a threshold expressed as

$$\hat{d} = \begin{cases} '1' & \text{if } \Lambda(z) > 1, \\ '0' & \text{if } \Lambda(z) < 1, \end{cases} \quad \text{where} \quad \Lambda(z) \triangleq \frac{\Pr(s_1|z)}{\Pr(s_2|z)}. \tag{16.4}$$

By using the Bayes's theorem [1,2], the likelihood ratio in equation (16.4) can be expressed as

$$\Lambda(z) \left(\triangleq \frac{\Pr(s_1|z)}{\Pr(s_2|z)} \right) = \frac{f_{z|s_1}(z)\Pr(s_1)}{f_{z|s_2}(z)\Pr(s_2)}, \tag{16.5}$$

where $\Pr(s_1)$ and $\Pr(s_2)$ are the prior probabilities of s_1, s_2, which correspond to, respectively, '1' and '0' being transmitted, and $f_{z|s_1}(z)$ and $f_{z|s_2}(z)$ denote the conditional PDFs of z conditioned on s_1 and s_2.

(a) Derive equation (16.5).
(b) The binary bits '1' and '0' are typically equally probable, that is, $\Pr(s_1) = \Pr(s_2) = 0.5$. In this case, the likelihood ratio expressed in equation (16.5) can be simplified as

$$\Lambda(z) = \frac{f_{z|s_1}(z)}{f_{z|s_2}(z)}. \tag{16.6}$$

By substituting the complete expressions of $\Pr(z|s_1)$ and $\Pr(z|s_2)$ derived in Problem 1.A into equation (16.6), show that $\Lambda(z)$ can be written as a function of a_1, a_2, and σ_0 as

$$\Lambda(z) = \exp\left(-\frac{a_1^2 - a_2^2}{2\sigma_0^2} + \frac{z(a_1 - a_2)}{\sigma_0^2} \right). \tag{16.7}$$

(c) For equally probable binary data, by using equations (16.4) and (16.6), show that the ML detection rule is equivalent to the decision rule below:

$$\hat{d} = \begin{cases} '1' & \text{if } f(z|s_1) > f(z|s_2), \\ '0' & \text{if } f(z|s_2) > f(z|s_1). \end{cases} \tag{16.8}$$

1.C Assume $a_1 > a_2$. By substituting equation (16.7) into (16.4) and rearranging, we can derive two regions (ranges) of z, according to which the transmitted bit is detected as '0' and '1', respectively, as expressed in equation (16.9):

$$\hat{d} = \begin{cases} '1' & \text{if } z > \frac{a_1 + a_2}{2}, \\ '0' & \text{if } z < \frac{a_1 + a_2}{2}. \end{cases} \tag{16.9}$$

1.C-1 [T]Derive equation (16.9).

1.C-2 [T]We can interpret equation (16.9) as shown below. Specify the two decision regions R_1 and R_2.

- If the decision variable z falls in region R_1, that is, $z \in R_1$, then we decide that '1' is transmitted.
- If the decision variable z falls in region R_2, that is, $z \in R_2$, then we decide that '0' is transmitted.

R_1 and R_2 are called the "decision regions."

1.C-3 [T]From equation (16.9), explain that ML detection of binary signals over an AWGN channel reduces to comparing z with a certain threshold.

1.D We can find the probability that the ML detection expressed by equation (16.9) generates an incorrect bit decision.

1.D-1 [T]Complete the following problems.

(a) Assume that '1' is transmitted. From equation (16.9), a bit error ($\hat{d}$ = '0') occurs if z is '**?**' (smaller than or larger than) '**?**' (threshold). Determine the two quantities marked by '**?**'.
(b) By using the PDF $f_{z|s_1}(z)$ derived in 1.A, show that the probability of the condition in (a) that leads to an erroneous bit decision can be written in the form of the Q-function as

$$\text{BER when } '1' \text{ is transmitted} = \Pr(\hat{d} = '0' | s_1) = Q\left(\frac{a_1 - a_2}{2\sigma_0}\right). \tag{16.10}$$

Note that the Q-function expressed in terms of the Gaussian PDF is studied in Section 1.A of Chapter 15.

1.D-2 [T]Similar to 1.D-1, show that the BER when '0' is transmitted is also equal to $Q\left(\frac{a_1 - a_2}{2\sigma_0}\right)$.

1.D-3 [T]If $\Pr(s_1) = \Pr(s_2) = 0.5$, show that the average BER is given as

$$p_b = Q\left(\frac{a_1 - a_2}{2\sigma_0}\right). \tag{16.11}$$

1.E By using the BER formula in equation (16.11) and the **erfc()** function in MATLAB, calculate the BERs for the following cases:

1. $a_1 = 8, a_2 = -8, \sigma_0^2 = 4$
2. $a_1 = 2, a_2 = -2, \sigma_0^2 = 0.25$
3. $a_1 = 1, a_2 = -1, \sigma_0^2 = 0.5$
4. $a_1 = 4, a_2 = 0, \sigma_0^2 = 0.25$
5. $a_1 = 10, a_2 = -6, \sigma_0^2 = 4$

1.F

(a) Compare the BERs of the five cases in 1.E and identify the cases that have the same BER.
(b) From the BER formula in 1.D-3, derive the conditions on a_1, a_2, and σ_0^2 that result in the same BER.
(c) Are the results in (a) and (b) consistent with each other?

1.G [A]Under the fixed energy constraint, that is, $a_1^2 + a_2^2 = $ constant, find the conditions on a_1 and a_2 that minimize the BER.

16.2   BER SIMULATION OF BINARY COMMUNICATIONS OVER AN AWGN CHANNEL

2.A [WWW]The m-file below simulates ML detection of binary data communication systems in an AWGN channel environment. In **PART1** of the MATLAB code, the decision variable **z** in equation (16.3) is generated; in **PART2**, the decision is made by employing the ML detection rule expressed by equation (16.9). For generating the decision variable in equation (16.3), we consider the simple case with $a_1 = 1, a_2 = -1$, and, $\sigma_0^2 = 1.5$.

```
clear
%%%% PART 1 %%%%%%%%%%%%
a1=1;  % = a₁ in (16.3)
a2=-1;  % = a₂ in (16.3)

vn=1.5;  % = σ₀² of n₀ in (16.3)
n0=randn*sqrt(vn);  % = n₀ in (16.3)

d = (rand > 1/2); % = d, transmitted binary data ('1' or '0'), i.e., bit

if d==1  z_nonoise = a1;  end  % z_nonoise implies z for the case when there is no
noise.
if d==0  z_nonoise = ? ;  end

z=z_nonoise+n0 %  = z in (16.3)
```

```
%%%%%%%%%%%%%%%%%%%%%%

%%%%% PART 2 %%%%%%%%%%%%%%%
if z> ? d_estimate=1; end    % Implement ML rule in (16.9)
if z< ? d_estimate=?; end    % Implement ML rule in (16.9)
decision_check=(d_estimate==?) %If the decision is correct, decision_check
will be 1, otherwise, it will be 0.
%%%%%%%%%%%%%%%%%%%%%%
```

2.A-1   Explain the following settings:

    (a) Why should **n0** be set to **randn*sqrt(vn)**?

    (b) Why should **d** be set to **(rand > 1/2)**?

2.A-2   Complete the quantities marked by '?' in the MATLAB code and then capture the m-file. Make sure that these parameters are formulated in such a way that they can be changed in relation to the settings of a_1, a_2, and σ_0^2 for simulating different cases.

2.A-3   Execute the completed m-file at least three times. Check whether the bit decision is made correctly each time.

2.A-4   [WWW]Modify the m-file as shown below. We repeat the steps for generating data and noise and the ML detection in the m-file completed in 2.A-2 until the number of incorrect bit decisions reaches Nid. The simulated BER is calculated as the ratio of the number of erroneous bit decisions to the total number of bits simulated.

    Complete all the quantities marked by '?' and capture the completed m-file.

```
clear
Nid=1XXX; %XXX are the last three digits of your student ID number.
a1=1;
a2=-1;
vn=1.5;

errcnt=0;
bitcnt=0;
while errcnt <Nid

    d= (rand > 1/2);
    if d==1 z_nonoise=a1; end
    if d==0 z_nonoise=a2; end

      n0=randn*sqrt(vn);

    z=z_nonoise+n0;
```

```
    if z> ?  d_estimate=1; end
    if z< ?  d_estimate=?; end

    if d_estimate~=?
       errcnt=errcnt+1;
    end

    bitcnt=bitcnt+1;
  end

BER= ?/?;
```

2.A-5   Execute the m-file and capture the BER result.

2.B   [WWW]In the m-file below, we consider the antipodal signaling, that is, $a_2 = -a_1$ and use an outer loop to simulate the BER for each of the following cases: $a_1 = 0.25, 0.5, 0.75, 1, 1.25, 1.5, 1.75, 2, 2.5, 3, 3.5$. This allows us to plot the simulated BERs as a function of a. In generating the BER plot, use **semilogy()**, instead of **plot()**, to display the y axis in log scale.

```
clear
Nid=1XXX;
vn=1.5;

a1_vector=[0.25, 0.5, 0.75, 1, 1.25, 1.5, 1.75, 2, 2.5, 3, 3.5];
for n=1:length(a1_vector)
   a1=a1_vector(?);
   a2=-a1;
   errcnt=0;
   bitcnt=0;
   while errcnt <Nid
      ... % Copy the body inside the while loop of m-file in 2.A-4
   end
   BER(n)= ?/?;
end
figure
semilogy(a1_vector,?)
xlabel('a_1'); ylabel('BER');
grid on
```

2.B-1   Complete all the quantities marked by '**?**' and capture the completed m-file.

2.B-2   Execute the completed m-file and capture the BER graph. Save the figure in .fig format for use in the problem in 2.C.

2.B-3   The BER graph captured in 2.B-2 should show that the BER decreases as a_1 increases. Explain the reason.

2.C   In the MATLAB command window as shown below, we calculate the theoretical BER values using equation (16.11) for a_1 = 0.25, 0.5, 0.75, 1, 1.25, 1.5, 1.75, 2, 2.5, 3, 3.5, and overlay the theoretical BER curve on the simulated BER curve obtained in 2.B-2.

   (a) Execute the following and capture the figure.

```
>> a2_vector=-a1_vector;
>>BER_exact=0.5*erfc(((a1_vector-(a2_vector))./(2*sqrt(vn)))/sqrt(2));
   %Calculate the BER vector at once by using vector operation.
>> hold on
>> semilogy(a1_vector,BER_exact, 'r');
```

   (b) Compare theoretical and simulated BER curves. Do they match each other?

2.D.   [A]Consider binary data d with nonequal prior probabilities described as

$$Pr\,(d = 1) = Pr(s_1) = \frac{1}{10}, \quad Pr\,(d = 0) = Pr(s_2) = \frac{9}{10}. \qquad (16.12)$$

2.D-1   [T]A bit sequence satisfying the priors described in equation (16.12) can be generated by replacing the line 'd=(rand > 0.5);' in the m-file of 2.B with 'd=(rand > 0.9);'. Justify this approach.

2.D-2   Complete the revision of the line described in 2.D-1 and execute the m-file. Capture the simulated BER graph.

2.D-3   [T]According to the ML decision rule in equation (16.9), the bit decision is made by comparing z with the threshold $(a_1 + a_2)/2$. However, the likelihood function in equation (16.7) and ML decision rule in equation (16.9) are derived for equally probable binary data, that is, $Pr(s_1) = Pr(s_2) = 1/2$.

   (a) For bits with unequal probabilities as expressed in equation (16.12), derive again the likelihood function and the ML decision rule.
   (b) Compared with the decision threshold for the case with equally probable bits, that is, $(a_1 + a_2)/2$, is the decision threshold derived in (a) larger or smaller?
   (c) Intuitively explain why for the case expressed in equation (16.12) the threshold is not $(a_1 + a_2)/2$ anymore.

2.D-4   In the m-file of 2.D-2, modify the lines 'if z > ? ....' and 'if z < ? ...' on the basis of the new ML decision threshold derived in 2.D-3(a). Execute command **hold on** in the command window and run the m-file.
   Capture the simulated BER graph.

2.D-5   Compare the BERs with the new decision threshold and the previous BERs on the basis of the decision threshold $(a_1 + a_2)/2$.

16.3 [A]ML DETECTION IN NON-GAUSSIAN NOISE ENVIRONMENTS

In equation (16.3), the term n_0 denotes the noise component of the decision variable z and is assumed to be Gaussian. Consider an environment where the PDF of n_0 is not Gaussian but is expressed as

$$p_{n_0}(x) = \begin{cases} |x|, & -1 \le x \le 1, \\ 0, & \text{elsewhere.} \end{cases} \tag{16.13}$$

Also a_1 and a_2 denote the signal component in the decision variable z when $s_1(t)$ (bit '1') and $s_2(t)$ (bit '0') are transmitted, respectively. In this section we assume $a_1 = 0.75$ and $a_2 = -0.75$.

3.A Complete the following problems.

3.A-1 Sketch the PDF of n_0 in equation (16.13). Draw the x axis by a thin but solid line and the PDF curve by a solid and thick line from $x = -3$ to $x = 3$.

3.A-2 Show that the area under the PDF curve equals 1.

3.B For the case where the noise component n_0 has the PDF as expressed in equation (16.13), sketch the following two conditional PDFs of z:

3.B-1 $f_{z|s_1}(z)$: The conditional PDF of z when $s_1(t)$ is transmitted, that is, $z = a_1 + n_0$.

3.B-2. $f_{z|s_2}(z)$: The conditional PDF of z when $s_2(t)$ is transmitted, that is, $z = a_2 + n_0$.
  Draw the x axis by a thin line and the PDF curve by a solid thick line from $x = -3$ to $x = 3$. Use the fact that if $Y = X + C$, where X is a random variable and C is a constant, then the PDF of Y is a shifted version of the PDF of X.

3.C Recall from equation (16.8) that for equally probable binary data, the ML rule is equivalent to making a decision on the basis of which one of $f_{z|s_1}(z)$ and $f_{z|s_2}(z)$ is bigger. Based on the conditional PDF curves sketched in 3.B, complete the following problems.

3.C-1 Find the region (range) of z that satisfies $f_{z|s_1}(z) > f_{z|s_2}(z)$.

3.C-2 Find the region (range) of z that satisfies $f_{z|s_2}(z) > f_{z|s_1}(z)$.

3.C-3 According to equation (16.8), if the decision variable z falls in the region obtained in 3.C-1, how should the transmitted bit, '0' or '1', be estimated?

3.D The BER of ML decision in a Gaussian noise environment was derived in 1.D. Using a similar approach, calculate the BER of the ML decision derived in 3.C for the system considered in this subsection.

3.E.
  (a) We can generate a random variable that has a PDF in equation (16.13). Execute the following in the command window and capture the histogram.

```
>>n0_samples=sign(rand(1,1e5)-0.5).*sqrt(rand(1,1e5));
>>hist(n0_samples,100); axis([-2 2 0 5e3])
```

(b) Is the histogram shape consistent with the sketch in 3.A-1?

(c) Revisit the m-file created in 2.A-4 and change the line '**n0=randn*?;**' to '**n0=sign(rand-0.5)*sqrt(rand)**' to generate the noise sample n_0 that has a PDF expressed in equation (16.13). In addition, in the second and third lines, set '**a1=0.75**', '**a2=-0.75**'. Do not change other parts of the m-file.

Execute the modified m-file and show the BER result.

3.F   In addition to the modifications made in 3.E(c), modify the lines '**if z>?** …' and '**if z <?**…' of the m-file properly according to the decision regions derived in 3.C-1 and 3.C-2.

3.F-1   Capture the two modified lines.

3.F-2   Execute the modified m-file and capture the BER result.

3.F-3   Is the simulated BER in 3.F-2 consistent with the theoretical BER derived in 3.D? Note that the limited number of bits simulated could result in a BER value that is not very precise. If they are not consistent, your derivation or simulation (or both) may be incorrect. In that case, try to correct one or both.

3.G   Compare the simulated BERs in 3.E and 3.F-2. Explain the reasons that have caused the difference.

REFERENCES

[1] A. Papoulis, *Probability, Random Variables, and Stochastic Processes*, New York: McGraw-Hill, 1965.

[2] A. Leon-Garcia, *Probability and Random Processes for Electrical Engineering*, 2nd ed., North Reading, MA: Addison-Wesley, 1994.

[3] S. M. Kay, *Fundamentals of Statistical Signal Processing: Estimation Theory*, Upper Saddle River, NJ: Prentice Hall, 1993.

[4] H. V. Poor, *An Introduction to Signal Detection and Estimation*, New York: Springer, 1998.

[5] H. L. Van Trees, *Detection, Estimation, and Modulation Theory, Part 1*, Hoboken, NJ: Wiley-Interscience, 2001.

17

SIGNAL VECTOR SPACE AND MAXIMUM LIKELIHOOD DETECTION I

- Map an M-ary symbol into a point in the vector space.
- Implement maximum likelihood detection (MLD) by using the Euclidean distance in the vector space and difference energy in the waveform domain in the additive white Gaussian noise (AWGN) environment.

17.1 [T]ORTHOGONAL SIGNAL SET

Consider the orthogonal signal set $\{x_1(t), x_2(t), x_3(t), \dots\}$ over the time interval $[0\ T]$ expressed as

$$x_n(t) = A\cos(2\pi n\Delta_f t), \quad \text{where } \Delta_f = \frac{1}{T}. \tag{17.1}$$

1.A What is the difference of the frequencies of the two signals, $x_n(t)$ and $x_{n+1}(t)$?

1.B Calculate the correlation between $x_n(t)$ and $x_m(t)$ over the time interval $t = [0\ T]$ and show that these two signals are orthogonal when $n \neq m$.

1.C Calculate the energy E_n of $x_n(t)$.

1.D Let us define another signal set $\{\psi_1(t), \psi_2(t), \psi_3(t), \dots\}$, where $\psi_n(t)$ is defined as $x_n(t)/\sqrt{E_n}$. Show that the signal set $\{\psi_1(t), \psi_2(t), \psi_3(t), \dots\}$ is an orthonormal set [1–4] over the time interval $[0\ T]$.

Problem-Based Learning in Communication Systems Using MATLAB and Simulink, First Edition.
Kwonhue Choi and Huaping Liu.
© 2016 The Institute of Electrical and Electronics Engineers, Inc. Published 2016 by John Wiley & Sons, Inc.
Companion website: www.wiley.com/go/choi_problembasedlearning

17.2 [T]MAXIMUM LIKELIHOOD DETECTION IN THE VECTOR SPACE

2.A Consider a case with $A = 1$ and $T = 1$ in equation (17.1). Substitute $A = 1$ and $T = 1$ into the expressions of $\psi_1(t)$, $\psi_2(t)$, and $\psi_3(t)$ in the answer of 1.D and write them as a function of time t only:

$$\psi_1(t) = ?, \ \psi_2(t) = ?, \ \psi_3(t) = ?. \tag{17.2}$$

2.B Consider an 8-ary transmission system using eight waveforms $s_i(t)$, $i = 1, 2, 3, ..., 8$. Each waveform represents 3 bits. The eight waveforms are constructed by linearly combining the three orthonormal basis functions $\psi_1(t)$, $\psi_2(t)$, and $\psi_3(t)$. Table 17.1 shows the mapping between the index i of $s_i(t)$, and the three data bits represented by $s_i(t)$. The table also shows how each of each of the eight waveforms is constructed from the linear combination of $\psi_1(t)$, $\psi_2(t)$, and $\psi_3(t)$.

For example, consider the following data bit stream to be transmitted:
01000111001101010010101010110101 . . ., for which $s_3(t)$, $s_2(t)$, $s_7(t)$? ? ? ? ? ?. . . will be transmitted in order. Complete the quantities marked by ? ? ? ? ?. . . with a proper signal sequence.

2.C. Suppose that one of the 8-ary signals $\{s_1(t), s_2(t), ..., s_8(t)\}$ is transmitted. In the receiver, it is received together with the noise component $n(t)$. Let $r(t)$ denote the received signal. To make the analysis mathematically tractable, we use a single function to represent the sum of signal plus noise expressed as

$$r(t) = s_i(t) + n(t) = \cos(1.532\pi t), \quad 0 \le t \le 1. \tag{17.3}$$

Although this simplification is unrealistic in a practical system because, in general, a single sinusoidal signal does not approximate well the signal plus a random noise term, it does not affect the description of the maximum likelihood detection (MLD) procedure. In the following problems, we demodulate the three data bits from the received signal $r(t)$ given in equation (17.3) by using MLD.

TABLE 17.1 Construction of $s_i(t)$ From $\psi_1(t)$, $\psi_2(t)$, and $\psi_3(t)$ and the Three Bits Each Waveform Represents.

Signal index i	Three data bits	Construction of $s_i(t)$
1	000	$s_1(t) = -2\psi_1(t) - \psi_2(t) - 5\psi_3(t)$
2	001	$s_2(t) = \psi_1(t) - 2\psi_2(t) - 3\psi_3(t)$
3	010	$s_3(t) = 0.5\psi_1(t) + 4\psi_2(t) + 2\psi_3(t)$
4	011	$s_4(t) = 3\psi_1(t) - \psi_2(t) - \psi_3(t)$
5	100	$s_5(t) = -\psi_1(t) - 4\psi_2(t) + 3\psi_3(t)$
6	101	$s_6(t) = \psi_1(t) + 3\psi_2(t) - 0.5\psi_3(t)$
7	110	$s_7(t) = -\psi_1(t) - 6\psi_2(t) + 3\psi_3(t)$
8	111	$s_8(t) = -2\psi_1(t) - 3\psi_2(t) + \psi_3(t)$

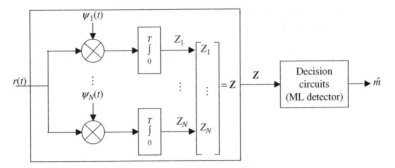

FIGURE 17.1   Block diagram of the two-stage MLD in a vector space.

2.C-1   Express each element of the 8-ary signal set $\{s_1(t), s_2(t), \ldots, s_8(t)\}$ as a corresponding point in the 3-D vector space where $\psi_1(t)$, $\psi_2(t)$, and $\psi_3(t)$ are orthonormal vectors, representing the x, y, and z axes, respectively. For example, in Table 17.1, $s_1(t)$ is mapped as a point $(-2, -1, -5)$ in this 3-D vector space. Following the convention, we express this mapping as $s_1(t) \Rightarrow \mathbf{s}_1 = (-2, -1, -5)$. Complete the quantities marked by '?' in equation (17.4) below:

$$s_2(t) \Rightarrow \mathbf{s}_2 = (?, ?, ?), \; s_3 \Rightarrow \mathbf{s}_3 = (?, ?, ?), \ldots, s_8 \Rightarrow \mathbf{s}_8 = (?, ?, ?). \quad (17.4)$$

2.C-2   Fig. 17.1 shows the block diagram of the two-stage MLD for M-ary signaling in the vector space. The block enclosed by the left-side rectangular box corresponds to the step to convert the received signal $r(t)$ into a vector $\mathbf{Z}$ (the coordinates of a point) in the vector space as

$$r(t) \Rightarrow \mathbf{Z} = (Z_1, Z_2, \ldots, Z_N). \quad (17.5)$$

Note that we have chosen symbol $\mathbf{Z}$ (rather than $\mathbf{r}$) to represent the vector corresponding to $r(t)$, since by convention symbol $\mathbf{Z}$ is often used to represent the decision variable.

(a) From Fig. 17.1, complete the expressions on the right-hand side of each '=' sign in equation (17.6) that computes $Z_1, Z_2, \ldots, Z_N$, the elements of the vector $\mathbf{Z}$, from $r(t)$, $\psi_1(t)$, $\psi_2(t)$, and $\psi_3(t)$.

$$Z_1 = \int ?dt, \; Z_2 = ?, \; Z_3 = ?, \ldots, Z_N = ? \quad (17.6)$$

(b) The integration interval T in Fig. 17.1 should be set equal to the duration of the considered M-ary signal. For the signal given in equation (17.3), $T = 1$. The dimension of the vector space is $N = 3$, since the 8-ary signals considered are generated from three orthonormal basis vectors (see Table 17.1). Substitute $r(t)$ expressed in equation (17.3), and $\psi_1(t)$, $\psi_2(t)$, and $\psi_3(t)$ expressed in equation (17.2) into the expressions completed in (a) above and calculate the values of Z_1, Z_2, and Z_3. Record the calculated values.

(c) In the noiseless case, $r(t)$ equals $s_i(t)$, which is the ith signal of the 8-ary signal set in Table 17.1. Repeat the calculation in (b) for this case and show that Z is equal to s_i obtained in 2.C-1.

2.C-3   Calculate the Euclidean distances between $r(t)$ and each of $s_i(t)$ ($i = 1, 2, 3, ..., 8$) in the vector space.

2.C-4   Answer the following questions:

(a) Which of the eight signals $s_i(t)$ ($=1, 2, 3, ..., 8$) is closest to $r(t)$ in terms of the Euclidean distance?

(b) From the answer to question (a), estimate the three data bits transmitted, that is, the MLD result for the three data bits transmitted.

2.C-5   Repeat the MLD steps in 2.C-2 to 2.C-4 for the received signal $r(t)$ below:

$$r(t) = \cos(4.53\pi t), \quad 0 \le t \le 1. \tag{17.7}$$

17.3   MATLAB CODING FOR MLD IN THE VECTOR SPACE

3.A   [WWW]The m-file below creates a vector **p2t**, which is the sampled version of $\psi_2(t)$ without using the expression derived in equation (17.2). Note that the term "vector" in the context of the sampled version of a waveform differs from the term "vector" in the context of converting (mapping) a waveform into a "vector" (point) in a vector space. In the m-file, the energy of $x_2(t)$ is calculated via numerical integration, rather than the derivation in 1.C. Refer to Section 2.1 of Chapter 2 for numerical integration.

For each of the lines in bold, explain what the variable on the left-hand side represents and justify how the right-hand side expression is properly formulated accordingly.

```
Ts=1;
tstep=Ts/10000;
tvector=0:tstep:Ts;

Delta_f=1/Ts;
f2=2*Delta_f;

x2t=cos(2*pi*f2*tvector);
E2=sum(x2t.^2)*tstep; % numerical integration (refer to Section 2.1 of Chapter 2)
p2t=x2t/sqrt(E2); %the sampled version of ψ₂(t)
```

3.B [WWW]The m-file below implements the MLD steps in 2.C-1 to 2.C-4. The vectors **rt**, **p1t**, **p2t**, and **p3t** are the sampled versions of the received signal $r(t)$, the orthonormal basis functions $\psi_1(t)$, $\psi_2(t)$, and $\psi_3(t)$, respectively. Again, all the required integration processes such as calculating the signal energy or mapping the received signal into a point in the vector space (see equation (17.6)) are implemented using numerical integration.

```
clear
rand(1XXX);
% XXX=the last three digits of your student ID number. This is irrelevant to the main
goal of m-file, but be sure to include this.
Ts=1;
tstep=Ts/10000;
tvector=0:tstep:Ts;

Delta_f=1/Ts;
f2=2*Delta_f;

x2t=cos(2*pi*f2*tvector);
E2=sum(x2t.^2)*tstep;  % Numerical integration. Refer to Section 1 of Chapter 2 for
numerical integration.
p2t=x2t/sqrt(E2); %the sampled vector of ψ₂(t)

% Add the part to generate p1t (= the sampled version of ψ₁(t))
x1t=?;.
E1=?;
p1t=?;
% Add the part to generate p3t (= the sampled version of ψ₃(t))
x3t=?;.
E3=?;
P3t=?;

rt=cos(? ); % the equation (17.3) or ( 17.7)

% Create the vectors (the coordinates in the vector space) for each of 8-ary signals in the
equation (17.4).
s1t_in_vector_space=[?,?,?]; %=s₁
s2t_in_vector_space=[?,?,?];
...
s8t_in_vector_space=[?,?,?];

% Implement the expressions in the equation (17.6) by the numerical integration method.
Z1=sum(????)*tstep;
Z2=???;
```

```
Z3=???;
rt_in_vector_space=[?,?,?];  % = Z in the equation (17.5)

% Implement the calculation performed in 2.C-3.

ED_rt_s1t=sum(abs(rt_in_vector_space  -  ? ).^2); % Euclidean distance 'square'
between r(t) and s₁(t) in the vector space

% Add the part to generate ED_rt_s2t ~ ED_rt_s7t below.
ED_rt_s2t=sum(abs(?-?).^2);
...

...
ED_rt_s8t=sum(abs(?-?).^2);

% Implement the calculation carried out in 2.C-4.
[T1 T2]=min([ED_rt_s1t, ?,?,..., ED_rt_s8t]);
T2
```

3.B-1   Refer to the explanations provided in the comments and complete the unknown quantities marked by '**?**'. Capture the completed m-file.

3.B-2   Run the completed m-file for the received signal $r(t)$ in equation (17.3).

- (a) Record the value of **T2**.
- (b) Record the MLD result, that is, the three estimated data bits according to the result in (a).
- (c) Are the execution results consistent with the results in 2.C-4?
- (d) Repeat (a), (b), and (c) for the received signal $r(t)$ given in equation (17.7).

3.B-3   Note that in the m-file, we compare the square of the Euclidean distance, instead of the Euclidean distance itself.

- (a) Explain why it makes no difference in terms of the decision result of MLD.
- (b) What are the advantages of using the Euclidean distance square, rather than the Euclidean distance itself in terms of implementation.

17.4   MLD IN THE WAVEFORM DOMAIN

The energy of the difference between $r(t)$ and $s_i(t)$ can be calculated in the waveform domain as

$$E_{r(t)-s_i(t)} = \int_0^T |r(t) - s_i(t)|^2 dt. \tag{17.8}$$

In this section we write an m-file to compute the energy of the difference of two signals given in equation (17.8). Based on this quantity, we perform MLD for the two examples of $r(t)$ given in equations (17.3) and (17.7).

4.A [WWW]First, create vectors **s1t, s2t, s3t, ..., s8t**, which are the sampled versions of the eight basis waveforms $s_i(t)$ $(i = 1, 2, 3, ..., 8)$ in Table 17.1 as shown at the bottom of the m-file below. Refer to the explanations provided as comments and complete the unknown quantities marked by '**?**'. Capture the completed m-file.

```
clear
Ts=1;
tstep=Ts/10000;
tvector=0:tstep:Ts;

Delta_f=1/Ts;
f2=2*Delta_f;

x2t=cos(2*pi*f2*tvector);
E2=sum((x2t.^2)*tstep);
p2t=x2t/sqrt(E2);

% Add the part to generate p1t (= the sampled vector of ψ₁(t))
x1t=?;.
E1=?;
p1t=?;
% Add the part to generate p3t (= the sampled vector of ψ₁(t))
x3t=?;.
E3=?;
P3t=?;
% Add the part to generate the vectors s1t (= the sampled vector of s₁(t)), s2t,..., s8t.
% For example, 's1t' is generated as follows:
s1t=?*p1t+?*p2t+?*p3t;  %Refer to Table 17.1.
s2t=?
...
...
s8t=?
```

4.B [WWW]Add the following code fragment to the end of the m-file above to calculate the energy expressed in equation (17.8) using numerical integration. Capture the added part.

```
% Add the followings to the m-file in 4.A.
rt=cos(?);% equation (17.3) or equation (17.7)
```

E_rt_s1t= sum(abs(???).^2)*tstep; % Implement $E_{r(t)-s_i(t)} = \int_0^T |r(t) - s_i(t)|^2 dt$ by the

numerical integration.

...

...

...

E_rt_s8t= sum(????)*?tstep;

4.C   To the m-file in 4.B, further add the part to find $s_i(t)$ that minimize $E_{r(t)-s_i(t)}$ using the command '**[T1 T2]=min()**' as done in the m-file in 3.B. Capture the added part.

4.D   Run the completed m-file separately for the two cases of the received signal $r(t)$ given in equations (17.3) and (17.7).

  (a) Determine $s_i(t)$ that minimize $E_{r(t)-s_i(t)}$ for the two cases.
  (b) Are the results consistent with the results using the minimum Euclidean distance method in vector space, that is, the results in 3.B-2(a) and 3.B-2(d)?

4.E   Does the m-file completed in 3.B require more computation than the m-file in 4.D? Why?

REFERENCES

[1] D. C. Lay, *Linear Algebra and Its Applications*, 3rd ed., Boston: Addison-Wesley, 2006.

[2] G. Strang, *Linear Algebra and Its Applications*, 4th ed., Belmont, CA: Brooks Cole, 2006.

[3] S. Axler, *Linear Algebra Done Right*, 2nd ed., New York: Springer, 2002.

[4] W. Rudin, *Real and Complex Analysis*, New York: McGraw-Hill, 1987.

18

SIGNAL VECTOR SPACE AND MAXIMUM LIKELIHOOD DETECTION II

- Perform maximum likelihood detection (MLD) on M-ary signal sequence received with additive white Gaussian noise (AWGN) and detect the data bit sequence in the M-ary signal sequence.
- Generate a random data sequence received with AWGN and detect the data bits by MLD .

18.1 ANALYZING HOW THE RECEIVED SIGNAL SAMPLES ARE GENERATED

Consider a 4-ary transmission system where each signal $s_i(t)$ represents a bit pair as shown in Table 18.1.

1.A [T]For example, the bit sequence 0110110100011101001110 is mapped into the signal sequence $s_2(t)$, $s_3(t)$, $s_4(t)$, $s_2(t)$, $s_1(t)$, $s_2(t)$, $s_4(t)$?, ?, ?, ? for transmission. Complete the unknown quantities marked by '?' with the 4-ary symbols shown in Table 18.1.

1.B [WWW]From the companion website, download **st_and_rt.mat** to the work directory and execute **load st_and_rt.mat** and then **whos** in the command window. Capture the results showing the variables saved in the mat file.

Problem-Based Learning in Communication Systems Using MATLAB and Simulink, First Edition. Kwonhue Choi and Huaping Liu.
© 2016 The Institute of Electrical and Electronics Engineers, Inc. Published 2016 by John Wiley & Sons, Inc.
Companion website: www.wiley.com/go/choi_problembasedlearning

TABLE 18.1   Mapping Bit Pair and 4-Ary Signals.

Data bit pair	Transmitted signal $s_i(t)$
00	$s_1(t)$
01	$s_2(t)$
10	$s_3(t)$
11	$s_4(t)$

1.C [WWW]All the variables saved in **st_and_rt.mat** were created by executing the m-file **rt_gen.m** shown below. The vectors **s1t**, **s2t**, **s3t**, and **s4t** are the sampled versions of the 4-ary signals $s_1(t)$, $s_2(t)$, $s_3(t)$, and $s_4(t)$, respectively. The vector **rt** is the sampled version of the received signal $r(t)$ for the case where 100 data bits are transmitted according to the signal mapping in Table 18.1 over an additive white Gaussian noise channel. The details of **rt_gen.m** shown below are for your reference only. Do not run **rt_gen.m** because running it will update the variables in **st_and_rt.mat**.

```
clear
T=1; %M(4)-ary signal duration
L=32; % Number of samples per 4-ary symbol (length of the sampled versions of
4-ary symbols below)
tstep=T/L;
tvector=tstep:tstep:T;

%%%%%%%%%%%%%%%% 4-ary symbol set generation %%%%%%%%%%%%%%%%
p1t=sqrt(2)*cos(3*pi*tvector); % z-axis orthonormal basis for 2-D vector space
p2t=sqrt(2)*sin(3*pi*tvector); % y-axis orthonormal basis
a= ? ;b= ?;c= ?;d= ?;e= ?;f= ?;g= ?;h= ? ; %Intentionally not shown for the problem.
s1t=a*p1t + b*p2t;
s2t=c*p1t + d*p2t;
s3t=e*p1t + f*p2t;
s4t=g*p1t + h*p2t;
%%%%%%%%%%%%%%%%%%%%%%%%%%%%%%%%%%%%%%%%%%%%%%%

%%%%%% data bit  stream generation %%%%%%%%%%%%%%%%%%%%%%%%%%%
databits=(rand(1,100)>0.5);  % databits = data bit stream
%%%%%%%%%%%%%%%%%%%%%%%%%%%%%%%%%%%%%%%%%%%%%%%%%%%%

%%%%%%%%%%%%% Transmitter signal x(t) generation %%%%%%%%%%%%%%%
xt=[] %Initialize the transmit signal vector
for k=1:50   %% 100 bits = ? 4-ary symbols
    %%% 2 bits => 4-ary symbol Mapping %%%%%%%%%%%%%%%%%%%%%%
    if databits([2*k-1, 2*k]) == [0 0]
        st=s1t;
```

```
    elseif databits([2*k-1, 2*k]) == [0 1]
        st=s2t;
    elseif databits([2*k-1, 2*k]) == [1 0]
        st=s3t;
    else
        st=s4t;
    end
    %%%%%%%%%%%%%%%%%%%%%%%%%%%%%%%%%%%%%%%%%%%%%

    xt=[xt st]; %Concatenate 'st' to 'xt'. Consequently, 'xt' becomes a transmit signal
vector for 100 bit data stream.
end

%%%%%%%%%%%%%%%%%%%%%%%%%%%%%%%%%%%%%%%%%%%%%%%%

xt_len=L*50;  % or length(xt), i.e., vector length of 'xt'

%%%%%%%%% Received signal after AWGN channel %%%%%%%%%%%%%%%%%
noise_sample=6*randn(1,xt_len);  %AWGN
rt=xt+noise_sample;  %noise addition
%%%%%%%%%%%%%%%%%%%%%%%%%%%%%%%%%%%%%%%%%%%%%%%%%

save st_and_rt.mat L T tstep tvector  p1t p2t s1t s2t s3t s4t rt
save data_bits.mat databits
```

Examine **rt_gen.m** above and then complete the following problems.

1.C-1   Complete all unknown quantities marked by '?' in the following description.
(a) The vectors **s1t**, **s2t**, **s3t**, and **s4t**, the sampled versions of the 4-ary symbols,
are created from a linear combination of the two vectors '**?**' and '**?**', which are the
sampled versions of the two orthogonal basis functions $\psi_1(t)$ and $\psi_2(t)$ for a 2-D
vector space.

(b) In the two code lines that generate two orthogonal bases, **tvector** contains
the sampling time instants. The vectors **s1t**, **s2t**, **s3t**, **s4t** as well as **p1t** and **p2t**
have the same sample interval and signal duration in seconds as those of **tvector**.
Therefore the duration of the 4-ary symbols is '**?**' seconds, and the sampling interval
of their sampled versions is determined by the variable '**?**', which has a value of '**?**'
seconds.

1.C-2.   The vector **data_bits** is the transmitted bit stream with a length of '**?**' bits. It
is converted into '**?**' (how many) 4-ary symbols.

1.C-3   As commented in the m-file above, **p1t** and **p2t** are the sampled versions of
the two orthogonal basis functions $\psi_1(t)$ and $\psi_2(t)$, from which **s1t**, **s2t**, **s3t**, and **s4t**
are generated by linear combination.

(a) Execute the following lines and capture the results.

```
>>load st_and_rt.mat
>>sum(p1t.^2)*tstep
>>sum(p2t.^2)*tstep
>>sum(p1t.*p2t)*tstep
```

(b) The last three commands in (a) perform numerical integration (see Section 2.1 of Chapter 2). For example, the first numerical integration in (a) corresponds to $\int_0^1 (\psi_1(t))^2 dt$. Write the mathematical expressions that correspond to the other two numerical integrations.

(c) Based on the execution results in (a), explain why **p2t** and **p1t** are two orthonormal basis functions. Ignore the numerical integration error.

1.C-4   The vector **xt** is the sampled version of the 4-ary symbol stream to be transmitted. Note that **xt** is formed by concatenating the sampled vectors **s1t**, **s2t**, **s3t**, and **s4t** one by one, which are selected according to every two-tuple of data bits and the mapping rules in Table 18.1. In the final step of the m-file, the sampled version of the received signal **rt** is generated from **xt**. Explain what this final step does.

1.C-5   The first **L** samples of **rt**, that is, **rt(1:L)**, are the sampled received signal for the first transmitted 4-ary symbol; the next **L** samples, that is, '**rt((L+1):(L+L))**', are the sampled received signal for the second transmitted 4-ary symbol, and so on. Properly set the sample indexes in **r(?:?)** to obtain the sampled received signal for the *n*th transmitted 4-ary symbol.

18.2   OBSERVING THE WAVEFORMS OF 4-ARY SYMBOLS AND THE RECEIVED SIGNAL

2.A   We can plot $s_1(t)$, using the variables saved in **st_and_rt.mat** through the following three lines of code. Plot the graphs of $s_1(t)$, $s_2(t)$, $s_3(t)$, and $s_4(t)$ and capture them.

```
>>load st_and_rt.mat
>>plot(tvector, s1t);
>>grid on;
```

2.B   The command below plots the part of the received signal that contains the first six 4-ary symbols (12 data bits).

```
>>plot(tstep:tstep:(6*T), rt( 1 :( 6*L) ) ); xlabel('t [sec]')
```

(a) Execute the command above and capture the plot.
(b) Compare the received signal waveform over the first 4-ary symbol period with each of 4-ary waveforms in 2.A and determine which one of $\{s_1(t), s_2(t), s_3(t), s_4(t)\}$ is transmitted. Repeat this process for the last five 4-ary symbols.
(c) If a decision cannot be made using visual inspection, explain why.

18.3   MAXIMUM LIKELIHOOD DETECTION IN THE VECTOR SPACE

The transmitted bit stream **data_bits** is not saved in **st_and_rt.mat**. In this section we estimate **data_bits** from the received signal sample vector **rt** using the other sampled vectors, **p1t, p2t ,s1t, s2t, s3t, s4t**, saved in **st_and_rt.mat**. To this end, go through the following steps.

3.A   If M-ary signals $s_1(t), ..., s_M(t)$ are expressed as a linear combination of N orthonormal basis functions $\psi_n(t), n = 1, ..., N$, that is, as

$$s_i(t) = \sum_{k=1}^{N} a_{ik}\psi_k(t), \ i = 1, ..., M \quad \text{(generally } N \leq M), \tag{18.1}$$

then each of the M-ary signals can be mapped into a point in the N-dimensional vector space, which is spanned by the N orthonormal basis functions $\psi_n(t), n = 1, ..., N$ as

$$s_i(t) \Rightarrow \mathbf{s_i} = \left(a_{i1}, a_{i2}, ..., a_{iN}\right). \tag{18.2}$$

The value of a_{ij} can be calculated from $s_i(t)$ and the orthogonal basis functions $\psi_n(t), \ n = 1, ..., N$, as

$$\begin{aligned} a_{ij} &= \int_{t_1}^{t_2} s_i(t)\psi_j^*(t)dt \\ &= \int_{t_1}^{t_2} s_i(t)\psi_j(t)dt \text{ for real-valued } \psi_j(t), \end{aligned} \tag{18.3}$$

where t_1 and t_2 are the boundaries of the symbol period, which are common to all of $s_1(t), ..., s_M(t)$ and $\psi_n(t), n = 1, ..., N$.

3.A-1   [T]Substitute equation (18.1) into the right-hand side of equation (18.3) and show that it equals a_{ij}.

3.A-2   Next, using equation (18.3), we convert the 4-ary ($M = 4$) signals in **rt_gen.m** into some of the points in the vector space expressed in equation (18.2).

(a) Determine the value of N from the file **rt_gen.m**.
(b) The integral in equation (18.3) can be calculated by using numerical integration. The m-file below calculates a_{11} and a_{12} in equation (18.3), which correspond to the (x, y) coordinates of a point **s1** in the 2-D vector space

for the signal **s1t**. Complete the m-file by properly filling in the quantities marked by '**?**'. Revisit Section 18.1 to review the variables that correspond to the sampled versions of $s_1(t), \ldots, s_4(t)$ and $\psi_n(t)$, $n = 1, 2(=N)$.

Capture the completed m-file and the execution results.

```
clear
load st_and_rt.mat
a11=sum(s1t.*p1t)*tstep;%numerical integration  a11 = ∫₀¹ s₁(t)ψ₁(t)dt
a12=sum(????)*tstep;  %numerical integration
s1=[a11,a12]
```

3.A-3   Expand the m-file in 3.A-2 by adding the code lines that calculate the vector space coordinates of the remaining 4-ary signals **s2t**, **s3t**, and **s4t**. Capture the completed m-file and the execution results.

3.B   [WWW]The following steps complete the maximum likelihood detection (MLD) process. Go through the following steps and complete the code fragment below. Then append the completed code fragment to the end of the m-file in 3.A-3.

- Step 1. Set **n** = 1. This variable is used as the time index of the 4-ary symbol stream in **rt** (the sampled version of $r(t)$).
- Step 2. Recall that the variable **L** is the number of samples per 4-ary symbol in **rt**. Therefore, if **rt_nth** denotes the sampled vector for the nth received 4-ary symbol in **rt**, then it is extracted from **rt** as '**rt_nth = rt(((n-1)*L+1) : n*L);**'.
- Step 3. As done in 3.A-2 and 3.A-3, convert **rt_nth** into a point in the vector space denoted by **z**, whose (x,y) coordinates are denoted by **z1** and **z2**, respectively.
- Step 4. Calculate the Euclidean distances between z and all the points of the 4-ary symbols in the vector space. The 4-ary symbol that is closest to **z** will be chosen as the 4-ary symbol transmitted.
- Step 5. De-map the detected 4-ary symbol into 2 bits using Table 18.1.
- Step 6. Increase the value of **n** by 1 through '**n=n+1**' to proceed to detect the next received 4-ary symbol. Return to Step 2 and complete the process again.

```
% Add the following fragment to the m-file in 3.A-3
data_bits_hat=[];   % Initialize a vector to concatenate the demodulated bit steam.
for n=1:50
  rt_nth= rt( ((n-1)*L+1) : n*L );  % Step 2
  z1= sum(rt_nth.*p1t)*tstep; z2= ???; % Step 3
  z=[z1,z2];                % Step 3
  ED_z_s1=sum(abs(z-s1).^2);        %Step 4: Euclidean Distance between z and s1.
  ED_z_s2=???;
  ED_z_s3=???;
  ED_z_s4=???;
```

```
    [T1 T2]=min([ED_z_s1, ED_z_s2, ED_z_s3,ED_z_s4]); %Step 4: Find the clos-
est point to z.
    if T2==1
        twobits_hat= [0 0] ; % Step 5: Set twobits_hat by the detected two bits accord-
ing to Table 18.1.
    elseif T2==2
        ??? ;
    elseif T2==3
        ??? ;
    else
        ??? ;
    end

    data_bits_hat=[data_bits_hat twobits_hat];%Concatenate the detected bits until all of
    4-ary symbols are detected.
end
```

3.B-1   Capture the completed m-file. Be sure to save the completed m-file because it will be used in the next chapter.

3.B-2   Remove all the existing comments and add a comment to explain what each line does. Especially for the lines with '=', explain what the variable on the left-hand side represents and justify how the right-hand side expression is properly formulated accordingly.

3.C   Run the m-file above. Then execute **data_bits_hat(1:12)** in the command window to display the first 12 demodulated bits. Capture the results.

3.D   [WWW]Download **data_bits.mat** from the companion website to the work folder and execute **load data_bits.mat** in the command window. The variable **databits** saved in **data_bits.mat** is the actual transmitted data bits that were used to create **rt** in file **rt_gen.m**.

(a) Execute **databits(1:12)** and capture the results.
(b) Compare the transmitted bits with the detection results in 3.C. Are there any bit errors? To check the number of bit errors, execute **sum(data_bits_hat~=databits)** in the command window.

3.E   [WWW]As in Section 17.4 of Chapter 17, the m-file below performs MLD in the waveform domain; that is, it demodulates the transmitted data bit stream by calculating the energy of the difference between **rt_nth** and each of **s1t, s2t, s3t,** and **s4t**. Complete this m-file and then capture it. Be sure to save the completed m-file because it will be needed in the next chapter.

```
clear
load st_and_rt.mat
data_bits_hat=[];
```

```
for n=1:50
  rt_nth= rt( ((n-1)*L+1) : n*L );

  E_rt_s1t=sum(abs(?-?).^2)*?; %Numerical integration to calculate the energy of
  the difference between rt_nth and s1t.
  ...
  ...
  E_rt_s4t=??;

  [T1 T2]=min([E_rt_s1t,?,?,?]);
  if T2==1
   twobits_hat=? ;
  elseif ??
  ...
  elseif ??
  ...
  else
  ...
  end

  data_bits_hat=[data_bits_hat  twobits_hat];

end
```

3.F   Execute the m-file above and then execute **data_bits_hat(1:12)** in the command window to display the first 12 demodulated bits. (a) Capture the results. (b) Are the transmitted bits correctly demodulated?

19

CORRELATOR-BASED MAXIMUM LIKELIHOOD DETECTION

- Investigate the statistical properties of additive white Gaussian noise (AWGN) in the vector space.
- Implement a correlation-based maximum likelihood detector.

19.1 STATISTICAL CHARACTERISTICS OF ADDITIVE WHITE GAUSSIAN NOISE IN THE VECTOR SPACE

1.A We check the statistical characteristics of (AWGN) in the vector space through simulation.

1.A-1 [WWW]Recall that the file **st_and_rt.mat** used in Chapter 18 contains the sampled versions of the 4-ary signals **s1t**, **s2t**, **s3t**, and **s4t**, the sampled versions of the two orthogonal basis functions **p1t** and **p2t**, and the sampled version of the received signal **rt**.

In the m-file below, we generate **rt** for the case where only the AWGN is received and replace the original received signal **rt** saved in **st_and_rt.mat**. In this case the sample length of **rt** is set to 100,000 times **L**, which is the sample length of 4-ary symbols.

Problem-Based Learning in Communication Systems Using MATLAB and Simulink, First Edition.
Kwonhue Choi and Huaping Liu.
© 2016 The Institute of Electrical and Electronics Engineers, Inc. Published 2016 by John Wiley & Sons, Inc.
Companion website: www.wiley.com/go/choi_problembasedlearning

Create the m-file below.

```
clear
load st_and_rt.mat
noise_sample= 2*randn(1,L*100000);   %AWGN
rt=noise_sample;   % AWGN only received signal
```

1.A-2   [WWW]As done in the Section 3.B of Chapter 18, the code fragment below converts 70,000 received symbols in **rt** into their corresponding points in the vector space. The coordinates of the points in the vector space are calculated. Note that **rt** contains noise only.

It is informative to see whether or not **z1** and **z2** are correlated. As one of empirical ways to check the correlation between **z1** and **z2**, we collect **z2** when **z1** falls in two disjoint regions **R1** and **R2** and then check whether or not the conditional probability density function (PDF) of **z2** on the two regions of **z1** are the same. In the m-file, for illustration, **R1** and **R2** are set to [0 0.1] and [0.2 0.3], respectively. If **z1**∈**R1**, then **z2** is appended to the vector **z2when_z1inR1** as its new element; if **z2**∈**R2**, then **z2** is appended to **z2when_z1inR2** as its new element.

(a) Complete the two quantities marked by '**?**' in the code fragment below.
(b) Append this code fragment to the m-file completed in 1.A-1 and capture the completed m-file.

```
z1s=[]; z2s=[]; z2when_z1inR1=[]; z2when_z1inR2=[];
for n=1:70000

    rt_nth= rt( ((n-1)*L+1): n*L ); %Received signal sample vector during the n-th 4-ary
symbol duration.

    z1=sum(rt_nth.*?)*tstep; % Refer to (18.3) for converting a waveform into a point in
a vector space.
    z2=sum(rt_nth.*?)*tstep; % Refer to (18.3) for converting a waveform into a point in
a vector space.

    if 0<=z1 & z1<=0.1
        z2when_z1inR1=[ z2when_z1inR1 z2]; %Collect z2 samples only whey z1 falls in
the region R1.
    end

    if 0.2<=z1 & z1<=0.3
        z2when_z1inR2=[ z2when_z1inR2 z2];
    end

    z1s=[z1s z1]; z2s=[z2s z2];
end
```

1.B   Execute the m-file completed in 1.A-2 and then execute the following in the command window to plot the empirical distributions of z1 and z2.

```
>>figure
>>hist(z1s,100)
>>figure
>>hist(z2s,100)
```

1.B-1   Capture the resulting plots.

1.B-2   Do the histograms look like Gaussian distributions?

1.B-3   [T,A]Note that the elements of vector **rt_nth** are realizations of i.i.d. Gaussian random variables with a mean 0 and variance 4. Also, from the code lines that generate **z1** and **z2** in the m-file, **z1** and **z2** are the weighted sums of the elements of **rt_nth** with the weighting vectors **p1t** and **p2t**, respectively. From these observations, explain why **z1** and **z2** should follow the Gaussian distribution.

1.C   Execute the following in the command window to calculate the mean and variance of **z1s** and **z2s**.

```
>>mean(z1s)
>>var(z1s)
>>mean(z2s)
>>var(z2s)
```

1.C-1   Capture the execution results.

1.C-2   Are the means and variances of **z1** and **z2** approximately equal? Ignore small errors due to a finite number of samples simulated.

1.C-3   [T,A]From the note made in 1.B-3 and using equations (15.5) and (15.6), we can derive the theoretical mean and variance of **z1** and **z2**.
   (a) Prove that the theoretical means of **z1** and **z2** are both zero.
   (b) From equation (15.6), explain why the theoretical variances of **z1** and **z2** can be calculated by the following commands.

```
>>z1variance= sum((p1t*tstep).^2)*4
>>z2variance= sum((p2t*tstep).^2)*4
```

   (c) Execute the commands above and capture the results. Are the simulation results in 1.C-1 approximately equal to the theoretical results?
   (d) Using the fact that **z1** and **z2** are Gaussian and their theoretical mean and variance are known, write the exact PDFs of **z1** and **z2**.

1.D Execute the following in order to plot the histograms of the elements in **z2when_z1inR1** and **z2when_z1inR2** and calculate their variances.

```
>>figure
>>hist(z2when_z1inR1,30)
>>figure
>>hist(z2when_z1inR2,30)
>>var(z2when_z1inR1)
>>var(z2when_z1inR2)
```

1.D-1 Capture the plots and the calculated variances.

1.D-2 From the results in 1.D-1, determine the distributions of the elements of **z2when_z1inR1** and **z2when_z1inR2**. Are they Gaussian and are their mean and variance the same as those of the elements of **z2s**?

Note that compared with the histograms obtained in 1.B-1, here the number of bins along the x axis is decreased to 30 because the number of samples of **z2when_z1inR1** and **z2when_z1inR2** is much smaller than that of **z2s**.

1.D-3 Set different regions for **z1**, that is, **R1** and **R2**, for example, **R1** = [−2 0], **R2** = [3 6]. Then execute the m-file completed in 1.A-2 for each setting. After each simulation, plot the histograms of the elements of **z2when_z1inR1** and **z2when_z1inR2** and check their variances as done in 1.D-1. Note that if **R1** or **R2** is set further away from 0, then the number of samples of **z2** that fall into that region becomes smaller. Thus the histogram will become less accurate.

(a) Judged from the simulation results, are the distributions of the elements of **z2when_z1inR1** and **z2when_z1inR2** identical regardless of **z1**'s region?

(b) If the answer in (a) is yes, then it means that **z1** and **z2** are independent. Explain why this result can verify the independence of **z1** and **z2**.

1.E [T,A]In this subsection, we mathematically investigate the statistical properties of **z1** and **z2**. From equation (18.3), the coordinates denoted by (z_1, z_2) of the received noise in the vector space can be calculated as

$$
z_1 = \int_0^T n(t)\psi_1(t)dt \text{ for real-valued } \psi_1(t),
$$

$$
\tag{19.1}
$$

$$
z_2 = \int_0^T n(t)\psi_2(t)dt \text{ for real-valued } \psi_2(t),
$$

where $n(t)$ is the AWGN, T is the duration of the 4-ary signal, which is set to be 1 in the m-file, and the two orthogonal basis functions $\psi_1(t)$ and $\psi_2(t)$ considered in the m-file are given as

$$\begin{aligned}
\psi_1(t) &= \sqrt{2}\cos(3\pi t), \quad 0 \le t \le 1, \\
\psi_2(t) &= \sqrt{2}\sin(3\pi t), \quad 0 \le t \le 1.
\end{aligned} \tag{19.2}$$

Note that the vector **noise_sample** in the m-file is the sampled version of $n(t)$ in equation (19.1). Also, from the answer to 1.A-2(a), **p1t** and **p2t** in the m-file are the sampled versions of $\psi_1(t)$ and $\psi_2(t)$ in equation (19.2), respectively. Consequently **z1** and **z2** in the m-file are the numerically represented versions of z_1 and z_2 in equation (19.1).

1.E-1   Explain why z_1 and z_2 in equation (19.1) follow the Gaussian distribution.

1.E-2   The autocorrelation of AWGN $n(t)$ is given as $N_0/2\delta(\tau)$, where $N_0/2$ is the two-sided power spectral density of $n(t)$ [1]. Show that the mean and variance of both z_1 and z_2 are 0 and $N_0/2$, respectively.

1.E-3   Show that the correlation between z_1 and z_2, $E[z_1 z_2]$ is zero.

1.E-4   Can we say that z_1 and z_2 are independent Gaussian random variables?

1.E-5   Revise $\psi_2(t)$ as $\psi_2(t) = \sqrt{2}\,\sin(3\pi t + \pi/3)$ $(0 \le t \le 1)$, which is not orthogonal to $\psi_1(t)$. Show that the correlation between z_1 and z_2, $E[z_1 z_2]$, is 0.5 now.

1.F   In 1.C-3(c), we showed that if the received signal contains noise only, that is, '**rt = noise_sample**', then, the (x,y) coordinates in vector space, that is, **z1** and **z2**, are zero-mean Gaussian random variables with a variance 0.125. In 3.A-2 of Chapter 18, we converted the 4-ary signals into points in the vector space. For example, the sample vector of the first 4-ary signal, **s1t**, is mapped into (**a11, a12**).

1.F-1   Consider the case that **s1t** is received with noise. This can be implemented by inserting '**rt_nth=rt_nth+s1t**' below the line '**rt_nth= rt(((n-1)*L+1): n*L);**' in the m-file in 1.A-2. Then, **z1** will be a Gaussian random variable with mean **a11** and variance 0.125, and **z2** will be a Gaussian random variable with mean **a12** and variance 0.125. Justify these conclusions.

1.F-2   Now consider the case that **s3t** is received with noise. Determine the PDFs of **z1** and **z2** for this case.

1.G   [A]Recall from **rt_gen.m** in Section 1.C of Chapter 18 that **p2t** saved in **st_and_rt.mat** is the sampled version of $\psi_2(t)$, that is, **sqrt(2)*sin(3*pi*tvector)**. Let us modify **p2t** to be the sampled version of $\psi_2(t)$ given in 1.E-5. To this end, add the following line before the '**for**' statement in the m-file in 1.A-2.

```
p2t=sqrt(2)*sin(3*pi*tvector+pi/3);
```

1.G-1   Execute the modified m-file and then repeat 1.B-1, 1.B-2, 1.C-1, 1.C-2, 1.D-1, and 1.D-2. In which problem(s) is the result different from the original result obtained before **p2t** is modified? Ignore the small difference between the results due to a finite number of samples generated in the simulation.

Capture the results that are different from original ones.

1.G-2   Based on the result in 1.G-1, are **z1** and **z2** still independent after **p2t** is modified? Justify your answer using the result in 1.G-1.

1.G-3   For the problems that have the same results as the original ones, explain why they are not changed even after **p2t** is modified.

1.H   Here we investigate the effect of the orthogonal basis vectors on the noise vector.

1.H-1   If the basis vectors in the vector space are mutually orthogonal, then the elements of the Gaussian noise vector in the vector space are independent of one another. Among the problems completed in this chapter so far, which one and its answer empirically verify this? Also which problem and its answer theoretically verify this?

1.H-2   If the basis vectors in the vector space are not mutually orthogonal, then the elements of noise vector in the vector space will be dependent. Among the problems completed in this chapter so far, which problem and its answer empirically verify this? Also which problem and its answer theoretically verify this?

19.2   CORRELATION-BASED MAXIMUM LIKELIHOOD DETECTION

Consider an M-ary signal transmission system in an AWGN environment. Let T denote the M-ary symbol duration. Denote the complex-valued M-ary signals by $s_i(t)(i = 1, 2, ..., M)$ and the received signal by $r(t)$.

2.A   [T]The left-hand side of equation (19.3) is the index of the estimated signal that results in the minimum squared error with the received signal.

$$\arg\min_k \int_0^T |r(t) - s_k(t)|^2 dt = \arg\max_k \text{Re}\left[\int_0^T r(t)s_k^*(t)dt\right]. \quad (19.3)$$

Equation (19.3) holds if all the M-ary signals $s_i(t)$ ($i = 1, 2, ..., M$) have the same energy, that is, $E_{s_1(t)} = E_{s_2(t)} = ... = E_{s_M(t)}$. Note that the term $\int_0^T r(t)s_k^*(t)dt$ on the right-hand side is the correlation between $r(t)$ and $s_k(t)$. This equation shows that for signaling schemes such as MPSK and MFSK, for which all the M-ary signals $s_i(t)$ ($i = 1, 2, ..., M$) have the same energy, MLD could be implemented on the basis of correlation, rather than the Euclidean distance. Such implementation results in the so called "correlation-based maximum likelihood receiver."

2.A-1 Prove equation (19.3).

2.A-2 (a) Prove $\int_0^T r(t)s_k^*(t)dt = \langle z, s_k^* \rangle$, where z and s_k denote, respectively, the coordinates of $r(t)$ and $s_k(t)$ in the vector space, and $\langle z, s_k^* \rangle$ denotes the inner product of z and s_k.

(b) Using the identity in (a), show that equation (19.3) can be written as $\arg\min_k \int_0^T |r(t) - s_k(t)|^2 dt = \arg\max_k \text{Re}[< z, s_k^* >]$.

2.A-3. Describe the advantage of using $\arg\max_k \text{Re}[< z, s_k^* >]$ over using the left- or right-hand sides of equation (19.3) for detection in terms of computational complexity.

2.B [WWW]The m-file below demodulates a received signal using the correlation-based MLD. As done in Chapter 18, we first load the sampled version of $r(t)$, that is, **rt**, and the sampled version of $s_i(t)$, that is, **s1t, s2t, s3t,** and **s4t**, from **st_and_rt.mat**. Recall that in Chapter 18, we have already demodulated **rt** in **st_and_rt.mat** by using the Euclidean distance-based MLD.

In the line '**C_rt_nth_s1t=sum(rt_nth.*s1t)*tstep**', the right-hand side **sum(rt_nth.*s1t)*tstep** is the numerically calculated correlation of the received signal with one of the M-ary waveforms $s_1(t)$, that is, $\int_0^T r(t)s_1^*(t)dt$. Since $r(t)$ and $s_i(t)$ ($i = 1,2,..., M$) are all real-valued in the considered system, we do not need an additional operation to take the real part. In the next line '**[T1 T2]=max([C_rt_nth_s1t,?,?,?])**', we find which one among $s_i(t)$ ($i = 1,2,..., M$) has the highest correlation with $r(t)$.

```
clear
load st_and_rt.mat

data_bits_hat=[];   %Initialize the vector for saving the demodulated bits
for n=1:50
rt_nth= rt( ((n-1)*L+1) : n*L ); %Sampled version of received signal during n-th M-ary
symbol duration

  C_rt_nth_s1t=sum(rt_nth.*s1t)*tstep; %Correlation between the received signal and
s1(t) by the numerical integration.
...

...
  C_rt_nth_s4t=???;

  [T1 T2]=max([C_rt_nth_s1t,?,?,? ]); %Find the most highly correlated M-ary signal
with the received signal.

  if T2==1
    twobits_hat= ??  %Refer to Table 18.1 for mapping data bit pair and 4-ary symbol.
```

```
elseif T2==2

...

else

...

end


data_bits_hat=[data_bits_hat  twobits_hat]; % Connect the detected bit to bit stream
end
```

2.B-1   Complete the m-file and capture it.

2.B-2   Execute the following in the command window to check whether all of the 100 bits are correctly demodulated. (a) Capture the result. (b) How many bits are incorrectly demodulated?

```
>>load data_bits.mat
>>sum(data_bits_hat ~= databits)
```

2.B-3   In Chapter 18, we performed the Euclidean distance-based MLD on the same received sample vector **rt** and confirmed that all of the 100 data bits are correctly demodulated. The study in 2.A shows that the correlation-based detection is equivalent to the Euclidean distance-based MLD. However, the result in 2.B-2 should show that correlation-based detection results in some errors.

Identify the aspects in the considered case that have caused the correlation-based detection being not equivalent to the MLD.

2.C   [T]Let us simulate the case where the 4-ary symbols have the same energy. By properly setting the values of **a, b, c, ..., h** in **rt_gen.m** as was analyzed in Chapter 18, **s1t, s2t, s3t**, and **s4t** could have the same energy. Then we can regenerate a new **rt** on the basis of the modified **s1t, s2t, s3t**, and **s4t**.

2.C-1   Assume that a signal $s(t)$ is mapped into a vector **s** in a vector space. Calculate the energy of $s(t)$ from **s**.

2.C-2   [WWW]From the companion website, download **rt_gen.m** into your current MATLAB work folder.

(a) Open **rt_gen.m** and set a=2 and b=−1, which are the (x, y) coordinates of **s1t** in the vector space. Show that the energy of **s1t** equals 5 using the answer to 2.C-1.

(b) Properly set the values of **c, d, e, f, g**, and **h** so that the other three signals, **s2t, s3t**, and **s4t**, all have the same energy as **s1t**, that is, 5. Note that there are multiple sets of solutions. Select one set of them.

2.D   Run **rt_gen.m** that was completed in 2.C-2 to create the equi-energy 4-ary signals **s1t, s2t, s3t,** and **s4t,** and regenerate the received signal **rt** using these newly created 4-ary signals.

2.D-1   Identify and record the file creation date of **st_and_rt.mat**, which contains **s1t, s2t, s3t, s4t,** and **rt**. Is it updated?

2.D-2   Save the m-file created in Section 3.B of Chapter 18 in the current work folder where the updated version of **st_and_rt.mat** resides. Then execute the m-file to perform MLD on the updated **rt** with the updated **s1t, s2t, s3t,** and **s4t**. After this is done, execute **'data_bits_hat_by_MLD=data_bits_hat'** in the command window. Capture the execution result.

2.D-3   Execute the m-file completed in 2.B-1 again to perform the correlation-based detection on the updated **rt** with the updated **s1t, s2t, s3t,** and **s4t**.

(a) After running the m-file, execute **sum(data_bits_hat ~= data_bits_hat_by_ MLD)** in the command window. Capture the execution result. Explain what this command does.

(b) Does the result in (a) show that the correlation-based detector and MLD perform the same?

(c) Document the condition(s) that guarantees the equivalence between the MLD and the correlation-based detection.

REFERENCE

[1] A. Papoulis, *Probability, Random Variables, and Stochastic Processes*, 3rd ed., New York, NY: McGraw-Hill, 1991.

20

PULSE SHAPING AND MATCHED FILTER

- Perform raised cosine pulse shaping and plot the eye diagram.
- Investigate the spectrum and eye diagram for different roll-off factors of raised cosine pulse shaping.
- Perform matched filtering to the pulse-shaped signal and analyze the eye diagram.
- Convert a MATLAB signal into an actual electric signal and observe the eye diagram in an oscilloscope.

20.1 [T]RAISED COSINE PULSES

The built-in MATLAB function **rcosine(1,L,'normal',r)** outputs the sampled version of a raised cosine pulse [1–3]. The argument **r** is a desired roll-off factor and the argument **L** is the desired number of samples per symbol. For example, the raised cosine pulse with a roll-off factor of 0.5 can be plotted in MATLAB as follows.

```
>>Ts=1; % symbol duration
>>L=16; % number of samples per symbol
>>r=0.5; %roll-off factor
>>t=-3:Ts/L:3; % time vector for x-axis
>>pt =rcosine(1,L,'normal',r); % sampled version of the raised cosine pulse
```

Problem-Based Learning in Communication Systems Using MATLAB and Simulink, First Edition.
Kwonhue Choi and Huaping Liu.

```
>>plot(t,pt)
>>grid on
>>hold on
```

1.A   Execute the commands above. Then generate and plot **pt** for each of the following values of the roll-off factor: 0, 0.25, 0.75, and 1. Plot the four curves with different line colors and use **legend( )** to indicate the corresponding roll-off factor for each curve.

1.B   Note that **rcosine()** generates the truncated version of the raised cosine pulse centered at time zero. Record the length of the truncated raised cosine pulse using the symbol period as the unit.

1.C   (a) From the related literature or textbooks (see the content mapping table at the beginning of this book), review and summarize the conditions on the pulse for zero intersymbol interference (ISI) [2–4]. (b) Based on the shape of **pt**, determine whether or not ISI exists if **pt** is used for pulse shaping. Justify your answer?

1.D   If the parameter **normal** in **rcosine(1,Ns,'normal',r)** is replaced by **sqrt**, then this function creates the sampled version of the square-root raised cosine (SRRC) pulse. Replace **normal** by **sqrt** and repeat 1.A to plot the SRRC pulses for each of the following roll-off factors: 0, 0.25, 0.5, 0.75, and 1.

1.E   Based on the shape of **pt** in 1.D, determine whether ISI exists if the SRRC pulse is used for pulse shaping. Justify your answer.

NOTE: At this point we are checking the ISI condition with the pulse-shaped signal itself, not the signal after any additional processing such as matched filtering in the receiver (will be discussed in Section 20.3). The overall pulse shape is determined by filters applied in both the transmitter and the receiver.

20.2   PULSE SHAPING AND EYE DIAGRAM

2.A   Consider a system where the pulse shape $p(t)$ is a triangular waveform as shown in Fig. 20.1.

2.A-1   The pulse-shaping process is equivalent to performing convolution between the impulse-modulated data stream and the pulse $p(t)$. First, consider a case where signal $x_1(t)$ contains a single data symbol that is impulse modulated at time zero, as

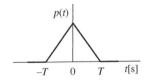

FIGURE 20.1   Triangular pulse $p(t)$.

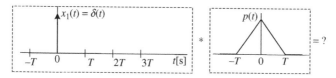

FIGURE 20.2   Signal $x_1(t)$ and pulse $p(t)$.

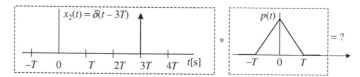

FIGURE 20.3   Signal $x_2(t)$ and pulse $p(t)$.

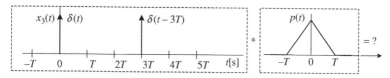

FIGURE 20.4   Signal $x_3(t)$ and pulse $p(t)$.

shown on the left of Fig. 20.2. Sketch the pulse-shaped signal, that is, the convolution of $x_1(t)$ and $p(t)$.

2.A-2   Second, consider a case where signal $x_2(t)$ contains a single data symbol that is impulse modulated at $t = 3T$, as shown on the left of Fig. 20.3. Sketch the pulse-shaped signal, that is, the convolution of $x_2(t)$ and $p(t)$.

2.A-3   Third, consider a case where signal $x_3(t)$ contains two data symbols that are impulse modulated, one is at $t = 0$ and the other at $t = 3T$, as shown on the left of Fig. 20.4. Sketch the pulse-shaped signal, that is, the convolution of $x_3(t)$ and $p(t)$.

2.A-4   Now consider the general case of data signal $x(t)$ as shown on the left of Fig. 20.5, where four consecutive bits are impulse modulated. One might think that by using the linearity property, the pulse-shaped signal should be easily obtained as shown in Fig. 20.6. Explain why the sketch in Fig. 20.6 is incorrect, and sketch the correct pulse-shaped waveform.

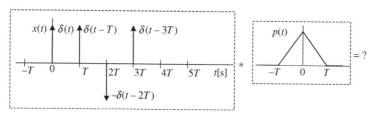

FIGURE 20.5   General signal $x(t)$ and the pulse $p(t)$.

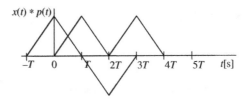

FIGURE 20.6   Incorrect sketch of $x(t)$ pulsed-shaped by $p(t)$.

2.B [WWW]From the companion website, download **tx_sig_gen.m** into your MATLAB work folder. The m-file **tx_sig_gen.m** shown below creates **tx_signal**, that is, the sampled version of the transmitted signal $s(t)$. This signal contains 100 randomly generated binary data bits that are pulse shaped by the raised cosine pulse with a roll-off factor of 0.25.

```
%tx_sig_gen.m
clear
rand(1,1XXX); % XXX= the last three digits of your student ID. Irrelevant to the goal
of this m-file but mandatory.
Ts=1;
L=16;
t_step=Ts/L;
%%%%%%%%%<1. Pulse waveform generation > %%%%%%%%%%%%%%%%%%%%%
pt=rcosine(1,L,'normal',?);

%%%%%%%%%<2. 100 bit (binary symbol) generation>%%%%%%%%%%%%%%%%%%%
Ns=?;
data_bit=(rand(1,Ns)>0.5); % You may alternatively use 'data_bit=randi([0 1],1,Ns)'
instead.

%%%%%%%%%<3. Unipolar to Bipolar (amplitude modulation)>%%%%%%%%%%%%%
amp_modulated=2*data_bit-1; % 0=> -1, 1=>1

%%%%%%%%%<4. Impulse modulation >%%%%%%%%%%%%%%%%%%%%%%%%%%%%%
impulse _modulated=[];
for n=1:Ns
  delta_signal= [amp_modulated(n) zeros(1, L-1)];
  impulse_modulated=[impulse_modulated  delta_signal];
end

%%%%%%%%%<5.Pulse shaping (Transmitter filtering)>%%%%%%%%%%%%%
tx_signal=conv(impulse_modulated, pt);
```

2.B-1   Complete the two quantities marked by '**?**' in the m-file. Save the completed m-file as **tx_sig_gen_Nid.m**. In the subsequent problems in this section, use **tx_sig_gen_Nid.m**, not **tx_sig_gen.m**. The original file **tx_sig_gen.m** will be needed again in Section 20.4 of this chapter.

Capture **tx_sig_gen_Nid.m**.

2.B-2   For each line that contains '=', provide a comment to explain what the variable at the left-hand side represents and justify how the right-hand side expression is formulated accordingly.

2.B-3   At the end of **tx_sig_gen_Nid.m**, append the code fragment below and execute the m-file.

(a) Capture the figure with two subplots.
(b) Based on the waveforms in the figure, explain the effects of impulse modulation and pulse shaping.

```
figure(100)
subplot(2,1,1)
stem(t_step:t_step:(Ns*Ts), impulse_modulated,'.');
axis([0  Ns*Ts  -2*max(impulse_modulated)  2*max(impulse_modulated)]);
grid on
title('impulse modulated')
subplot(2,1,2)
plot(t_step:t_step:(t_step*length(tx_signal)), tx_signal);
axis([0  Ns*Ts  -2*max(tx_signal)  2*max(tx_signal)]);
grid on
title('pulse shaped')
```

2.B-4.   By only visual inspection of the pulse-shaped waveform in B-3, is it easy to determine whether there exists ISI? Note that for the binary signaling simulated (a positive pulse represents a bit '1' and a negative pulse represents a bit '0'), ISI can be checked by determining whether **tx_signal** equals 1 or −1 at time instants that are equal to integer multiples of one bit duration.

2.C   In this section we generate the sampled waveform **tx_signal** using the m-file **tx_sig_gen_Nid.m** completed in 2.B and plot the eye diagram [2, 3].The window size for the eye diagram is set to three bit periods for the binary system considered.

Before we start, set the variable **Ns** to 1XXX, where XXX is the last three digits of your student ID.

2.C-1   Note that **tx_signal** is the sampled version of the transmitted signal $s(t)$ with **L** samples per symbol. Therefore, if we execute **tmp=tx_signal(1:(3*L))**, then **tmp** will be the sampled version for the first 3 symbol periods of $s(t)$.

The following command extracts the k-th three-bit period of $s(t)$ and stores this portion of the signal in variable **tmp**. For example, if **k** = 2, **tmp** will be the sampled version of $s(t)$, $3T_s < t \leq 6T_s$; if $k = 3$, **tmp** will be the sampled version of $s(t)$, $6T_s < t \leq 9T_s$, and so on. Test this out.

```
tmp=tx_signal( ((k-1)*3*L+1) : (k*3*L) );
```

2.C-2   [WWW]Now we overlay the waveform traces over every three-bit period on top of one another in the same graph. Append the code fragment below to **tx_sig_gen_Nid.m**. Note that in the **'for'** loop, we start with **k** = 1 to exclude the first three-bit portion because the peak of the raised cosine pulse for the first symbol (bit) appears at $t = 3T_s$ due to the delay in the convolution process with **pt**, whose peak is at $t = 3T_s$.

```
%Append the following lines to tx_sig_gen_Nid.m
figure(200)
for k=2:floor(Ns/3)   % k is the index of three consecutive symbol portion.

    tmp=tx_signal( ? : ? ); % k-th three consecutive symbol portion. Refer to 2.C-1.
    plot(t_step*(0:(3*L-1)), tmp);

    axis([0 3 min(tx_signal) max(tx_signal)]);
    grid on; hold on
    pause
end
hold off
```

(a) Complete the quantities marked by '?' and capture the completed line.
(b) Before you run the completed file **tx_sig_gen_Nid.m**, sketch what you expect to see when you run it.
(c) Execute the completed **tx_sig_gen_Nid.m** and press any key repeatedly. Whenever you press a key, the next three-bit period of the transmitted signal will be overlaid in the figure. Capture the resulting figure after you press the key for 3, 10, and 30 times, respectively.

2.C-3   Press Ctrl-C to stop executing the m-file. Inside the **'for'** loop of the m-file, comment out the line **pause**, and run the m-file again. (a) Capture the resulting eye diagram. (b) Does your sketch in 2.C-2(b) look similar to the eye diagram? (c) Do the curves on the eye diagram intersect exactly at 1 or −1 at integer multiples of the symbol period? (d) If the answer in (c) is "yes," then we say there is no ISI. Why?

2.C-4   Note that the same eye pattern repeats every symbol duration. In the eye pattern for one symbol duration, draw four lines connecting the pair of points where the curves intersect at one time instant to the next pair of points. For each of these four branches, specify the two corresponding bits.

2.C-5 Does the eye diagram captured in 2.C-3 verify your answer to 1.C?

2.D In this section we consider 4-ary PAM signals. To this end, modify the line that generates the vector **amp_modulated** in **tx_sig_gen_Nid.m** as follows.

```
amp_modulated=2*ceil(rand(1,Ns)*4)-5;  %Now Ns is not the number of data bits but
the number of 4-ary data symbols.
```

2.D-1 In the modified m-file, record all the possible values of the elements of **amp_modulated**. To verify the answer, repeatedly execute **2*ceil(rand*4)-5** in the command window.

2.D-2 Suppose that the first eight elements of **amp_modulated** are [−1 3 3 −3 −1 1 1 −3]. Sketch the waveform **tx_signal** after pulse shaping for this eight-symbol duration.

2.D-3 [WWW]Make sure that in **tx_sig_gen_Nid.m**, the line 'amp_modulated= 2*data_bit-1;' is modified as 'amp_modulated=2*ceil(rand(1,Ns)*4)-5;' for 4-ary PAM signals. Execute the modified 'tx_sig_gen_Nid.m' and capture the eye diagram.

2.E Let us compare the eye diagrams of 4-ary pulse amplitude modulation (PAM) signals with different roll-off factors.

2.E-1 Execute **tx_sig_gen_Nid.m** for each of the following roll-off factors: 0, 0.5, 0.75, and 1.

   (a) Capture the four corresponding eye diagrams.
   (b) Observe the changes in the eye diagram according to the roll-off factor value. Document the differences especially in terms of eye opening, that is, whether or not the eye is open wider as the roll-off factor increases.

2.E-2 The code lines below plot the one-sided PSD of **tx_signal** in dB scale.

```
figure(300)
pwelch(tx_signal, L*8, [ ], 2048,16);
axis([0  1 -10 15])
hold on
```

Append the code lines above to the end of **tx_sig_gen_Nid.m** and then execute the m-file for each of the following roll-off factor values: 0, 0.5, 0.7, and 1. After this is completed, set the four curves in different colors. This can be accomplished by using the edit plot icon in the menu bar. Finally, execute the following in the command window to add a legend for the PSD curves.

```
>>legend('r =0', 'r=0.5', 'r=0.7' , 'r=1')
```

Capture the resulting PSD plot.

2.E-3   Analysis of the spectrum of raised cosine pulse-shaped signals and the 6-dB bandwidth.

(a) From the captured PSD plot in 2.E-2, read and record the PSD value (y axis) at the center frequency of the spectrum, which is zero Hz.

(b) The 6-dB bandwidth is defined as the frequency B where the PSD value is 6 dB below the PSD value at the center frequency. Read and record the 6-dB bandwidths of the raised cosine pulse-shaped signals with the four different roll-off factors.

2.E-4   In 2.E-3(b), is the 6-dB bandwidth of the raised cosine pulse-shaped signals equal to 0.5, regardless of the roll-off factor? Go through the following steps and explain it.

(a) Denote R_s as the symbol rate in Hz and $H_{RC}(f)$ as the Fourier transform of the raised cosine pulse. Then $H_{RC}(f)$ decreases to a certain value at $f = R_s/2$ Hz, regardless of the roll-off factor. What is that value?

(b) Calculate the power reduction in dB as the amplitude decreases by 50%.

(c) From (a) and (b), explain why all PSDs in 2.E-2 have the same 6-dB bandwidth and why it equals 0.5.

2.E-5   The bandwidth of the raised cosine pulse-shaped signals.

(a) From the captured PSD plot in 2.E-2, measure the 20-dB bandwidths of the four raised cosine pulse-shaped signals.

(b) Establish a relationship between the roll-off factor and the bandwidth.

2.E-6   Based on the results in 2.E-1(b) and 2.E-5(b), summarize the advantages and disadvantages of a small or a large roll-off factor.

20.3   EYE DIAGRAM AFTER MATCHED FILTERING

If the pulse-shaped signal received with additive white Gaussian noise (AWGN) passes through a filter matched to the pulse applied at the transmitter, then the signal-to-noise ratio of the filter output is maximized. Such a filter is called a "matched filter" [5, 6]. The impulse response of the matched filter for a pulse $p(t)$ is written as $p^*(-t)$. Hence we can generate the matched filter output by using convolution between the received signal and $p^*(-t)$.

3.A   Suppose that the pulse $p(t)$ is a raised cosine pulse. For this case, explain why the matched filter's impulse response $p^*(-t)$ is equal to $p(t)$.

3.B   From 3.A, if the pulse-shaped signal **tx_signal** is received with noise, we can implement the matched filtering process by convolving **pt** with the received signal. Let us first perform matched filtering to the noise-free received signal to see how matched filtering changes the eye diagram.

Below the line '**tx_signal=conv(impulse_modulated, pt);**' in the m-file **tx_sig_gen_ Nid.m**, insert the following line to generate the matched filter output, **matched_out**.

```
matched_out=conv(tx_signal,pt);
```

Inside the second **'for'** loop, which plots the eye diagram, replace all **tx_signal** by **matched_out** to plot the eye diagram using signal **matched_out**, rather than **tx_signal**. Set the roll-off factor to 0.5 and execute the modified **tx_sig_gen_Nid.m** for the binary as well as the 4-ary PAM signaling cases. Capture the eye diagrams obtained using **matched_out** for the binary and the 4-ary PAM cases.

3.C (a) Check whether ISI exists or not at the matched filter output. In other words, at the time instants equaling integer multiples of the symbol duration, that is, at $t = 1, 2$, seconds, do the curves in the eye diagram intersect exactly at 1 and -1 for the binary case, and exactly at 3, 1, -1, -3 for the 4-ary PAM case? (b) If ISI exists, explain why.

3.D Here we investigate why the SRRC pulse, rather than the raised cosine pulse, should be used for pulse shaping at the transmitter if a matcher filter is applied at the receiver. To this end, in **tx_sig_gen_Nid.m**, modify the line **'pt=rcosine(1,L,'normal', ?);'** as **'pt=rcosine(1,L,'sqrt',r);'** to implement pulse shaping with the SRRC pulse.

First, let us check the eye diagram with signal **tx_signal**, not **matched_out**, for SRRC pulse shaping. To this end, replace again all **matched_out** inside the second **'for'** loop by **tx_signal**.

Execute the modified **tx_sig_gen_Nid.m** for both the binary and the 4-ary signaling cases and capture the corresponding eye diagrams obtained with **tx_signal**.

3.E Check whether or not ISI exists in **tx_signal**.
(a) Explain why ISI exists.
(b) It is important to note that for symbol detection, ISI in the transmitted signal before the matched filter does not matter. Design a method to remove ISI in the receiver and justify your approach.

3.F Now let us examine the eye diagram obtained using **matched_out**, assuming SRRC pulse shaping. To this end, replace again all **tx_signal** inside the second **'for'** loop by **matched_out**. Execute the modified **tx_sig_gen_Nid.m** for both the binary and the 4-ary signaling cases. Capture the corresponding eye diagrams obtained using **matched_out**.

3.G System with SRRC pulse shaping in the transmitter and a matched filter in the receiver.
(a) In such a system, does ISI exist in the matched filter output? Why?
(b) We have seen two ISI-free cases: SRRC pulse shaping, followed by matched filtering in the receiver and raised cosine (RC) pulse shaping with an ideal LPF at the receiver. Explain why at the symbol detection stage, no ISI for both cases.

3.H Execute **tx_sig_gen_Nid.m** for each of the following five roll-off factors: 0, 0.25, 0.5, 0.75, and 1. After each case is finished, capture the eye diagram obtained using **matched_out**. Do these for both the binary and the 4-ary PAM signaling cases.

3.I Theoretically, the filtered SRRC pulse by a matched filter is equivalent to an RC pulse. Thus ISI does not exist regardless of the roll-off factor. However, from the captured eye diagrams in 3.H, the **matched_out** at symbol decision instants is not exactly 1 or −1 for the binary case (−3, −1, 1, 3 for the 4-ary PAM case) as the roll-off factor decreases.

From 1.B and 1.D, you may find the difference between the ideal SRRC pulse and the practical SRRC pulse used in the m-file. Based on this, explain what causes ISI, especially for cases with a small roll-off factor.

3.J To investigate pulse shaping and matched filtering over an AWGN channel, insert the following line right below the line '**tx_signal=conv(impulse_modulated, pt);**' in **tx_sig_gen_Nid.m**.

```
rx_signal = tx_signal + 0.15* randn(1, length(tx_signal));
```

To see the eye diagram of the noisy received signal, replace all **matched_out** inside the second '**for**' loop by **rx_signal**. Set the roll-off factor to 1 and execute the m-file for the binary as well as the 4-ary PAM signaling cases. Capture the resulting eye diagrams.

3.K Effects of matched filtering when applied to noisy received signals.

3.K-1 To see the eye diagram of a noisy received signal after matched filtering, properly modify the line '**matched_out=conv(tx_signal, pt);**'. Capture your modified m-file.

3.K-2 Replace all **rx_signal** inside the second '**for**' loop by **matched_out** and execute the m-file for the binary and the 4-ary PAM signaling cases. Capture the resulting eye diagrams.

3.K-3 Compare the eye diagrams in 3.J and 3.K-2 and assess the eye opening after matched filtering.

(a) We have seen the eye opening with matched filtering for the noiseless case in 3.D and 3.F. Compared with the eye diagram over a noisy channel obtained in 3.K-2, the impact of match filtering to the eye opening is more significant for which environment—noiseless or noisy environment?

(b) From (a), document two of the main functions of matched filtering assuming SRRC pulse shaping in an AWGN channel.

20.4 GENERATING AN ACTUAL ELECTRIC SIGNAL AND VIEWING THE EYE DIAGRAM IN AN OSCILLOSCOPE

In this section we first convert the pulse-shaped signal generated in MATLAB into an actual electric signal through the PC's audio port. Then we observe the eye diagram of the actual electric signal in an oscilloscope.

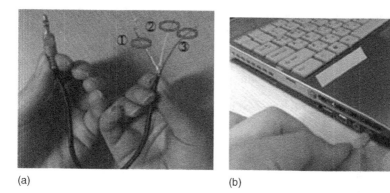

(a)   (b)

FIGURE 20.7   Audio cable after sheath removed (left) and connection to the audio out port of a PC (right).

Before we start the experiment, complete the following steps:

Step 1. Obtain a 3.5 ϕ audio cable, which is commonly used in any audio device. You may use an earphone cable instead, but it will be cut for this experiment.

Step 2. Remove the sheath of the 3.5 ϕ audio cable as shown on the left of Fig. 20.7, and you will see three wires. The uncovered wire ① will be connected to the ground port of the oscilloscope later in the experiment. Wire ② (typically white) is for the symbol clock output, and wire ③ (typically red) is for the raised cosine pulse-shaped waveform.

Step 3. Connect the audio cable to the audio output of a PC that is running MATLAB as shown on the right of Fig. 20.7.

Step 4. Connect wire ① of the cable to the ground clip of two oscilloscope probes for CH A and CH B, respectively, as shown on the left of Fig. 20.8. Then connect the two probes to wires ② and ③ of the cable as shown on the right of Fig. 20.8.

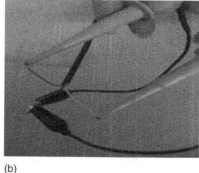

(a)   (b)

FIGURE 20.8   Connection to the ground clips of two probes (left) and the connection of the stereo audio signal wires to the probes (right).

Step 5. Set the PC audio output volume at around 75% of the maximum. NOTE: depending on the PC configuration, the left and right stereo outputs of the sound card might be mixed and the signal waveform may not appear correctly. In this case, change the audio sound setting on the control panel to separate the left and right stereo outputs.

NOTE: This audio cable will be reused in Chapter 23.

4.A   Now we check whether or not the preparation has been done correctly. Execute the following commands in the command window:

```
>>t=0:0.001:100;
>>x=sin(2*pi*t);
>>y=cos(4*pi*t);
>>z(1,:)=x;
>>z(2,:)=y;
>>soundsc(z', 1000)
```

It will take 100 seconds for the last command **soundsc(z', 1000)** to finish. While the command **soundsc(z', 1000)** is running, first push the autoscale (or autoplay) button of the oscilloscope for instant triggering. Then properly adjust the time axis scale by SEC/DIV dial so that two sine waveforms can be displayed for several periods as shown in Fig. 20.9. Also adjust the amplitude scale by VOTS/DIV dial of CH1 and CH2.

Capture the oscilloscope screen. It is recommended to save the screen in an image file if the oscilloscope supports it.

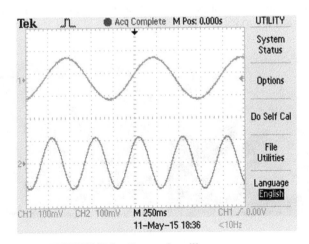

FIGURE 20.9   Captured oscilloscope screen.

4.B Observe the raised cosine pulse-shaped signal in the oscilloscope. Before we start this experiment, if the last command **soundsc(z', 1000)** in 4.A is still running, press Ctrl-C to stop it.

4.B-1 [WWW]From the companion website, download **tx_sig_gen.m** into your MATLAB work folder. Append the code fragment below to the end of the m-file and save it as **eye_diagram.m**. This m-file first creates a two-row matrix, **signal_out**: the first row of the matrix is the symbol clock signal; the second row of the matrix is the raised cosine pulse-shaped signal. Then **signal_out** is converted into a stereo audio signal through the PC's audio output. Note that the variable **Nrepeat** determines the number of repetitions of audio signal conversion, and the variable **n** controls the symbol clock rate.

In the m-file, set **Ns** = 10000 and **XX** = the last two digits of your student ID. Capture the completed m-file.

```
%%%%%%%%%%%%%%%%%%%%%% Creation of Symbol Clock %%%%%%%%%%%%%%
clk=[];
for di=1:ceil(length(tx_signal)/(L/2))
  clk=[clk (-1)^di*ones(1,L/2)];
  % No of samples per clock period (1 clock period)= L samples.
end
clk=clk(1:length(tx_signal)); %Match the length of clk with the length of tx_signal.

%%%%%%%Converting the symbol clock and pulse shaping waveform into an stereo
audio signal%%%%%%
signal_out=[clk' tx_signal']';
XX=??;  % Set XX = Last two digits of your student ID
n=ceil(40*(XX+1)/100);
Nrepeat=n;
for k=1:Nrepeat  %Repeat n times to output
  soundsc(signal_out',8000*n);
  % The sample rate is 8000*n Hz. Therefore the symbol(clock) rate is 8000*n/L Hz
  done=k
end
```

4.B-2 Under the same experimental setting in 4.A, execute **eye_diagram.m**. Note that the symbol clock signal and the raised cosine pulse-shaped waveform will appear in CH1 and CH2, respectively.

Now go through the following steps:

Step 1. Properly adjust the amplitude scale by using the VOTS/DIV dial of CH1 and CH2 to observe the signal waveforms clearly as shown in Fig. 20.10. Note that the clock signal will not be a rectangular pulse due to the distortions caused by the PC's sound card.

Step 2. Adjust the SEC/DIV dial to show 10–20 cycles of the clock signal as shown in Fig. 20.10.

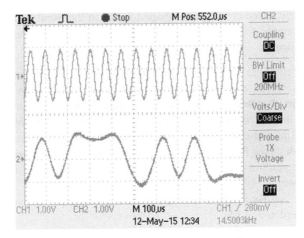

FIGURE 20.10   Oscilloscope screen still cut.

(a) Press the Run/Stop button repeatedly for 3–4 times and observe the still cuts of the pulse-shaped signal. If the setting is correct, you should see the still cut similar to what is shown in Fig. 20.10. Capture the still cuts.

(b) From the still cuts, is it possible to check whether ISI exists in the pulse-shaped signal? Justify your answer.

4.B-3 In order to see the eye diagram of the pulse-shaped signal, first trigger the symbol clock signal to hold still. If **eye_diagram.m** finishes running, execute it again and go through the following steps:

Step 1. Disconnect the probe whose CH displays the raised cosine waveform. Leave the other probe for the clock signal connected.

Step 2. Press the TRIG MENU button and select the clock signal as the triggering signal source.

Step 3. Adjust the TRIG Level control dial to keep the clock signal in the screen to hold still.

Step 4. Adjust the time axis scale by SEC/DIV dial so that 3–4 clock cycles are displayed in the screen.

Capture the oscilloscope screen.

4.B-4 Record the frequency of the clock signal in Hz measured in the oscilloscope.

4.C Observe the eye diagram of the raised cosine pulse-shaped waveform. First, complete the following steps:

Step 1. Reconnect the audio cable wire that carries the raised cosine pulse-shaped signal to the open probe, which was disconnected in Step 1 of 4.B-3.

Step 2. If **eye_diagram.m** has finished running, run it again and properly adjust the amplitude scale by VOTS/DIV dial of both CH1 and CH2 to observe the signal waveforms clearly as shown in Fig. 20.10.

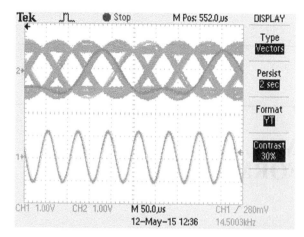

FIGURE 20.11   Illustration of eye diagram.

Step 3. Increase the "memory duration" in the "display mode" menu. Select a proper memory duration that gives the best view of the eye diagram. If everything has been done properly, the screen should display something like what is shown in Fig. 20.11, where the pulse-shaped signal is shifted up and the clock signal is shifted down for better presentation.

(a) Modify the roll-off factor in eye_diagram.m and rerun it. Observe the eye diagram with the modified roll-off factor. Capture the eye diagrams in the oscilloscope screen for roll-off factors of 0, 0.5, 0.75, and 1 for the binary signaling case.

(b) Change the line '**amp_modulated=2*data_bit-1;**' into '**amp_modulated=2* ceil(rand(1,Ns)*4)-5;**' to generate the 4-ary PAM signals. Capture the eye diagrams in the oscilloscope screen for roll-off factors of 0, 0.5, 0.75, and 1, for the 4-ary PAM signaling.

REFERENCES

[1] H. Nyquist, "Certain Topics in Telegraph Transmission Theory," *Transactions of the American Institute of Electrical Engineers*, Vol. 47, 1978, pp. 617–644.

[2] K. Feher, *Digital Communications : Microwave Applications*, Englewood Cliffs, NJ: Prentice Hall, 1980.

[3] J. Proakis, *Digital Communications*, 3rd ed., New York: McGraw-Hill, 1995.

[4] E. R. Kretzmer, "Generalization of a Technique for Binary Data Communication," *IEEE Transactions on Communications Technology*, Vol. 14, 1966, pp. 67–68.

[5] D. O. North, "An analysis of the factors which determine signal/noise discrimination in pulsed carrier systems," *Rep. PTR-6C*, Princeton, NJ: RCA Laboratories, 1963.

[6] G. L. Turin, "An Introduction to Matched Filters," *IRE Transactions on Information Theory*, Vol. 3, No. 6, 1960, pp. 311–329.

21

BER SIMULATION AT THE WAVEFORM LEVEL

- Design a binary phase shift keying (BPSK) system that includes pulse shaping and matched filtering, and perform bit error rate (BER) simulation.
- Set the variance of the noise samples according to the signal-to-noise ratio (SNR).
- Understand the effects of the roll-off factor on BER when there exists a symbol timing error.
- Perform passband modulation and demodulation, and observe the power spectral density (PSD) of the passband signal and the effects of phase error on BER.

21.1 E_B/N_0 SETTING IN BASEBAND BPSK SIMULATION

In evaluating the error performance of a digital communications system, the BER is often plotted as a function of E_b/N_0 in dB with a certain step size (e.g., 2 or 5 dB). If a fixed pulse with energy E_b is used in the simulation, the desired E_b/N_0 can be set by adjusting N_0, the single-sided noise power spectral density. For most commonly used signaling schemes such as M-ary PSK and M-ary QAM, BER simulation at the waveform level typically involves the following steps:

Step 1. Design an appropriate pulse shape. For example, the pulse should meet the bandwidth requirement while maintaining intersymbol interference (ISI)-free operation.

Step 2. Calculate the energy of the designed pulse, which is equal to the bit energy denoted by E_b for binary signaling.

Problem-Based Learning in Communication Systems Using MATLAB and Simulink, First Edition.
Kwonhue Choi and Huaping Liu.
© 2016 The Institute of Electrical and Electronics Engineers, Inc. Published 2016 by John Wiley & Sons, Inc.
Companion website: www.wiley.com/go/choi_problembasedlearning

Step 3. Set E_b/N_0 [dB] value.

Step 4. Convert the E_b/N_0 value in dB into the E_b/N_0 value in the linear scale.

Step 5. Calculate N_0 using E_b/N_0 in the linear scale obtained in Step 4 and E_b obtained in Step 2.

Step 6. Set the variance of the noise samples according to the value of N_0.

Step 7. Generate the sampled transmitted signal vector, the sampled noise vector according to the value of N_0, and the sampled received signal vector. Then estimate the transmitted bit (or bits) from the sampled received signal vector. Count the number of erroneously detected bits and store this information.

Step 8. Repeat Step 7 with independently generated transmitted bits and received noise until the total number of bits simulated is sufficiently large to generate a statistically meaningful average BER.

Step 9. Change the E_b/N_0 [dB] value and perform Steps 4–8 again to generate the corresponding BER; this is repeated until BERs for all target E_b/N_0 [dB] are simulated.

Step 10. Plot the BER curve as a function of E_b/N_0 [dB].

1.A [WWW] The following m-file first designs the pulse **pt** and then performs Step1– Step 5 mentioned above to calculate N_0 for each of the given E_b/N_0 [dB] values. The variable **EbN0dB** is E_b/N_0 in the dB scale, and **EbN0** is E_b/N_0 in the linear scale. We consider a square-root raised cosine pulse with a roll-off factor of 0.5. We set the symbol (which is a bit for BPSK considered in this chapter) duration at **Ts**=1 second, and the number of samples per symbol at **L**=16.

Complete all quantities marked by '**?**' in the m-file. Capture the completed m-file.

```
clear
%%%Signal pulse design %%%%%%%%%%%%%%%%%%%%%%%%%%%%%%%%%%%%%
Ts=1; L=16;
t_step=Ts/L;
pt=rcosine(1,?,?,?); N=length(pt);
%We consider a square-root raised cosine pulse with a roll-off factor of 0.5. Refer to 1.D
   of Chapter 20 for the use of the function rcosine().

%%% Calculation of Signal (bit in case of BPSK) Energy %%%%%%%%%%%%%%%%%%%
Eb=sum(?)*t_step ;
% We will go through the same process in tx_sig_gen.m in 2.B of Chapter 20 to
   generate the transmitted signal. A bit is converted into a discrete unit impulse
   if bit is '1' and its inverted version if bit is'0'. The discrete unit impulse (or its inverted
   version) is convolved with the shaping pulse pt obtained above. Summing it up, this is
   equivalent to transmitting pt if bit is '1' and -pt if bit is '0'. Consequently,
   the bit energy Eb is equal to the energy of pt. Refer to 3.A of Chapter 17 regarding
   how to numerically calculate the energy of a sampled signal.
```

%%%% N0 setting part %%%%%%%%%%%%%%%%%%%%%%%%%%%%%%%
EbN0dB=100; % Suppose Eb/N0 [dB] = 100dB.

EbN0=10^(?); %Conversion into linear scale.
N0= ? ; % Eb and EbN0 have been set already and EbN0 means Eb/N0.
% In the two lines above, do not hard code the values of EbN0 and N0 for the specific
    case of EbN0dB=100; instead, complete EbN0 and N0 as a function of EbN0dB
    so that if we change the value of EbN0dB to any different value, EbN0 and N0 will
    be changed accordingly.

1.B   Execute the m-file completed in 1.A and then execute (type and enter) **N0** in the command window to see the calculated value of **N0**. Capture the execution result.

1.C   [T]Fig. 21.1 illustrates the digital filtering process performed on the sampled signal after noise rejection by the analog filter. The received signal $r(t)$ is the sum of the transmitted signal $s(t)$ and the AWGN $n(t)$ with a spectral density $N_0/2$. Set the bandwidth of the noise rejection filter to a half of the filter output sampling frequency, that is, bandwidth = **0.5*(1/t_step)**.

1.C-1   The left-hand side of Fig. 21.2 shows the PSD of the noise $n(t)$. Draw the exact PSD of the noise after it passes through the noise rejection filter.

1.C-2   How is the signal power calculated from the signal PSD? Write an equation that establishes this relationship.

1.C-3   Let $P_{n,\mathrm{NRF}}$ denote the noise power at the noise rejection filter output before sampling. It can be calculated as

$$P_{n,\mathrm{NRF}} = \frac{N_0}{2 \times \mathbf{t\_step}}. \tag{21.1}$$

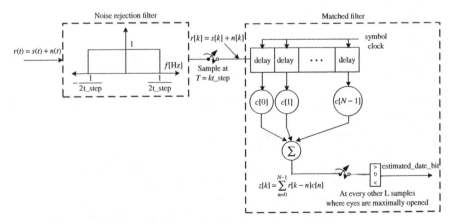

FIGURE 21.1   Noise rejection and matched filter.

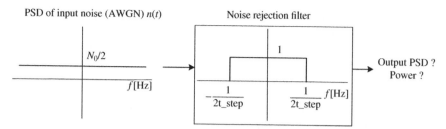

FIGURE 21.2   PSD and signal power before and after passing through the noise rejection filter.

Prove equation (21.1) using the exact PSD function obtained in 1.C-1 and the equation obtained in 1.C-2.

1.C-4   Let us examine the distribution and correlation model of the noise samples at the noise rejection filter output. Since the noise rejection filter is a linear system, the noise samples after passing through the noise rejection filter ..., $n[k-1]$, $n[k]$, $n[k+1]$,... remain to be zero mean and identically distributed Gaussian random variables. We can show that the noise samples ..., $n[k-1]$, $n[k]$, $n[k+1]$,... are independent.

(a) First, find the noise autocorrelation $R_n(\tau)$, that is, the inverse Fourier transform of the noise PSD at the noise rejection filter output.
(b) Then show that the correlation values of the noise samples are zero except at $k = 0$, that is, $R_n(k*\textbf{t\_step}) = 0, \forall$ integer $k \neq 0$. Therefore these uncorrelated samples are also independent of one another because they are Gaussian.

1.C-5   Next, let us find out the variance of the noise samples at the noise rejection filter output. Although we cannot say, in general, that the variance (ensemble average of the squared samples) of any zero-mean sequence equals the signal power (time average of the squared samples), the variance of noise samples is equal to the noise power given in equation (21.1). This is due to a certain statistical property that the noise signal has. What is this property?

1.D   As mentioned in 1.C, the noise samples ..., $n[k-1]$, $n[k]$, $n[k+1]$,... are zero mean independent Gaussian random variables whose variances can be calculated as equation (21.1). Let **v_n** denote the variance of the noise samples. Add the following line that generates **v_n** to the end of the code in 1.A and properly complete the quantity marked by '**?**'.

```
%Add to the m-file in 1.A
v_n=?;  %As a function of N0 and t_step
```

1.E   Now we can generate the sampled vector of the received signal in Fig. 21.1, including both the signal and the noise ..., $r[k-1]$, $r[k]$, $r[k+1]$, ... (the corresponding MATLAB variable is **r_samples**). Since the noise rejection filter bandwidth

1/(2*t_step) is greater than the bandwidth of data signal $s(t)$, the data signal term at the noise rejection filter output remains to be $s(t)$.

The code fragment below, first, generates the sampled signal vector ..., $s[k - 1]$, $s[k]$, $s[k + 1]$, ... (the corresponding MATLAB variable is **tx_signal**) through similar steps used in **tx_sig_gen.m** in Chapter 20. Then, it generates the sampled noise vector ..., $n[k - 1]$, $n[k]$, $n[k + 1]$, ... of the same length (the corresponding MATLAB variable is **n_samples**). Finally, it adds the two sampled vectors together. Add this code fragment to the end of the m-file completed in 1.D.

```
% Append the following to the m-file in 1.D

% First, copy and paste the parts 2, 3, 4, and 5 (as listed below) from tx_sig_gen.m in
    Chapter 20.
<2. Ns bit binary symbol generation>
<3. Unipolar to Bipolar (amplitude modulation)>
<4. Impulse modulation>
<5. Pulse shaping (Transmitter filtering)>

%%%%%%< Noise sample generation >%%%%%%%%%%%%%%%%%%%%%%%%%
n_samples=sqrt(v_n)*randn(1,length(tx_signal));%Generate n[k]

%%%%%%<Generation of received signal sample>%%%%%%%%%%%%%%
r_samples = tx_signal + n_samples;
```

1.E-1   Complete the m-file and capture it.

1.E-2   Note that in the line that generates the noise samples, there is a scaling factor **sqrt(v_n)**. Justify why it is needed there.

21.2   MATCHED FILTER AND DECISION VARIABLES

The m-file in Section 21.1 generates the sampled vector ..., $r[k - 1]$, $r[k]$, $r[k + 1]$, ... of the received signal modeled in Fig. 21.1. The right-hand side of Fig. 21.1 (after sampling) shows matched filtering applied to the sampled signal.

2.A   Add a line to the m-file in 1.E to generate the sampled matched filter output vector ..., $z[k - 1]$, $z[k]$, $z[k + 1]$, ... (the corresponding MATLAB variable is **z_samples**) as follows. Complete the two quantities marked by '**??**', one with a proper function name and one with a proper variable name. The materials in Section 3.B of Chapter 20 can help answer this question.

Capture the completed line.

```
z_samples=??(r_samples,?? );
```

2.B   In order to examine the waveforms of **z_samples** and **impulse_modulated**, execute the m-file and then execute the following commands in the command window. Capture the resulting plot.

```
>>figure
>>subplot(2,1,1)
>>stem(impulse_modulated(1:20*L),'.');
>>grid on
>>subplot(2,1,2)
>>plot(z_samples(1:20*L));
>>grid on
```

2.C   Note that in the line 'N=length(pt)' of the m-file, the variable **N** is set to the length of **pt**, which is equal to the pulse-shaping filter length and the matched filter length as well. Type **N** in the command window to see its value. Capture the execution result.

2.D   The plot obtained in 2.B shows that there is a delay of **N** samples between **z_samples** and **impulse_modulated**. Explain why this delay exists. The following steps will help answer this question.

Step 1. Execute **plot(pt)** and find the sample index of the peak of **pt**, which is the center of even symmetry of **pt**.

Step 2. The right-hand side of Fig. 21.1 (after sampling) shows the matched filtering step, which is the convolution of the input and **pt**. Therefore the filter coefficients are $c[0] =$ **pt(N)**, $c[1] =$ **pt(N-1)**, ..., $c[N-2] =$ **pt(2)**, $c[N-1] =$ **pt(1)**. Suppose that the initial values of the delay line inside the filter are set to all 0s and the sequence 1, 0, 0, 0, 0, 0..., is a trial input to the delay line. Then the filter output sequence will be **pt(N)**, **pt(N-1)**, **pt(N-1)**, ... Find out the sample index of the peak of the filter output sequence. How large a delay (in samples) does the convolution operation, **conv([1 0 0 0 0 ...]**, pt), introduce?

Step 3. In the transmitter and the receiver, how many times of convolution between **impulse_modulated** and **pt** are performed to obtain **z_samples**?

2.E   Based on the number of samples per bit and the delay between **z_samples** and **impulse_modulated**, **z_samples(N+(n-1)*L)** is the decision variable of the nth transmitted bit, **data_bit(n)**. In other words, we estimate **data_bit(n)** on the basis of the sample **z_samples(N+(n-1)*L)**. Explain why the sample index of **z_samples** should be set to **N+(n-1)*L**.

2.F   Using the decision variable in 2.E, we can detect the first 5 bits of **data_bit** by executing the commands below in the command window. Explain why the decision variable for the **nth** bit should be set to **(z_samples(N+(n-1)*L) > 0)**. The part of the m-file under **<3. Unipolar to bipolar (amplitude modulation)>** will help answer this question.

```
>>estimated_data_bit(1)= (z_samples(N+(1-1)*L) >0 ) ;
>>estimated_data_bit(2)= (z_samples(N+(2-1)*L) >0 ) ;
>>estimated_data_bit(3)= (z_samples(N+(3-1)*L) >0 ) ;
>>estimated_data_bit(4)= (z_samples(N+(4-1)*L) >0 ) ;
>>estimated_data_bit(5)= (z_samples(N+(5-1)*L) >0 ) ;
>>estimated_data_bit
>>data_bit(1:5)
```

2.G (a) Execute the commands listed in 2.F and capture the result. (b) Is **estimated_data_bit** identical to **data_bit(1:5)**?

2.H   All **Ns** bits in the vector **data_bit** can be detected by the code fragment below. Append this fragment to the m-file in 2.A.

```
for k=1:Ns
 estimated_data_bit(k)= (z_samples(N+(k-1)*L) >0 ) ;
end
```

2.H-1   Execute the m-file and then execute **sum(estimated_data_bit~=data_bit)** in the command window. Capture the result.

2.H-2.   Does the result in 2.H-1 imply no error? Why?

21.3   COMPLETING THE LOOP FOR BER SIMULATION

Now we are ready to implement Step 8 and Step 9 explained at the beginning of Section 21.1. After this, we complete the whole simulation loop and obtain the BER curve.

3.A   [WWW] Implementation of Step 8. After incorporating the code fragment below, the m-file will be capable of simulating the BER for the case of $E_b/N_0 = 0$ dB.

```
[ Step 1~Step6]: Copy and paste the m-file completed in 1.D. Be sure to set EbN0dB=0.

sum_Ne=0; % Initialize the number of total errors
N_frame=20; % Number of frames for simulation
```

```
[ for loop in order to implement Step 8 ]
for frame_number=1:N_frame

    [ Step 7] Copy and paste the added parts in 1.E, 2.A and 2.H as follows
    < Ns bit binary symbol generation>
    < Unipolar to Bipolar (amplitude modulation)>
    < Impulse modulation>
    < Pulse shaping (Transmitter filtering)>
    %%%%%%< Noise sample generation>%%%%%%%%%%%%%%%%%%%%%%%%%%
    n_samples=sqrt(v_n)*randn(1,length(tx_signal));%Generate ..., n[k-1], n[k],
    n[k+1], ...

    %%%%%%<Generation of received signal sample>%%%%%%%%%%%%%%%
    r_samples=tx_signal + n_samples;

    z_samples=conv(r_samples, pt); %Matched filter

    estimated_data_bit= (z_samples(N+((1:Ns)-1)*L) >0 ); % Data bit estimation in 2.H

    [Newly added part for Step 8]
    Ne=sum(estimated_data_bit~=data_bit); % ?
    sum_Ne=sum_Ne +Ne; % ?
end

%Calculate the BER by the definition, i.e., BER= Total number of bit errors /Total num-
ber of transmitted bits
Total_bit=?*?;
BER=sum_Ne/Total_bit
```

3.A-1   How many bits are generated in one frame and how many frames are used to calculate the BER?

3.A-2   The following two lines are added in the '**for**' loop in Step 8. Explain what these two lines do.

```
...
Ne=sum(estimated_data_bit~=data_bit); % ?
sum_Ne=sum_Ne +Ne; % ?
...
```

3.A-3   Complete the m-file and execute it.

   (a) Complete the quantities marked by '**?**' in the line '**Total_bit=?*?**' so that **Total_bit** is a function **Ns** and **N_frame** (i.e., do not hard code the value of variable **Total_bit** for a specific number of bits and frames). Justify your answer.

(b) Execute the completed m-file and capture the simulated BER results.

3.B   Let us check whether or not the simulated BER result obtained in 3.A-3 matches the theoretical BER of BPSK through the following steps.

3.B-1   The theoretical BER of BPSK is given in the form of the Q-function. The Q-function can be calculated by using the MATLAB built-in function **erfc( )** as below.

```
BER_exact=0.5*erfc(?)
```

(a) Complete the quantity marked by '**?**' in the command above; this quantity is a function of **EbN0** so that **BER_exact** is the theoretical BER.
(b) Execute the completed command above in the command window and capture the result.

3.B-2   Check whether or not the simulated BER approximately matches the theoretical BER.

3.C   Change **EbN0dB** to 9 and execute the m-file again. After this, execute the command in 3.B-1.

3.C-1   Capture the simulated BER result and the theoretical BER result. Check whether or not the simulated BER matches the theoretical BER.

3.C-2   If these BERs do not match, investigate what might have caused this mismatch.

3.D   [WWW]In the m-file in 3.C, replace the '**for**' loop with a '**while**' loop. Modify the m-file as shown below.

```
[Maintain all the part before the 'for' loop of the m-file simulated in 3.C and modify the
   line 'N_frame=20' as follows]
N_frame=0; % Initialize the variable which counts the number of frames

[Change the 'for' loop as follows]
while ? < 25
[Maintain all the parts inside the for loop and add the following line to count the num-
ber of the simulated frames]
N_frame=N_frame+1;
end
```

```
Total_bit=?*?;  %As you completed in 3.A-3.
BER=(sum_Ne/( Total_bit))
BER_exact=0.5*erfc(?) %As you completed in 3.B-1(a).
```

3.D-1   With the modified m-file, iteration will stop when the total number of bit errors reaches 25.

   (a) What is the proper variable for '**?**' in the line '**while ? < 25**'.
   (b) Execute the modified m-file again with '**EbN0dB=9**', and examine whether the simulated BER approximately equals the theoretical BER.

3.D-2   The issue of the inaccurate simulated BER in 3.C should have been resolved in the modified m-file in 3.D-1, since the number of frames simulated is sufficiently large. Note that in the m-file in 3.D-1, the number of frames to simulate is not explicitly specified. Explain how the number of frames simulated automatically increases to a sufficiently large value when a '**while**' loop is used.

3.E   [WWW]Now add the last steps (Steps 9 and 10) to the m-file in 3.D. Modify the m-file as below to repeatedly simulate the BERs for $E_b/N_0 = 1, 3, 5, 7, 9$ dB. Make a copy of this m-file for use in Section 21.5.

   Execute the modified m-file and capture the resulting BER figure.

```
clear
EbN0dB_vector=[1 3 5 7 9];
for snr_i=1:length(EbN0dB_vector)
  EbN0dB=EbN0dB_vector(snr_i);

[Copy and paste whole contents of the m-file in 3.D and delete the first line 'clear' and the
  line 'EbN0dB=9' in the middle]

  BER_vector(snr_i)=BER;
  BER_exact_vector(snr_i)=BER_exact;
end

figure
semilogy(EbN0dB_vector, BER_vector);
hold on
semilogy(EbN0dB_vector, BER_exact_vector, 'r');
xlabel('Eb/N0 [dB]');ylabel('BER');legend('Simulated', 'Theory');grid on; hold off;
```

3.F   The plot captured in 3.E shows both the theoretical BER curve and the simulated BER curve. Compare these two curves. Does the simulated BER curve match the theoretical curve?

21.4 [A]EFFECTS OF THE ROLL-OFF FACTOR ON BER PERFORMANCE WHEN THERE IS A SYMBOL TIMING ERROR

4.A   If the actual sampling instants to obtain the decision variables are not exactly at the optimal sampling instants, then we say that there is a symbol timing error [1]. This problem studies how symbol timing errors affect the BER performance.

4.A-1   In the m-file in 3.E, perfect symbol timing is assumed in forming the decision variables from the received signal **z_samples**. Recall that the number of samples per bit is **L** and that in 2.D we examined that there is a delay of **N** samples between **z_samples** and **impulse_modulated**. Hence **z_samples(N+(n-1)*L)** corresponds to the decision variable of the **n**th bit at the maximum eye-opening instant.

Execute the following commands in the command window and capture the result that shows the sample indexes for the first 10 symbols.

```
>>sample_index=N+((1:Ns)-1)*L;
>>sample_index(1:10)
```

4.A-2   Now let us introduce a symbol timing error. Execute the following commands in the command window.

```
>>t_offset=2;
>>sample_index=N+t_offset+((1:Ns)-1)*L;
>>sample_index(1:10).
```

Note that for simplicity, here we introduced a fixed timing error. In practical systems, the timing error is caused by many factors such as clock jitter, which could result in both periodic and random timing errors that might be different for different symbols.

   (a) Capture the execution result.
   (b) Compare the results with those in 4.A-1 and explain the role of the parameter **t_offset**; that is, explain what it implies when **sample_index** in 4.A-2(a) is used for obtaining the symbol decision variables.

4.A-3   In the m-file in 3.E, replace the line 'estimated_data_bit=(z_samples (sample_index) >0);' by the following three lines. Now the sampling instants where the decision variables are obtained will be the two samples off the maximum eye-opening points.

```
t_offset=2;
sample_index=N+t_offset+((1:Ns)-1)*L;
estimated_data_bit= (z_samples(sample_index) >0 ) ;
```

TABLE 21.1   BERs According to the Symbol Timing Errors with Different Roll-Off Factors.

Timing offset [samples]	0	2	4	8
Roll-off factor				
0.25				
0.5				
0.75				

(a) Run the m-file and capture the BER plot.

(b) Compare the simulated BER with the theoretical BER that assumes perfect symbol timing.

4.B   Replace the line '**EbN0dB_vector=[1 3 5 7 9];**' with '**EbN0dB_vector =4;**' and modify '**while sum_Ne<25**' into '**while sum_Ne<500**'. Run the modified m-file for the cases of t_offset=0, 2, 4, 8 and the roll-off factor = 0.25, 0.5, 0.75, a total of 12 different combinations. Record the simulated BER in Table 21.1.

4.C   BER performance versus timing error and roll-off factor.

(a) Describe how the BER changes as the timing error increases.

(b) For nonzero timing error cases, for example, **t_offset** = 2, 4, or 8, describe how the BER changes as the roll-off factor increases.

(c) Using the result in 3.H of Chapter 20, explain why the BER performance improves as the roll-off factor increases when there exists a symbol timing error.

4.D   Summarize the advantages and disadvantages of using a large and a small roll-off factor in terms of bandwidth efficiency and robustness against symbol timing errors.

21.5   PASSBAND BPSK BER SIMULATION AND EFFECTS OF CARRIER PHASE ERRORS

The goal of this section is to simulate the performance of BPSK in the passband and investigate the effects of carrier phase errors on the BER performance. The passband BPSK signal will be generated by multiplying the pulse-shaped baseband signal with a sinusoidal signal, called "carrier."

Open the m-file completed in 3.E. Make sure that the roll-off factor is set to 0.5 and no symbol timing error is introduced.

5.A   [WWW]Insert the following code fragment right below the line that generates **r_samples** in the m-file. For each of the inserted lines, add a comment to explain what the variable on the left-hand side of the '=' sign represents and how the right-hand side expression is formulated accordingly.

```
carrier_wave=cos(2*pi*3.2*(1:length(tx_signal))*t_step); % ??
passband_tx_signal=tx_signal.*carrier_wave;       % ??
r_samples_pass = passband_tx_signal + n_samples; % ??
```

5.B Let us first observe the spectrum of the transmitted passband signal (MATLAB variable **passband_tx_signal**). Modify the line '**EbN0dB_vector=[1 3 5 7 9];**' into '**EbN0dB_vector =10;**' and modify '**while sum_Ne<25**' into '**while sum_Ne<1**'. Then execute the modified m-file and the following in the command window.

```
>>figure
>>BW_v=fft(passband_tx_signal,1024);
>>plot([-512:511]/1024*L, abs(fftshift(BW_v)))
```

5.B-1 Capture the resulting plot.

5.B-2 Is the spectrum what you expected to see and why?

5.C [WWW]The transmitted bit sequence can be detected after converting the passband signal **r_samples_pass** into the baseband. Right below the line '**r_samples_pass = passband_tx_signal + n_samples;**' insert the following two lines, which convert the received passband signal, **r_samples_pass**, into the baseband by multiplying it with a local carrier wave. The local carrier has the same frequency and phase as the carrier wave used in the transmitter.

```
local_carrier=cos(2*pi*3.2*(1:length(tx_signal))*t_step);
mult_out=r_samples_pass.*local_carrier;
```

Execute the revised m-file and then the following in the command window. The resulting plot shows the PSD of **mult_out** (not **passband_tx_signal**).

```
>>figure
>>BW_v=fft(mult_out,1024);
>>plot([-512:511]/1024*L, abs(fftshift(BW_v)))
```

5.C-1 Capture the plot.

5.C-2 Is the spectrum what you expected to see and why?

5.D The PSD of **mult_out** obtained above contains the spectral components at 6.4 Hz (=3.2*2 Hz). In order to get the baseband signal only, the 6.4 Hz components should be eliminated by passing **mult_out** through an LPF as done in AM demodulation. However, let us skip this low pass filtering process and directly perform the matched filtering operation.

5.D-1 [WWW]Replace **r_samples** by **mult_out** in the line '**z_samples=conv (r_samples, pt);**' to perform matched filtering directly on **mult_out**. Execute the m-file and then the following in the command window. Capture the resulting spectrum plot.

```
>>figure
>>BW_v=fft(z_samples,1024);
>>plot([-512:511]/1024*L, abs(fftshift(BW_v)))
```

5.D-2   Explain why the baseband signal can still be restored although the low pass filtering step is skipped.

5.E   [WWW]Restore the line '**EbN0dB_vector =10;**' back to '**EbN0dB_vector= [1 3 5 7 9];**' and restore the line '**while sum_Ne<1**' back to '**while sum_Ne<25**'. Execute the m-file and capture the resulting BER curve.

5.F   If all the processes discussed so far have been followed properly, the simulated BER in 5.E should not match the theoretical BER.

5.F-1   In general, the symbol energy (bit energy in the case of BPSK) should be measured at the stage right before noise is added. If the signal is processed (e.g., up-conversion) before noise is added, then the energy per symbol should be recalculated. The simulated BER curves obtained in 5.E should show that, to achieve the same BER, the E_b/N_0 required is 3 dB (a factor of 2 in linear scale) higher than the theoretical value. This is caused by the up-conversion process in which **tx_signal** is multiplied by **carrier_wave**. Mathematically explain why the up-conversion process reduces the symbol energy by a half.

5.F-2   To resolve the above problem, the line '**carrier_wave = cos(2*pi*3.2* (1:length(tx_signal))*t_step);**' should be modified into '**carrier_wave = sqrt(2)*cos (2*pi*3.2*(1:length(tx_signal))*t_step);**'. Explain how this modification resolves the problem.

5.F-3   Modify the m-file as described in 5.F-2 and execute the modified m-file. Capture the BER versus E_b/N_0 plot.

5.F-4   Also explain why it is unnecessary to multiply **local_carrier** by **sqrt(2)** as done to **carrier_wave**.

5.G   Consider the case where carrier phase synchronization is imperfect. Modify the line that generates **local_carrier** as follows.

```
local_carrier= cos(2*pi*3.2*(1:length(tx_signal))*t_step+pi/4);
```

This leads to a $\pi/4$ phase error between the transmitted carrier wave (**carrier_wave**) and the local carrier (**local_carrier**).

5.G-1   Execute the modified m-file and capture the BER plot.

5.G-2   Compare the simulated BER curve with the phase error and the theoretical BER curve that assumes perfect phase synchronization [1]. Measure the SNR gap at a BER of 10^{-3}, that is, the difference in the required E_b/N_0 in dB to achieve a BER of 10^{-3} between the two cases.

5.G-3   Mathematically explain the SNR gap measured in 5.G-2.

5.G-4   Calculate the SNR gap if the phase error is $\pi/8$.

5.G-5   Modify the m-file properly and verify the answer to 5.G-4 in simulation.

5.G-6   Summarize why phase synchronization is essential to ensure a good BER performance.

REFERENCE

[1] S. Bregni, *Synchronization of Digital Telecommunications Networks*, Hoboken, NJ: Wiley, 2002.

22

QPSK AND OFFSET QPSK
IN SIMULINK

- Design a passband quadrature phase shift keying (QPSK) transmitter and receiver in Simulink.
- Obtain bit error rate (BER) curves by using a Simulink design in conjunction with an m-file.
- Generate an offset QPSK (OQPSK) waveform and investigate its characteristics.
- Observe the eye diagrams, constellation, and the signal trajectories of QPSK and OQPSK.

22.1 CHARACTERISTICS OF QPSK SIGNALS

A passband QPSK signal (non–pulse-shaped) is expressed as [1]

$$s_i(t) = \sqrt{\frac{2E_s}{T_s}} \cos(\omega_c t - \phi_i), \qquad 0 \le t \le T_s, i = 1, 2, 3, 4, \qquad (22.1)$$

where E_s denotes the symbol energy, T_s denotes the symbol duration, and the phase ϕ_i is determined according to two data bits b_I and b_Q, which form one 4-ary (or quaternary) symbol as shown in Table 22.1.

Problem-Based Learning in Communication Systems Using MATLAB and Simulink, First Edition.
Kwonhue Choi and Huaping Liu.
© 2016 The Institute of Electrical and Electronics Engineers, Inc. Published 2016 by John Wiley & Sons, Inc.
Companion website: www.wiley.com/go/choi_problembasedlearning

TABLE 22.1   Phase-Bit Mapping for 4-Ary Symbols.

4-Ary symbol index i	$b_I\, b_Q$	ϕ_i
1	11	$\pi/4$
2	10	$-\pi/4$
3	01	$3\pi/4$
4	00	$-3\pi/4$

Let the two orthonormal basis functions be $\psi_1(t) = \sqrt{2/T_s}\cos(\omega_c t)$, $\psi_2(t) = \sqrt{2/T_s}\sin(\omega_c t)$. The QPSK signal $s_i(t)$ can be written in terms of the basis functions as

$$s_i(t) = \sqrt{E_s}\cos(\phi_i)\psi_1(t) + \sqrt{E_s}\sin(\phi_i)\psi_2(t), \qquad i = 1, \ldots, 4. \qquad (22.2)$$

1.A [T]Prove equation (22.2) by using the two methods to be discussed next.

1.A-1   Directly expand equation (22.1) by using the trigonometric identity for $\cos(A\text{-}B)$.

1.A-2.   Convert $s_i(t)$ into a point in the two-dimensional (2-D) vector space spanned by the orthonormal basis functions $\psi_1(t)$ and $\psi_2(t)$.

   (a) The x coordinate of $s_i(t)$ in the vector space, that is, the projection of $s_i(t)$ onto the basis function $\psi_1(t)$, is obtained as $\int_0^{T_s} s_i(t)\,\psi_1^*(t)\,dt$. Calculate the (x, y) coordinates of $s_i(t)$ in the vector space.
   (b) Use the result in (a) to express $s_i(t)$ as a linear combination of $\psi_1(t)$ and $\psi_2(t)$.

1.B [T]Substitute the values of ϕ_i in Table 22.1 to equation (22.2) and rewrite $s_1(t)$, $s_2(t)$, $s_3(t)$, and $s_4(t)$.

1.C [T]Consider a 2-D vector space where $\psi_1(t)$ is the x axis (called in the in-phase channel or the I channel) unit vector and $\psi_2(t)$ is the y axis (called the quadrature channel or the Q channel) unit vector.

1.C-1   From the results in 1.A-2(a) and 1.B, determine the (x, y) coordinates of $s_1(t)$, $s_2(t)$, $s_3(t)$, and $s_4(t)$ in the vector space.

1.C-2   Draw the x and y axes of the 2-D space, and in the space, identify the points corresponding to $s_1(t)$, $s_2(t)$, $s_3(t)$, and $s_4(t)$. Also specify the two data bits that correspond to each point.

1.D [T]In the answer to 1.C-2, read the phase of each symbol point, that is, the angle of the line from the origin to the symbol point in the 2-D space, and record them. Are they consistent with the values of ϕ_i listed in Table 22.1?

1.E [T]Generally, an MPSK signal is expressed as $s_i(t) = \sqrt{2E_s/T_s}\cos(\omega_c t - 2\pi(i-1)/M), i = 1, 2, \ldots, M$, where E_s denotes the energy per M-ary symbol.

1.E-1   For the case of $M = 8$, calculate the coordinates of each $s_i(t)$, $i = 1$, $2, \ldots, 8$, in the vector space spanned by $\psi_1(t)$ and $\psi_2(t)$. Recall that $\psi_1(t) = \sqrt{2/T_s}\cos(\omega_c t)$, $\psi_2(t) = \sqrt{2/T_s}\sin(\omega_c t)$ in this chapter.

1.E-2   Establish the relationship between the M-ary symbol energy E_s and bit energy E_b in an equation.

1.F   From 1.B and 1.E-2, we can express $s_i(t)$ in equation (22.2) as

$$s_i(t) = x(t) + y(t), \tag{22.3}$$

where

$$x(t) = \begin{cases} \sqrt{E_b}\psi_1(t) & \text{if } b_I = 1, \\ -\sqrt{E_b}\psi_1(t) & \text{if } b_I = 0, \end{cases} \qquad y(t) = ?. \tag{22.4}$$

Complete the expression marked by '**?**' in equation (22.4). According to equation (22.4), $x(t)$ and $y(t)$ can be viewed as follows.

- $x(t)$ is an antipodal (BPSK) signal, which transmits the bit b_I using the basis function $\psi_1(t)$, that is, the dimension along the x axis.
- $y(t)$ is an antipodal (BPSK) signal that transmits the bit b_Q using the basis function $\psi_2(t)$, that is, the dimension along the y axis.

Consequently the QPSK signal $s_i(t)$ can be viewed as the sum of two independent BPSK signals, one carrying b_I and one carrying b_Q.

22.2   IMPLEMENTATION OF THE QPSK TRANSMITTER

In this section we design a QPSK modulator that generates $s_i(t)$ in Simulink. The main parameters of the modulator are set as follows.

- QPSK symbol duration $T_s = 1$ second
- Carrier frequency $f_c(= \omega_c/2\pi) = 30$ Hz
- Bit energy $E_b = 16$
- Number of samples per symbol in passband QPSK waveform = 256

2.A   Recall from 1.F that the QPSK signal $s_i(t)$ is the sum of two orthogonal BPSK signals, one carrying b_I and one carrying b_Q. Based on equations (22.3) and (22.4), design an mdl/slx file for the QPSK transmitter as shown in Fig. 22.1.

2.A-1   Identify the blocks whose outputs correspond to b_I, b_Q, $\psi_1(t)$, $\psi_2(t)$, $x(t)$, $y(t)$, and $s_i(t)$ in equations (22.3) an (22.4), respectively.

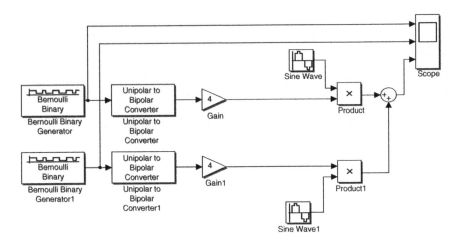

FIGURE 22.1 QPSK transmitter design.

2.A-2 Set the parameters of the blocks as specified in Table 22.2 and then save the design file as **QPSK_TX_SEC6.mdl/slx**. This file will be used again in Section 22.6. For the problems in this section, save the design file in a different name, **QPSK_TX.mdl/slx**, and use it; leave **QPSK_TX_SEC6.mdl/slx** untouched until Section 22.6.

Justify the parameter settings in Table 22.2.

TABLE 22.2 Parameter Settings for the QPSK Transmitter Design in Simulink.

Block	Parameter setting	Reason
Bernoulli Binary Generator	**Sample time** = 1	QPSK symbol duration $Ts = 1$
Bernoulli Binary Generator1	**Output data type** = **Boolean**	To convert into Boolean type
Bernoulli Binary Generator	**Initial seed** = Your student ID number	
Bernoulli Binary Generator1	**Initial seed** = Default setting (do not change)	
Unipolar to bipolar convertor,		
Unipolar to bipolar convertor1	**M-ary number** = 2	
Gain, Gain1	**Gain** = 4	
Sine wave, Sine wave1	**Amplitude** = sqrt(2) **Frequency (rad)** = 60*pi **Sample time** =1/256	
Sine wave	**phase** = pi/2	
Sine wave1	**phase** = 0	

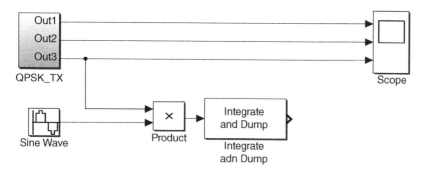

FIGURE 22.2   Correlation step to extract the in-phase component.

2.B   In this problem, we create $s_i(t)$ that consists of 20 QPSK symbols.

2.B-1   Determine the setting of **Simulation time** to simulate 20 QPSK symbols and justify the answer.

2.B-2   Run the simulation and capture the **Scope** display window that shows the waveforms of b_I, b_Q, and $s_i(t)$.

2.C   Compare the signal bandwidths of $x(t)$, $y(t)$, and $s_i(t)$ in equation (22.3). Explain how $s_i(t)$ and $x(t)$ as well as $y(t)$ have the same bandwidth even though $s_i(t)$ carries two separate information signals $x(t)$ and $y(t)$ without mutual interference.

2.D   Create a subsystem for the QPSK modulator as follows.

- Select all blocks except the **Scope**.
- Right-click in the selected area and choose **Create subsystem**.
- Change the subsystem name to QPSK_TX.

Capture the mdl/slx design.

22.3   IMPLEMENTATION OF THE QPSK RECEIVER

The Simulink design **QPSK_TX** completed in Section 22.2 outputs the transmitted signal QPSK $s_i(t)$, which is equivalent to the sum of two orthogonal BPSK signals; the I-channel carries b_I and the Q-channel carries b_Q, as shown in Section 1.F. In this section, we design the QPSK receiver that estimates b_I and b_Q from $s_i(t)$. Because of the orthogonality between the I and Q channels, the QPSK receiver is effectively two BPSK receivers running in parallel.

3.A   Fig. 22.2 illustrates the correlation step between $s_i(t)$ and $\psi_1(t)$ used to extract the in-phase component. Complete an mdl/slx file to extract the quadrature component. Then complete the whole receiver that will demodulate both the in-phase and quadrature components.

The following notes will help in this design:

- Create the two basis functions $\psi_1(t)$ and $\psi_2(t)$. The Simulink subsystem **QPSK_TX** created in 2.D has two sinusoidal blocks, **Sine Wave** and **Sine Wave1**, that generate the two basis functions. Since the receiver needs the same pair of basis functions, you can copy these two blocks for the receiver or create them the new blocks but set their parameters the same as these two blocks.
- The two **Product** blocks and the two **Integrate and dump** blocks, one for the in-phase component and one for the quadrature component, are identical.

3.A-1   The integration interval should be equal to the bit duration. Set the internal parameter **Integration period (Number of samples)** of the two **Integrate and dump** blocks to 256. Justify this setting.

3.A-2   Complete the design and capture it.

3.A-3   Capture the parameter setting windows of the two sinusoidal blocks and the two **integrate-and-dump** blocks.

3.B   To verify the operation of the **Integrate and dump** blocks, connect the input and output of **Integrate and dump** for the I channel to the **Scope** block. Set the simulation time to 20 seconds and run the simulation.

3.B-1   Capture the display window of the **Scope** window.

3.B-2   From the results, determine whether the **Integrate and dump** block operates as expected. Focus on the overall shape rather than the signal level.

3.C   Search for the **Compare To Constant** block in the library and add it to the mdl/slx design. To detect the binary bits using the decision variable, connect the output of **Integrate and dump** to the input of **Compare To Constant**. Set the parameters of **Compare To Constant** as follows:

- **Operator:** >
- **Constant value:** 0

Justify these settings.

3.D   Recall that the port **Out1** of the transmitter subsystem, **QPSK_TX**, corresponds to the I-channel binary data b_I. Connect this port as well as the output of **Compare To Constant** to the **Scope** block.

3.D-1   Run the simulation and capture the waveforms of the I-channel binary data b_I (transmitted bit stream) and the corresponding detected bit stream.

3.D-2   The result in 3.D-1 should show that relative to the transmitted waveform, the detected bit stream is delayed by one bit duration. Justify this.

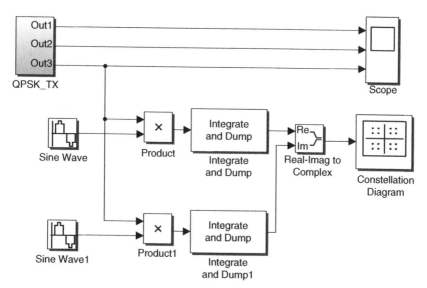

FIGURE 22.3 QPSK receiver design with a constellation diagram scope.

3.E Repeat the modifications to the mdl/slx design made in 3.C and 3.D for the Q channel; that is, let the output of the **Integrate and dump** block in the Q channel pass through **Compare To Constant** and connect the port **Out2** of **QPSK_TX** as well as the output of **Compare To Constant** to the **Scope** block.

Run the simulation and capture the waveforms of the Q-channel binary data b_Q (transmitted bit stream) and its corresponding detected bit stream.

3.F Like the eye diagram, the constellation diagram [1], which provides a visual insight on the performance, is also an important diagram for digital communication systems. In order to observe the constellation diagram of the received QPSK symbols, modify the mdl/slx file as shown in Fig. 22.3. The major changes are that we have added the **Real-Imag to complex** block and the **Constellation Diagram** block to the outputs of the **Integrate and dump** block for both the I and Q channels.

3.F-1 Capture the modified mdl/slx file.

3.F-2 Open the display window of **Constellation Diagram** and click the **Configuration Properties** icon in the menu bar. Set the parameter **Symbols to display** to 1000. In the mdl/slx design window, set the **Simulation stop time** to 2000 and run the simulation. Click the icon named **Scale X & Y Axes Limits** and then capture the display window of **Constellation Diagram**.

22.4 SNR SETTING, CONSTELLATION DIAGRAM, AND PHASE ERROR

Double-click to open the subsystem **QPSK_TX** in the mdl/slx file modified in 3.F. Insert a **Gaussian Noise Generator** block, connect it as shown in Fig. 22.4, and set its parameter **Sample time** to 1/256. If you are using a version that does not have a

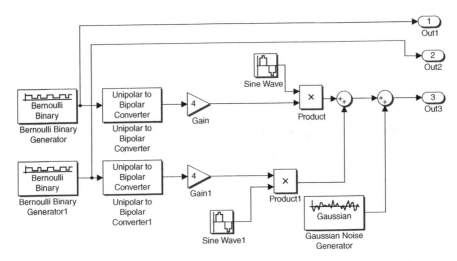

FIGURE 22.4 Model of the received QPSK signal over an AWGN channel.

Gaussian Noise Generator block in the library, then use the completed design file on the companion website. Now the output port **Out3** of **QPSK_TX** is the received QPSK signal over an AWGN channel.

4.A The following steps verify the setting of the parameter **Variance** of the **Gaussian Noise Generator** block, assuming $E_b/N_0 = 13$ dB. Document the calculation details.

- Convert E_b/N_0 [dB] into the linear scale value.
- Find N_0 from E_b/N_0 and the current E_b set in the mdl/slx file.
- Substitute N_0 into the noise variance setting expression, that is, the variance of noise sample $= N_0/(2 \times$ **t_step**), that was derived in 1.C-3 of Chapter 21 for the waveform level simulation over AWGN channels. The parameter **t_step** denotes the sample interval of the sampled waveforms. Thus it is set to be equal to 1/256 in this chapter.

4.B Set the parameter **Variance** of the **Gaussian Noise Generator** block as derived in 4.A and run the simulation. Capture the display window of **Constellation Diagram**. Prior to capturing, select the automatic axis scaling option in the display window menu.

4.C Repeat 4.A and 4.B for $E_b/N_0 =1$, 15, and 30 dB. Describe the changes in the constellation diagram according to the SNR setting and justify this change.

4.D In this subsection, we will observe the constellation diagram when there is a phase error.

4.D-1 Set the parameter **Variance** of the **Gaussian Noise Generator** block to 0 to simulate the noiseless case. Open the parameter setting windows of **Sine Wave** and

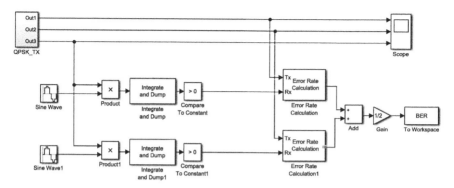

FIGURE 22.5   Completed mdl/slx file for QPSK BER simulation.

Sine Wave1 and add **pi/4** to the current setting of their parameter **Phase (rad)**. Run the simulation and capture the display window of **Constellation Diagram.**

4.D-2   Repeat 4.D-1 for phase errors of 15°, 30°, and 60°, and justify the results generated in the simulation.

22.5   BER SIMULATION IN SIMULINK USING A BUILT-IN FUNCTION sim( )

Obtaining a statistically meaningful BER in simulation requires simulating a number of independently generated bits and noise samples. How many bits to be simulated are sufficient? Over an AWGN channel, at a specific SNR, 10 to 100 times the inverse of the BER is typically acceptable.

For the systems designed in Simulink, we can simulate the BER curve using the **Error Rate Calculation** block and the built-in MATLAB function **sim( )**. The **Error Rate Calculation** block can be conveniently used to simulate the BER for each specific SNR value; the function **sim( )** provides an efficient way to simulate the Simulink model at different SNR values.

Modify the mdl/slx file in Section 22.4 as follows. Insert two **Compare To Constant** blocks as done in 3.C. To calculate the BERs of both channels, insert an **Error Rate Calculation** block for each channel. The output ports of the **Error Rate Calculation** blocks are configured as follows. These ports are connected to an **Add** block a **Gain** stage and finally a **To Workspace** block as shown in Fig. 22.5.

5.A   Set the parameters of the following blocks:

1. Two **Bernoulli Binary Generator** blocks in the subsystem **QPSK_TX**
    - **Output data type: Boolean** (select)
2. **Gaussian Noise Generator** in the subsystem **QPSK_TX**
    - **Variance**: v_n
    - **Sample time**: 1/256
3. Two **Compare to Constant** blocks: As done in 3.C.

4. Two **Error Rate Calculation** blocks:
   - **Receive delay**: 1 (Refer to 3.D-2 for reason.)
   - **Output data: Port** (select)
   - **Stop simulation: On** (enable)
   - **Target number of errors: Nerrs**
   - **Maximum number of symbols: inf**
5. **To Workspace**
   - **Variable name: BER**
   - **Limits data points to last:** 1
   - **Save format: Array**
6. **Gain**
   - **Gain** : 1/2 (in order to average I and Q channels)

Capture the parameter setting windows of all the blocks mentioned above.

5.B Set the **Simulation stop time** to **inf** and save the completed mdl/slx file as **QPSK_BER.slx** (or **QPSK_BER.mdl** in old Simulink versions).

Capture the completed mdl/slx design. Do not use **File/Copy** menu because this method does not capture the window frame, which shows the file name and the **Simulation stop time** settings. Instead, click the design window to select it and then press alt + print screen.

5.C The MATLAB built-in function **sim( )** enables us to simulate an mdl/slx model within an m-file. In this section we will write an m-file that repeatedly calls **QPSK_BER.slx (QPSK_BER.mdl)** to simulate the system BER performance at different E_b/N_0 values. The parameters not set in 5.A such as **v_n** and **Nerrs** will be set in this m-file. In particular, **v_n** will be set within each loop when **QPSK_BER.slx** is invoked to simulate the BER at a certain E_b/N_0.

5.C-1 The m-file below uses function **sim()** to invoke **QPSK_BER.slx/mdl** for each of the following E_b/N_0 [dB] values: 0, 1, 3, 5, and 7, and plots the BER versus E_b/N_0 curve. Complete the places marked by '?' and capture the completed m-file.

```
clear
Nerrs=50; %For fast simulation, decrease Nerrs but the simulation error will increase.
EbN0dB_vector=[0 1 3 5 7];
Eb=16;t_step=1/256;
for n=1:length(EbN0dB_vector)
  EbN0dB=EbN0dB_vector(n);
  EbN0=?;  % Convert EbN0dB into the linear scale value.
  N0=Eb/EbN0;
  v_n= ?;    % Set noise variance as a function of N0 and t_step.  Refer to the for-
mula in the equation (21.1)
  sim('QPSK_BER') % Simulate QPSK_BER.slx/mdl file
  BER_vector(n)=BER(1)
```

```
%'Error rate calculation' outputs 3-element array 'BER' in the form of [BER, num-
ber of bit errors, number of total bits].
    BER_theory(n)=0.5*erfc(sqrt(EbN0));
end
figure
semilogy(EbN0dB_vector, BER_vector ) %Plot BER curve.
hold on;
semilogy(EbN0dB_vector, BER_theory,'r')
legend('Simulated','Theory');grid on; hold off
```

5.C-2   Execute the completed m-file. The parameter **Nerrs** denotes the number of bit errors to be accumulated until simulation stops. Decreasing the value of **Nerrs** will speed up the simulation, but insufficient number of bit errors accumulated will result in statistically nonrepresentative BER results.

(a) Capture the simulated BER curve.

(b) Does the simulated BER curve match the theoretical BER curve of BPSK?

5.C-3   [T]Note that in the m-file in 5.C-1, **BER_theory** is the theoretical BER of BPSK. Explain why the simulated BER of QPSK should be identical to the theoretical BER of BPSK.

5.C-4   In the file **QPSK_BER.slx/mdl**, modify the phase settings of **Sine Wave** and **Sine Wave1** to introduce a phase error of pi/6. Save the modified **QPSK_BER.slx/mdl** and then execute the m-file in 5.C-1. Capture the simulated BER curve. Repeat the simulation for a phase error of pi/2.

5.C-5   Complete the following steps.

(a) [T]Derive the theoretical BER of QPSK as a function of the phase error θ.

(b) Modify the line '**BER_theory(n)=0.5*erfc(sqrt(EbN0));**' of the m-file in 5.C-1 to calculate the theoretical BER of QPSK derived in (a) assuming θ = pi/6.

(c) Execute the modified m-file and capture the simulated BER curve. Do the simulated and theoretical BER curves match?

22.6   PULSE SHAPING AND INSTANTANEOUS SIGNAL AMPLITUDE

6.A   Open the **QPSK_TX.mdl** file saved in 2.A-2. Note that the baseband signal multiplied by the output of **Sine Wave** becomes the passband signal whose spectrum is centered at the carrier frequency of 30 Hz. Thus the outputs of **Gain** and **Gain1** correspond to the baseband signals of the I channel and Q channel, respectively. We investigate the pulse shape of the baseband signals in the current design.

6.A-1   Connect the outputs of **Gain** and **Gain1** to the **Scope** and run the simulation for 10 seconds. Capture the display window of **Scope**.

6.A-2   Based on the captured plot, describe the shape of the current pulse $p(t)$.

6.B   The rectangular pulse in the time domain corresponds to a sinc function in the frequency domain. Thus pulse shaping using a rectangular pulse is bandwidth inefficient. Let us perform the raised cosine pulse shaping.

To this end, insert a **Raised cosine Transmit Filter** block between the **Gain** and **Product** blocks and another one between the **Gain1** and **Product1** blocks. Set the parameters as follows. Note that the parameter names could differ in the different versions of Simulink.

- **Roll-off factor**: 0.75
- **Output samples per symbol**: 256
- **Input processing**: **Element as channels (sample-based)**

Capture the modified mdl/slx design.

6.C   Let $I(t)$ and $Q(t)$ denote, respectively, the pulse shaped I- and Q-channel signals, that is, the outputs of the **Raised cosine Transmit filter** blocks. The passband signal $s(t)$ is written as

$$
\begin{aligned}
s(t) &= I(t) \sqrt{\frac{2}{T_s}} \cos\left(\omega_c t\right) + Q(t) \sqrt{\frac{2}{T_s}} \sin\left(\omega_c t\right) \\
&= \sqrt{\frac{2}{T_s} \left(I(t)^2 + Q(t)^2\right)} \cos\left[\omega_c t - \tan^{-1}\left(\frac{Q(t)}{I(t)}\right)\right].
\end{aligned}
\tag{22.5}
$$

Note that the instantaneous amplitude of $s(t)$ is $\sqrt{2/T_s \left(I(t)^2 + Q(t)^2\right)}$, and that the term $\sqrt{\left(I(t)^2 + Q(t)^2\right)}$ equals the magnitude of a vector $(I(t), Q(t))$ on the X-Y plane. Therefore the fluctuation of the amplitude of $s(t)$ can be predicted by observing how the magnitude of the vector $(I(t), Q(t))$ changes as a function of t. The plot showing the trajectory of the time-varying vector $(I(t), Q(t))$ is called the "signal trajectory" plot.

6.C-1   In order to observe the eye diagram of the pulse-shaped baseband signal $I(t)$ and $Q(t)$, and the signal trajectory of the passband signal $s(t)$, further modify the mdl/slx file in 6.B as shown in Fig. 22.6. If you are using a version that does not have the **Discrete-Time Signal Trajectory Scope** block in the library, then use the completed design file on the companion website.

Configuration Property setting for the **Discrete-Time Signal Trajectory Scope** block: **Samples per symbol** = 256

Configuration Property setting for the **Discrete-Time Eye Diagram** block: **Samples per symbol** = 256, **Symbols per trace** = 2.

Capture the modified mdl/slx file and the **Configuration Property** setting windows of **Discrete-Time Signal Trajectory Scope** and **Discrete-Time Eye Diagram**.

6.C-2   Before observing the signal trajectory and eye diagram, check the waveform of the pulse-shaped passband QPSK signal $s(t)$. Set the simulation stop time to 18 seconds and execute the simulation.

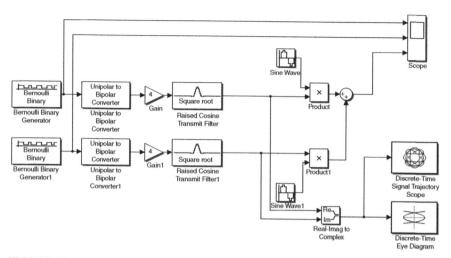

FIGURE 22.6   Pulse-shaped QPSK system with signal trajectory and eye diagram observation blocks.

(a) Right-click the pulse-shaped passband QPSK signal at the bottom of the **Scope** display window to autoscale it and then capture it.

(b) Does the amplitude of the signal change substantially?

6.C-3   In order to observe the signal trajectory over a sufficiently long time period, set the simulation stop time to **inf** and execute the simulation. Right-click the display windows of **Discrete-Time Signal Trajectory Scope** and **Discrete-Time Eye Diagram Scope** to autoscale them. After the **Autoscale** feature is turned on, capture the display windows of **Discrete-Time Signal Trajectory Scope** and **Discrete-Time Eye Diagram Scope**.

6.C-4   Based on the captured signal trajectory plot,

(a) Describe how significantly the instantaneous amplitude $\sqrt{I(t)^2 + Q(t)^2}$ changes.

(b) In the captured eye diagrams of $I(t)$ and $Q(t)$, find the time instants when the signal trajectory of $\sqrt{I(t)^2 + Q(t)^2}$ crosses the origin. Are these instants occurring at the maximum eye-opening points or occurring at the middle between the maximum eye-opening points?

6.D   Here we investigate the reason that causes the zero instantaneous amplitude of $s(t)$ to occur at the middle of two adjacent maximum eye-opening points.

6.D-1   Focus on the time instants when the signal trajectory crosses the origin. At these time instants, the signal trajectory moves from which quadrant and into which quadrant?

6.D-2   Write all possible QPSK bit pair (b_I, b_Q) transitions that cause the signal trajectory to cross the origin.

22.7 OFFSET QPSK

7.A If the instantaneous amplitude of the transmitted signal fluctuates over a wide range, then the linear operation range of the analog circuits (especially the power amplifier) should be large accordingly. This is not desirable in terms of hardware cost and power efficiency. The fluctuation of the instantaneous amplitude of the transmitted pulsed-shaped QPSK signals can be reduced by simply modifying one of the QPSK signal generation steps above.

In the mdl/slx design in 6.C-1, insert a **Delay** block between the pulse-shaping filter and the product block (**Product1**) in the Q channel only (or the I channel only) as shown in Fig. 22.7. Set **Delay Length** of the **Delay** block to 128, which corresponds to half of the QPSK symbol period. The resulting signal $s(t)$ is called the "Offset QPSK (OQPSK signal)" [1, 2].

7.A-1   Complete the mdl file as shown in Fig. 22.7 and run the simulation.

(a) Capture the signal trajectory.
(b) Compare the signal trajectories of OQPSK and the conventional QPSK signals. Examine both the regular QPSK and the OQPSK signals and then discuss how you expect the instantaneous amplitudes of both signals to change. Which one will likely change over a smaller range?

7.A-2   (a) Capture the eye diagram. (b) The half a symbol delay in the OQPSK transmitter places the maximum eye-opening points of the I and Q channels maximally apart from each other. Explain how this characteristic reduces the amplitude fluctuation of the OQPSK signals. In other words, how does this avoid the simultaneous bit pair transition?

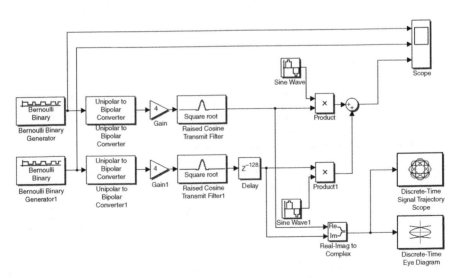

FIGURE 22.7   Offset QPSK transmitter.

7.B   Set the simulation stop time to 18 seconds and run the simulation again.

7.B-1   (a) Capture the display window of the **Scope** block. (b) Does the instantaneous amplitude change of OQPSK verify your observations made in 7.A-1?

7.B-2   Compare the captured waveform in 7.B-1 with the result captured in 6.C-2.

REFERENCES

[1] J. Proakis, *Digital Communications*, 3rd ed., New York: McGraw-Hill, 1995.

[2] S. Pasupathy, "Minimum Shift Keying: A Spectrally Efficient Modulation," *IEEE Communications Magazine*, Vol. 17, No. 4, 1979, pp. 14–22.

23

QUADRATURE AMPLITUDE MODULATION IN SIMULINK

- Design the quadrature amplitude modulation (QAM) transmitter and receiver in Simulink and obtain the bit error rate (BER) graphs through simulation.
- Compare the BER graphs obtained through simulation with the theoretical BER.
- Convert the pulse-shaped QAM signal into an electrical signal and observe the signal trajectory using an oscilloscope.

23.1 CHECKING THE BIT MAPPING OF SIMULINK QAM MODULATOR

The block **Rectangular QAM Modulator Baseband** provided in Simulink takes the input of a row vector composed of $\log_2 M$ bits and outputs the two-dimensional coordinates of QAM symbols in the two-dimensional vector space [1].

1.A In order to check the mapped points of four data bits in the vector space of 16-QAM, let us design a test system as shown in Fig. 23.1.

Set the parameters of the three blocks as follows.

1. **Constant**
    - **Constant value:** [0 0 0 0]' (Be sure to include the transposition operation.)
    - **Interpret vector parameters as 1-D:** Uncheck
    - **Sampling mode: Frame based**
    - **Frame period:** 1

Problem-Based Learning in Communication Systems Using MATLAB and Simulink, First Edition.
Kwonhue Choi and Huaping Liu.
© 2016 The Institute of Electrical and Electronics Engineers, Inc. Published 2016 by John Wiley & Sons, Inc.
Companion website: www.wiley.com/go/choi_problembasedlearning

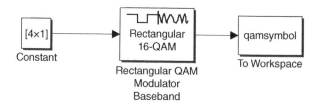

FIGURE 23.1   Test system for checking the 16-QAM bit mapping.

2. **Rectangular QAM Modulator Baseband**
   - **Input type**: Bit
   - **M-ary number**: 16
3. **To Workspace**
   - **Variable name: qamsymbol**
   - **Decimation**: 1
   - **Save format: Array**

Capture the parameter setting windows of each block.

1.B   Set **Simulation time stop time** to 0.5 seconds and run the simulation. This will generate the QAM-modulated symbol (MATLAB variable: **qamsymbol**) in the MATLAB work space.

1.B-1   Execute the following line in the command window and record the result.

```
>>qamsymbol
```

1.B-2   For every four bits set in the parameter **Constant value** of the **Constant** block, a complex **qamsymbol** will be generated in the work space after the simulation. Assign the 16 combinations of the 4-bit streams from [0 0 0 0]' to [1 1 1 1]' to the parameter **Constant value** of the **Constant** block one by one, execute the simulation for each combination, and record the value of **qamsymbol**. Note that since column vectors are required, you must be sure to transpose the 4-bit vector (to make it a column vector) for setting **Constant value** of the **Constant** block.

1.B-3   (a) Based on the results in 1.B-2, record the corresponding 4 bits on each of the constellation points in Fig. 23.2, for example, 0000 for point (−3.3). (b) Click **View Constellation** in the parameter window of the **Rectangular QAM Modulator Baseband** block. Is it the same as your mapping in (a)?

1.B-4   [T]The energy of each QAM symbol in the constellation equals the square of the distance between the point and the origin. For example, the energy of the symbol corresponding to the 0000 bit stream is $3^2 + 3^2 = 18$. So the symbol energy might vary depending on its location in the constellation.

   (a) Determine the symbol energy of each of the 16 symbols and calculate the average symbol energy of 16-QAM E_s.

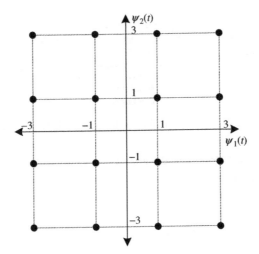

FIGURE 23.2   Bit mapping of the Rectangular QAM block.

(b) The theoretical average symbol energy formula for QAM with lattice distance of 2 is given as $E_s = 2(M\text{-}1)/3$. Compare the result in (a) with the theoretical average symbol energy.

1.B-5   (a) Write the relationship between bit energy E_b and M-ary symbol energy E_s. (b) By using this relationship, show that the bit energy E_b of 16-QAM with a constellation as shown in Fig. 23.2 is equal to 2.5.

1.B-6   The mapping confirmed in 1.B-3 is called "Gray mapping" [1,2]. Summarize the general conditions of Gray mapping and the advantages of Gray mapping.

1.C   A QAM signal received through an AWGN channel can be converted into a point in the 2-D vector space as shown in Fig. 23.2. Let (z_1, z_2) denote the point (coordinate) of the received signal. Because of the Gaussian noise, (z_1, z_2) will not fall exactly on one of 16 constellation points in Fig. 23.2, and both z_1 and z_2 can be any real values in the range of $[-\infty \ \infty]$.

In this subsection, we estimate the four transmitted bits using MLD on the received signal vector (z_1, z_2).

1.C-1   For 16-QAM, MLD is equivalent to calculating the distance between (z_1, z_2) and each of the 16 constellation points, and the one closest to (z_1, z_2) is chosen as the estimated transmitted symbol. For rectangular QAM like the one in Fig. 23.2, the estimate can be done by comparing each of the values of z_1 and z_2 with a set of thresholds.

For example, with 16-QAM, if z_1 is positive, then the estimated transmitted symbol for (z_1, z_2) will be one of the eight points in the first and fourth quadrants. Because of the mapping structure, the first bit (MSB) of these eight constellation points is always '1'. Thus the first bit can be estimated as '1' without having to determine which of the eight constellation points is closest to the received symbol.

TABLE 23.1   Decision Boundaries of 16-QAM With Gray Mapping.

Data bit	Decision condition
b4	IF $z_1 > 0$ **b4_estimate**= 1, ELSE b4_estimate= 0.
b3	
b2	
b1	

Simple decision rules can be formulated for the other three data bits in a similar way. Complete the decision rules to estimate the four transmitted bits from (z_1, z_2) and record them in Table 23.1, in which the four data bits are denoted as **b4**, **b3**, **b2**, and **b1** in the order from MSB to LSB, and **b4_estimate**, **b3_estimate**, **b2_estimate**, and **b1_esimate** denote their estimates.

1.C-2   From the table completed in 1.C-1, explain whether or not calculating the actual distance for MLD is required.

1.D   [T, A]Derivation of the BER of 16-QAM as a function of E_b/N_0. For simplicity, let γ represent E_b/N_0.

1.D-1   Substitute E_b (which equals 2.5 as obtained in 1.B-5) into the equation $E_b/N_0 = \gamma$ and then express N_0 as a function of γ.

1.D-2   With white Gaussian noise added to the signal component, z_1 follows the Gaussian distribution with variance $N_0/2$ and mean equaling the real part of the corresponding constellation point in Fig. 23.2. Similarly, z_2 also follows the Gaussian distribution with variance $N_0/2$ and mean equaling the imaginary part of the corresponding constellation point.

Consider the example that 0000 are transmitted. Then z_1 becomes a Gaussian random variable with a mean -3 and variance $N_0/2$. From this distribution of z_1 and using the Q function, express the error probability of **b4**, that is, the probability that **b4_estimate** is different from **b4**, according to the decision rules given in Table 23.1. For the considered example that 0000 are transmitted, the error probability of **b4** is the probability that **b4_estimate** is equal to 1 since **b4** is 0.

Repeat this process to obtain the error probability of **b4** for the other 15 constellation points. Then derive an expression of the average BER of **b4**.

1.D-3   Similarly, derive the BER expressions for **b3**, **b2**, and **b1**, also using the Q-function.

1.D-4   Are the BERs of **b4**, **b3**, **b2**, and **b1** identical? If not, compare these BER values and explain why.

1.D-5   Derive the average BER expression of **b4**, **b3**, **b2**, and **b1** as a function of only γ. Make sure that the answer to 1.D-1, which established N_0 as a function of γ, is applied in the average BER expression, if this has not been done yet.

1.D-6 The MATLAB function **berawgn()** can be used to calculate the theoretical BER of various modulation and demodulation methods. Execute the following lines of code to calculate the BER of 16-QAM assuming $\gamma = 10$. Compare the result with the value calculated from the answer to 1.D-5.

```
>> EbN0=10;
>> EbN0dB=10*log10(EbN0);
>> berawgn(EbN0dB,'qam',16,'gray');
```

23.2 RECEIVED QAM SIGNAL IN AWGN

2.A Design a Simulink block as shown in Fig. 23.3. You can search and import each block by typing in the block name in the **Simulink library browser**. Capture the completed design.

2.B Set all the parameters of each block as follows. Attention should be paid to **Sample time**.

1. **Bernoulli Binary Generator**
   - **Initial seed**: Your student ID number
   - **Sample time**: 1
   - **Frame-based outputs**: Check the box.
   - **Samples per frame**: 4
   - **Output data type: Boolean**

2. **Frame Conversion**
   - **Sampling mode of output signal: Sample-based**

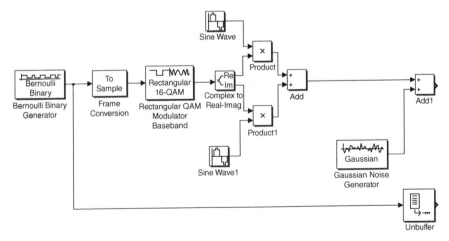

FIGURE 23.3 Block diagram of **16QAM_AWGN** for generating the received QAM signal over an AWGN channel.

3. **Rectangular QAM Modulator Baseband**
   - **Input type: Bit**
   - **M-ary number**: 16
4. **Sine Wave**
   - **Amplitude**: sqrt(2)
   - **Frequency (rad/s): 2*pi*20**
   - **Phase (rad): pi/2**
   - **Sample time**: 1/100
5. **Sine Wave1**
   - **Amplitude**: sqrt(2)
   - **Frequency (rad/s): 2*pi*20**
   - **Phase (rad): 0**
   - **Sample time**: 1/100
6. **Gaussian Noise Generator**
   - **Variance (vector or matrix): 0**
   - **Sample time**: 1/100

2.B-1   Capture the parameter setting windows of each block.

2.B-2   In the menu of the design window in 2.A, select **Edit/Select all** to select all the blocks in the design. Then left-click the selected blocks and choose **Edit/Create subsystem** to create a subsystem. Change the name of the created subsystem into **16QAM_AWGN**. Capture the subsystem **16QAM_AWGN**.

2.C   Make sure that the **Out1** port of **16QAM_AWGN** is the sum of the passband QAM signal and AWGN, and **Out2** port of **16QAM_AWGN** is the four data bits of the QAM signal.

2.C-1   Double-click the subsystem **16QAM_AWGN** to open it. Import a **Scope** block and connect the two outputs of the **Complex to Real-Imag** block to the **Scope** block and run the simulation for 50 seconds. Capture the **Scope** display window.

2.C-2   Based on the block parameters set in 2.B for the **16QAM_AWGN** block, what is the carrier frequency of the passband QAM signal at **Out1** port?

2.C-3   Based on the block parameters set in 2.B for the **16QAM_AWGN** block, explain why the in-phase basis function $\psi_1(t)$ and quadrature basis function $\psi_2(t)$ used for generating the passband QAM signal at **Out1** port are expressed $\psi_1(t) = \sqrt{2}\cos(2\pi \times 20t)$, $\psi_2(t) = \sqrt{2}\sin(2\pi \times 20t)$, respectively.

2.C-4   Consider the example that the first four data bits are 0110, that is, the output of the **Bernoulli Binary generator** block inside the **16QAM_AWGN** block is 0110, and assume that there is no noise. Then the passband QAM waveform $r(t)$ at **Out1** port is expressed as $r(t) = -\psi_1(t) - 3\psi_2(t)$. Justify this.

23.3   DESIGN OF QAM DEMODULATOR

3.A   Let $r(t)$ represent the signal at the **Out1** port of the **16QAM_AWGN** block. Complete the following equations that generate z_1 and z_2, the real and imaginary components of the nth received QAM symbol:

$$z_1 = \int_{(n-1)T}^{nT} r(t) \times \psi_1(t)dt, \quad z_2 = \int_{(n-1)T}^{nT} ? \times ?dt. \quad (23.1)$$

3.B   Design a Simulink file (mdl/slx file) to generate z_1 and z_2 from $r(t)$; that is, implement (23.1) in a Simulink design. Some useful guides for this implementation:

1. The basis functions $\psi_1(t)$ and $\psi_2(t)$ can be generated the same way as the ones generated in **16QAM_AWGN**.
2. The integration over one symbol period can be implemented using the **Integrate and dump** block. But the following settings are critical:
   - Set the internal variable **Integration period** of the **Integrate and dump** block to the number of samples per 16-ary symbol (=100).
   - The output of the **Integrate and dump** block should be multiplied by the sample interval (=1/100). This is similar to the numerical integration discussed in Section 2.1 of Chapter 2, in which **sum( )** should be multiplied by **t_step**.
3. Fig. 23.4 shows the design to generate z_1.

3.B-1   Complete the Simulink file that generates z_1 and z_2 from $r(t)$. Capture the completed design window.

3.B-2   Add the following two blocks in Fig. 23.5 on the right-hand side of the design in 3.B-1 and connect the generated z_1 and z_2, respectively, to the **Re** and **Im** ports of the **Real-Imag to Complex** block. In the configuration parameters of the **Constellation Diagram** block, set the parameter **Symbols to display** to 1000. Set the **Simulation stop time** to inf and run the simulation. You may stop the simulation after the constellation diagram has converged properly. Click the icon named **Scale X & Y Axes Limits** and then capture the **Constellation Diagram**.

3.B-3   Judged from visual inspection of the constellation diagram obtained in 3.B-2, have z_1 and z_2 been generated correctly?

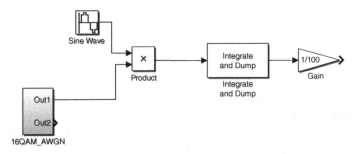

FIGURE 23.4   Part of the design that generates z_1.

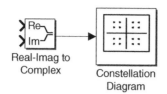

FIGURE 23.5 Connection for the **Constellation diagram** block.

3.B-4 Set the parameter **Variance (vector or matrix)** of the **Gaussian Noise Generator** block inside the **16QAM_AWGN** block to 10 and run the simulation again. Capture the constellation diagram.

3.B-5 Change the frequency of the two **Sine Wave** (I, Q basis) blocks used for generating z_1 and z_2 to 2*pi*20.00001 rad/s and run the simulation. Capture the constellation diagram.

3.B-6 With the current parameter settings, does the frequency error or the noise variance degrade more the BER performance? Explain the reason.

3.B-7 From the answers in B-5 and B-6, explain why QAM signals require coherent detection.

3.C In the design completed in 3.B-1, add the part that generates the estimates of **b4**, **b3**, **b2**, and **b1** denoted by **b4_estimate**, **b3_estimate**, **b2_estimate**, and **b1_esimate**, respectively. Apply the decision rules established in Table 23.1 to z_1 and z_2. Fig. 23.6 shows the incomplete design that generates only **b4_estimate**. Complete the remaining part to generate **b3_estimate**, **b2_estimate**, and **b1_esimate**. Properly use the **abs** and **compare to constant** blocks. Capture the completed design.

3.D To convert the estimated bits **b4**, **b3**, **b2**, and **b1** that each 16-QAM symbol represents into a serial bit sequence, connect the related blocks as shown in Fig. 23.7. To this end, go through the following steps.

1. Select the demodulation part, that is, all the blocks added in 3.B-1 and 3.C, and create a subsystem.
2. Add all the other required blocks as shown in Fig. 23.7.

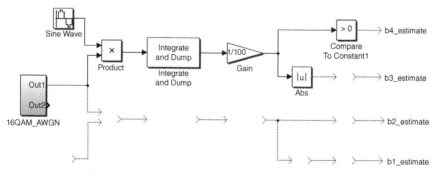

FIGURE 23.6 Incomplete design that detects only **b4**.

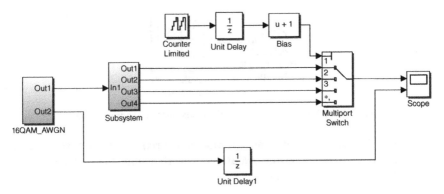

FIGURE 23.7   System that compares the transmitted bits with the estimated bits.

3. Set the following parameters:
   - **Upper limit** of the **Counter Limited** block: 3
   - **Bias** of the **Bias** block: 1
   - **Number of data ports** in the **Multiport Switch** block: 4
   - **Sample time** of the two **Unit Delay** blocks: 1

Capture the competed design.

3.E   In the design completed in 3.D, make sure that the **Variance** of the **Gaussian Noise Generator** block inside the **16QAM_AWGN** block is set to 0, and the frequency of the two Sine Wave (I and Q basis functions) blocks inside the demodulation subsystem is set to 2*pi*20 rad/s.

Set the **Simulation time** to 50 and set the variable **Initial Seed** of the **Bernoulli Binary Generator** block inside the **16QAM_AWGN** block to your student ID. Run the simulation and capture the display window of the **Scope** block.

3.F   Are the transmitted and received bits identical?

23.4   BER SIMULATION

In this section we simulate the BER performance of 16-QAM using a method similar to the one used in Chapter 22: employing a combination of a Simulink design and a MATLAB script file.

4.A   We start the simulation for a fixed E_b/N_0 and take $E_b/N_0 = 10$ dB as a test case. Continue the following steps to calculate the variance of the noise samples, that is, the value for the parameter **Variance** of the **Gaussian Noise Generator** block inside the **16QAM_AWGN** block, and set $E_b/N_0 = 10$ dB.

4.A-1   Convert this E_b/N_0 value in [dB] to a value in linear scale.

4.A-2   Use the answers to 4.A-1 and 1.B-5 to determine the value of N_0.

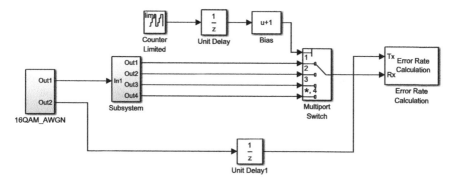

FIGURE 23.8   BER simulation-ready Simulink design.

4.A-3   In the simulations in Chapter 21, the sampled versions of the signal and noise waveforms with an interval **t_step** were used. Thus the variance of the noise samples should be calculated as

$$\text{Variance of the noise samples} = N_0/(2 * \text{t\_step}). \tag{23.2}$$

The details of this relationship can be found in 1.A–1.D of Chapter 21.

In the Simulink design of this chapter, the noise sampling interval (i.e., **sample time** of **16QAM_AWGN/Gaussian Noise Generator**) is set to 1/100; thus, **t_step**= 1/100. The parameter **Variance** of **16QAM_AWGN/Gaussian Noise Generator** can be found by substituting **t_step** and the calculated value of N_0 in 4.A-2 into equation (23.2). Show that the value for the **Variance** of **Gaussian Noise Generator** is 12.5.

4.B   Fig. 23.8 shows the final Simulink design, which is nearly ready for BER simulation in conjunction with an m-file except the settings of some parameters as follows.

1. Set **Simulation stop time** to **inf** and remove all **Scope** blocks if any (e.g., waveform, eye diagram, constellation), which are irrelevant to BER simulation.
2. Set the **Variance** of **Gaussian Noise Generator** inside the subsystem **16QAM_AWGN** as
   - **Variance (vector or matrix): v_n** (The variable **v_n** will be set in the m-file later.)
3. Set the parameters of **Error Rate Calculation** block as follows:
   - **Stop simulation**: Check (select)
   - **Output data: Workspace.**
4. Save the modified design as **QAM_BER.mdl/slx**.

4.B-1   Capture the completed design file **QAM_BER.mdl/slx**.

4.B-2   Execute the following line in the command window. The number 12.5 corresponds to the value of the **Variance** of **16QAM_AWGN/Gaussian Noise Generator** to set $E_b/N_0 = 10$. Refer to 4.A-3.

```
>> v_n=12.5
```

Then run the simulation using **QAM_BER.mdl/slx**. After the simulation is finished, the **Error Rate Calculation** block returns a vector **ErrorVec** whose first element corresponds to the simulated BER. Execute **ErrorVec(1)** in the command window and capture the simulated BER.

4.C Compare the simulated BER with the theoretical BER of 16-QAM (derived in 1.D-5).

4.D The m-file below simulates the BER curve of 16 QAM in conjunction with the Simulink design **QAM_BER.mdl/slx**.

```
clear
EbN0dB_vector=2:2:12;
Eb=?; %Answer to 1.B-5
t_step=1/100; % Sample time of 'out1' port in '16QAM_AWGN' block which is also
same as the noise sample interval.
for n=1:length(EbN0dB_vector)
   EbN0dB=EbN0dB_vector(n);
   EbN0=?;  % Convert EbN0dB [dB] into a value in linear scale.
   N0=Eb/EbN0; % Calculate N0 value
   v_n=?;     % Variance setting. Refer to (23.2)
   sim('QAM_BER') % Execute QAM_BER.mdl/slx file
   BER_vector(n)=ErrorVec(1) %BER of workspace is saved as a vector.
end
figure
semilogy(EbN0dB_vector , BER_vector ) %Plot the BER as the function of  Eb/N0[dB]
BER_theory=berawgn(EbN0dB_vector,'qam',16,'gray');
hold on
semilogy(EbN0dB_vector,BER_therory,'r')
grid on; xlabel('EbN0 [dB]');ylabel('BER');legend('Simulated','Theory');
```

4.D-1 Complete the places marked by '**?**' in the m-file and capture the completed m-file.

4.D-2 Execute the m-file and capture the simulated BER curve. Does the simulated BER match the theoretical result?

4.D-3 Insert a **Gain** block between **16QAM_AWGN** and the demodulator subsystem, and set **Gain** to 10. Then execute a following line in the command window to set E_b/N_0 to infinity.

```
>> v_n=0
```

Run the simulation using **QAM_BER.mdl/slx** first. Then execute a following line of command in the command window to see the simulated BER.

```
>> ErrorVec(1)
```

(a) Capture the simulated BER result. Justify why the BER is not 0 even though E_b/N_0 is set to infinity.

(b) Note that the simulated BER is approximately equal to 1/4. This is because the decision rule in Table 23.1 should be adjusted according to the received signal scaling factor but it was not. Explain why the current decision rule in Table 23.1 results in a BER equal to 1/4, not 1/2, which is the worst_case BER.

(c) In order to obtain the correct BER while the received signal is scaled by a factor of 10, the parameters of which blocks of QAM_BER.mdl/slx should be modified and to what values? Just provide the answers but do not modify these parameters in the mdl/slx file.

(d) In contrast to the QAM system, in the QPSK system, even if the received signal is scaled, the detection thresholds do not need to be adjusted accordingly. Justify this.

4.D-4 Summarize the disadvantages of M-ary QAM ($M > 4$) over QPSK in terms of receiver complexity and signal processing overhead.

4.E In the **Simulink Library Brower**, search for the **Spectrum Analyzer** block in the **DSP System Toolbox**. To see the spectrum of 16QAM signals, connect the **Spectrum Analyzer** to the **Out1** port of the **16QAM_AWGN** block in the design file.

4.E-1 Execute the mdl file and stop the simulation when the spectrum appears properly. (a) Capture the spectrum displayed in the window. (b) Record the passband center frequency. Comment on whether or not it is what you expected.

4.E-2 Zoom into the main lobe of the passband signal in the positive frequency range and measure the 20-dB bandwidth. For accurate measurement, zoom into the range of 18 Hz to 22 Hz along the x axis. (a) Capture the zoomed-in window. (b) Measure and record the 20-dB bandwidth of the main lobe. Is the measured bandwidth equal to the theoretical bandwidth?

4.F In this problem we analyze the trade-off between bandwidth and BER performance for a given modulation order M in QAM systems.

4.F-1 Examine again the BER graph of QPSK (equivalent to 4-QAM) obtained in Chapter 22. Compare the BER performances between 4-QAM and 16-QAM. Explain why there is a performance gap in terms of the required E_b/N_0 to achieve the same BER.

4.F-2 [T]Assuming pulse shaping with a roll-off factor of 0, data bit rate R_b, and QAM alphabet size M, show that the QAM symbol rate is $R_s = R_b / \log_2 M$ and the passband bandwidth is $B = R_b/(2 \log_2 M)$.

4.F-3 [T]The bandwidth efficiency is defined as [1]:

$$\text{Bandwidth efficiency} = \frac{\text{Data bit rate}}{\text{Bandwidth}}$$

$$= \frac{R_b[\text{bits/s}]}{B[\text{Hz}]}. \tag{23.3}$$

Calculate and compare the bandwidth efficiencies of QPSK (=4-QAM), 16-QAM, and 64-QAM.

4.F-4 Explain how bandwidth efficiency and BER performance change (degrade or improve) depending on the QAM constellation size M.

23.5 OBSERVING QAM SIGNAL TRAJECTORY USING AN OSCILLOSCOPE

5.A [T]The baseband signal in the **16QAM_AWGN** block refers to the signals right before the mixer, that is, the two outputs of the **Complex to Real-Imag** block. Connect the I and Q components of the baseband signals to the **Scope** blocks and observe their waveforms. Based on the waveforms observed, write an expression of the pulse shape of the baseband signal $p(t)$.

5.B [T]The pulse in 5.A is not suitable for practical systems. Why?

5.C Download the m-file **tx_sig_QAM.m** from the companion website to your work folder. This file generates the baseband I and Q signals of an M-ary QAM signal and plays the baseband I and Q signals as a stereo audio signal. The vector variables **tx_signal_I** and **tx_signal_Q** in the m-file correspond to the I and Q signals, respectively. Note that the pulses used to generate **tx_signal_I** and **tx_signal_Q** are a raised-cosine pulse.

For the lines in bold, explain what the variables on the left-hand side represent and justify how the right-hand side expression is properly formulated accordingly.

```
clear
a=1; b=1;
M=16;Ts=1;L=16;
t_step=Ts/L;
%%%%%%%%%<1. Pulse waveform generation > %%%%%%%%%%%%%%%%%%%%
pt=rcosine(1,L,'normal',0.75);

%%%%%%%%%<2. Generation of Ns number of M-ary symbols >%%%%%%%%%%
```

```
Ns=5000;
dI=2*ceil(rand(1,Ns).^a*sqrt(M))-(sqrt(M)+1);%Ns is the number of M-ary data symbols
dQ=2*ceil(rand(1,Ns).^b*sqrt(M))-(sqrt(M)+1);
%%%%%%%%<3. Impulse modulation >%%%%%%%%%%%%%%%%%%%%%%%%%
impulse_modulated_I=[ ];
impulse_modulated_Q=[ ];
for n=1:Ns
   impulse_signal_I=[dI(n) zeros(1, L-1)];
   impulse_modulated_I=[impulse_modulated_I impulse_signal_I];
   impulse_signal_Q=[dQ(n) zeros(1, L-1)];
   impulse_modulated_Q=[impulse_modulated_Q impulse_signal_Q];
end
%%%%%%%%<4.Pulse shaping (Transmitter filtering)>%%%%%%%%%%%
tx_signal_I=conv(impulse_modulated_I, pt);
tx_signal_Q=conv(impulse_modulated_Q, pt);
signal_out=[tx_signal_I' tx_signal_Q']';
n=10;
Nrepeat=n*4;
for k=1:Nrepeat  %Output repeated for Nrepeat times
   soundsc(signal_out',8000*n);
   %Sample rate is 8000*n Hz, Symbol rate is 8000*n/L Hz
   done=k
end
```

5.D Here we go through the following steps to see the signal trajectory of QAM in an oscilloscope.

   Step 1. Similar to the procedure of observing the eye diagram with an oscilloscope in Section 20.4 of Chapter 20, connect the stereo (left/right) audio output signals of a PC to the two probing ports of the oscilloscope (left channel to port A and right channel to port B). You may reuse the audio cable made for experiments in Section 20.4 of Chapter 20.

   Step 2. Set the display mode of the oscilloscope to XY plot and adjust the shift dial of A and B ports to place the curser in the center of the screen.

   Step 3. Set the PC audio volume to maximum and execute the m-file **tx_sig_QAM.m**. Then the pulse-shaped I and Q signals will appear in the oscilloscope as shown in Fig. 23.9. You may need to adjust the scale of the amplitude of A and B ports to observe the signal shape clearly. Be sure to set the same amplitude for both A and B ports.

5.D-1   Repeat Step 3 above for each of the following values of M: 4, 16, and 64 (change the value of **M** in the m-file **tx_sig_QAM.m**). Capture the oscilloscope screens for each case.

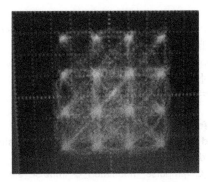

FIGURE 23.9   Illustration of the signal trajectory observed in an oscilloscope.

5.D-2   For $M = 16$, capture the oscilloscope screens when the roll-off factor is set to 0, 0.5, and 1.

5.D-3   Replace the second line of the m-file **tx_sig_QAM.m** by the two lines below. Then repeat 5.D-1 with the roll-off factor set to 0.7 and capture the resulting screen.

---

a=0.5+0.015*XX; % XX=Last 2 digits of your student ID number.
b=1/a;

5.E   Analyze the captured signal trajectory results in 5.D-1, 5.D-2, and 5.D-3.

5.E-1   From the result in 5.D-1, describe the difference in the signal trajectories according to M and justify what you observe.

5.E-2   From the result in 5.D-2, describe the difference in the signal trajectories according to the roll-off factor and justify what you observe.

5.E-3   [A]From the result in 5.D-3, describe the difference in the signal trajectories according to **a** and **b**, and justify what you observe.

REFERENCES

[1] J. Proakis, *Digital Communications*, 3rd ed., New York: McGraw-Hill, 1995.

[2] F. Gray, *Pulse communication*, US Patent 2,632,058.

24

CONVOLUTIONAL CODE

- Implement the encoding and decoding algorithms of convolutional codes [1–4].
- Simulate the bit error rate (BER) of convolutional codes.
- Observe the changes in the coded BER according to the parameter of the convolutional code.

24.1 ENCODING ALGORITHM

1.A Review 4.A and 4.B of Chapter 1 and solve the problems in these sections again to get refreshed on creating a used-defined MATLAB function.

1.B [T]Fig. 24.1 shows the structure of the convolutional encoder considered in this section. The variable m(k) denotes the kth input data bit and the pair (u1(k), u2(k)) denotes the kth encoded bit pair; that is, if m(1), m(2), m(3), … are the encoder input bits, then the encoder outputs u1(1), u2(1), u1(2), u2(2), u1(3), u2(3), … in order.

1.B-1   Determine the code rate [1–4] of the encoder shown in Fig. 24.1.

1.B-2   Determine the constraint length [2–4] of the encoder shown in Fig. 24.1.

1.B-3   Create a 5-bit vector **m** randomly in the MATLAB command window as shown below and capture the result. Do not clear the vector **m** created since it will be used in 1.C.

Problem-Based Learning in Communication Systems Using MATLAB and Simulink, First Edition.
Kwonhue Choi and Huaping Liu.
© 2016 The Institute of Electrical and Electronics Engineers, Inc. Published 2016 by John Wiley & Sons, Inc.
Companion website: www.wiley.com/go/choi_problembasedlearning

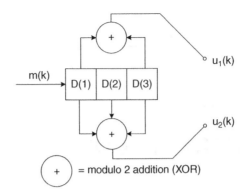

FIGURE 24.1 Convolutional code considered.

```
>>rand(1,XXX); %XXX=the last three digits of your student ID.
>>m=(rand(1,5)>0.5) %Do not append ';' to display the result. Alternatively, you may
   use m=randi([0 1],1,5) .
```

1.B-4   Suppose that the data vector **m** created in 1.B-3 is passed to the input of the encoder that is in the all-zero initial state. Manually encode **m** and record the encoder output bit sequence.

1.C [WWW]The following user-defined MATLAB function implements the encoder in Fig. 24.1. Executing **enc(m)** for any input bit stream **m** will return the corresponding encoding result.

```
% Save the m-file as same as the function name, i.e., save it as enc.m.
function coded = enc(data_bit)
Nb=length(data_bit);
D=zeros(1,3); % Initialize the states.
coded=[];
for n=1:Nb % n = clock index. Since one bit comes in at each clock period, Nb clock
    periods will be required.
  D(1)=data_bit(n);
  %%%%%%output (u1, u2) generation logic part %%%%%%
  u1=xor(D(1),D(3));
  u2=xor(xor(D(1),D(2)),D(3));

  %%%%%Coded bit concatenation %%%%%%%
  coded=[coded u1 u2];

  %%%%%%% memory updating %%%%%%%
  D(3)=D(2); %During each clock period, the bit in the memory shifts, i.e., D(2) =>D(3).
  D(2)=D(1); %During each clock period, the bit in the memory shifts, i.e., D(1) =>D(2).
end
```

1.C-1 The values (outputs) of state memories (D(2), D(3)) in Fig. 24.1 are synchronized for bit-by-bit operation. Each time the clock cycle changes, the output of the previous register will be applied as the input of the next shift register. The memory updating part at the bottom of the 'for' loop in **enc.m** implements this. Explain why the execution orders of these two lines should not be switched.

1.C-2 Execute **enc(m)** in the command window and check whether the result is same as the manually obtained answer in B-4.

1.D The free distance [2–4] of this encoder.

1.D-1 Execute **enc([0 0 0 0 0])** in the command window to obtain the encoder output for the input [0 0 0 0 0]. Capture the result.

1.D-2 Using the linearity property of convolutional codes, explain why the encoder output for an all-zero input sequence is also an all-zero sequence as observed in 1.D-1.

1.D-3 Execute **enc([1 0 0 0 0])** in the command window to obtain the encoder output for the input [1 0 0 0 0]. (a) Capture the result. (b) Count the number of bit '1' in the encoder output sequence.

1.D-4 Execute the following lines to regenerate **m** and another 5-bit sequence **m_diff**, which is equal to **m** except one of the first three bits is inverted. Capture the result to show the generated sequences **m** and **m_diff**.

```
>>m=(rand(1,5)>0.5)   %Do not add ';' at the end to display the result. Alterna-
tively you may use m=randi([0 1],1,5).
>>m_diff=m;
>>m_diff(1)=not(m(1)) %You may execute m_diff(2)=not(m(2)) or m_diff(3)=not(m(3))
instead. Do not append ';' either.
```

1.D-5 Execute the following lines of code to see the encoded outputs for the input **m** and **m_diff**, respectively. Capture the result.

```
>>enc(m)    %Do not append ';' to display the result.
>>enc(m_diff) %Do not append ';' to display the result.
```

1.D-6 The two input sequences **m** and **m_diff** in 1.D-4 differ by only one bit. Suppose that such a pair of sequences is each applied at the input of the encoder, generating two corresponding output sequences. The number of bits that these two output sequences differ from each other (Hamming distance) is called the "free distance." From the encoded output results in 1.D-5, determine the free distance of the convolutional code shown in Fig. 24.1.

1.D-7   Repeat 1.D-4 and 1.D-5 several times to check the free distance for other random input bit sequence pair **m** and **m_diff** (which differs from **m** by only one bit).

(a) Does the free distance depend on the specific encoder input sequence **m**?

(b) Is the free distance equal to the answer obtained in 1.D-3(b)?

1.D-8   Execute **xor(enc(m), enc(m_diff))** in the command window and capture the result.

(a) Note that the sequence captured in 1.D-3(a) (which is the encoded output for the input [1 0 0 0 0]) has the pattern "110111," followed by a sequence of 0s. Does this 5-bit pattern appear in the execution result of **xor(enc(m), enc(m_diff))**?

(b) Using the linearity property of convolutional encoder, explain the observation made in (a), that is, why **xor(enc(m), enc(m_diff))** has the 5-bit pattern of the encoded output for the input [1 0 0 0 0].

(c) Extend the discussion in (b) to explain the results in 1.D-7(a) and 1.D-7(b).

1.D-9   Denote the free distance of a convolutional code by d_{free} [2–4], then e_{min}, the correcting capability, that is, the number of errors that can be corrected is equal to $\lfloor (d_{free} - 1)/2 \rfloor$, where $\lfloor x \rfloor$ denotes the floor function, for example, $\lfloor 3.7 \rfloor = 3$. Calculate e_{min} of the convolutional code in Fig. 24.1.

1.E   Consider another convolutional encoder as shown in Fig. 24.2. For the input bit sequence m(1), m(2), ... m(k), ..., the encoder outputs u1(1), u2(1), u3(1), u1(2), u2(2), u3(2), ..., u1(k), u2(k), u3(k), ... in order.

1.E-1   Determine the code rate and constraint length.

1.E-2   Properly modify **enc.m** in 1.C to implement this encoder. Change the function name **enc()** into **enc2()** and save the modified m-file as **enc2.m**. Capture the completed m-file.

1.E-3   Execute **enc2([1 0 0 0 0])** in the command window. (a) Capture the result. (b) Identify the free distance d_{free} and the correcting capability e_{min}.

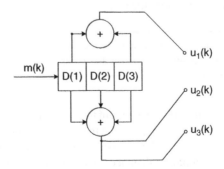

FIGURE 24.2   Convolutional encoder example 2.

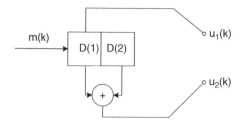

FIGURE 24.3   Convolutional encoder example 3.

1.E-4   Execute **m=(rand(1,5)>0.5)** in the command window to generate a 5-bit sequence **m**. (a) Capture the generated **m**. (b) Let **m** be the input sequence. Manually perform the encoding process and record the encoder output.

1.E-5   Execute **enc2(m)** in the command window and capture the result. Is the result consistent with your manual encoding result in 1.E-4?

1.F   Let us consider another convolutional code shown in Fig. 24.3.

1.F-1   Determine the code rate and constraint length.

1.F-2   Properly modify **enc.m** in 1.C to implement this encoder. Change the function name **enc()** into **enc3()** and save the modified m-file as **enc3.m**. Capture the completed m-file.

1.F-3   Execute **enc3([1 0 0 0 0])** in the command window. (a) Capture the result. (b) Identify the free distance d_{free} and the correcting capability e_{min}.

1.F-4   Execute '**m=(rand(1,5)>0.5)**' in the command window to generate a 5-bit sequence **m**. (a) Capture the generated **m**. (b) Let **m** be the input sequence. Manually perform the encoding process and record the encoder output.

1.F-5   Execute **enc3(m)** in the command window and verify the result in 1.F-4.

24.2   IMPLEMENTATION OF MAXIMUM LIKELIHOOD DECODING BASED ON EXHAUSTIVE SEARCH

2.A   [WWW]The m-file below performs channel encoding on a 4-bit stream **m** by using **enc()** for transmission employing antipodal signaling. It also creates the sampled received signal, a vector **r**, in an AWGN channel. Add a comment to each line to explain what it does or means. Capture the m-file with comments.

```
clear
rand(1,1XXX); % XXX=Last three digits of your student ID number. This is irrele-
vant of the goal of this code, but be sure to include it here anyway.
Nb=4;
m=rand(1,Nb)>0.5;
coded_m=enc(m);
s=2*coded_m-1;
r=s+randn(1,length(s));
```

2.B [WWW]We estimate **m** using maximum likelihood decoding, which consists of the following steps.

Step 1. Create a matrix **m_set** whose rows consist of all possible **m**, that is, the 16 combinations for 4 bits.

Step 2. Perform the following processes for each of the 16 4-bit sequences in **m_set**:

Step 2-1. Encode the 4-bit sequence to create its corresponding encoded bit sequence **m_k**.

Step 2-2. Form the antipodal modulated transmitted sequence **s_k** for **m_k**.

Step 2-3. Calculate the distance (or distance square) between the received vector **r** and **s_k**.

Step 3. Among the 16 distances calculated, identify the smallest one. From the **s_k** that is closest to **r** we identify the corresponding 4-bit message in **m_set**, which is the ML decoding result.

The code fragment below implements Step 1 and Step 2.

```
% Add the following to the m-file in 2.A
m_set(1,:)=[0,0,0,0];
m_set(2,:)=[0,0,0,1];
m_set(3,:)=[0,0,1,0];
.... % Complete all possible 4-bit sequences in order.
m_set(16,:)=[1,1,1,1];

for k=1:16
  m_k=m_set(k,:);
  coded_m_k=enc(m_k);
  s_k=2*coded_m_k-1;
     D_k(k)=sum((r-s_k).^2 ); % Use sum(abs(r-s_k).^2 ) for complex signal-
ing such as QPSK.
end
D_k %Incomplete Step 3.
```

2.B-1   Complete the m-file. Identify the lines in the m-file that implement each of the two steps. Add a comment to mark the lines (e.g., add **% Step 1** or **% Step 2** to the corresponding lines). Capture the completed m-file.

2.B-2   The m-file above implements decoding. However, the encoding function **enc( )** is used inside the '**for**' loop. Discuss what this encoding step does for ML decoding.

2.B-3   Make sure that **enc.m** is saved in your MATLAB work folder. Execute the m-file.

   (a) Display **D_k** and capture the result.
        For example, **D_k(4)** is the Euclidean distance between the received signal **r** and the encoded and transmitted signal for the 4-bit sequence [? ? ? ?]. Complete the 4 bits marked by '**?**'.
   (b) From **D_k**, determine the ML decoding result of the transmitted bit sequence **m**.
   (c) Execute (type) **m** in the command window to see the transmitted data sequence. Check whether or not the ML decoding in (c) has generated correct results.

2.B-4   Add the following two lines to the m-file to complete Step 3. The vector **m_hat** is the decoding result, that is, the estimate for **m**.

```
[T1 T2]=min(D_k);
m_hat=m_set(?,:);
dec_err=sum(m~=m_hat) % number of error bits after decoding.
```

Complete the quantity marked by '**?**'.
   HINT: The command '**[T1 T2] = max(a)**' returns T2 as the index of the maximum element of vector **a**.
   (a) Execute the m-file and capture the result.

2.B-5   The '**for**' loop does Step 2 for all of the 16 4-bit data sequences. If the data bit sequence length is increased from 4 to 100, then how many times Step 2-related processes need to be repeated within the '**for**' loop?

2.B-6   The preceding implementation of the ML decoding relies on the exhaustive search approach. Explain why such an approach is impractical.

2.C   [WWW]The vector **r** in the m-file of 2.B contains both the signal and the noise components. This m-file implements soft decision ML decoding.
   As shown in the simulation code below, add the two lines in bold to the m-file to implement hard decision decoding [2–5]. In hard decision decoding, the bit detection is made by using '**z=(r>0)**', where **r** is the received signal. Then the Euclidean distance **D_k** is calculated by comparing **z** with **coded_m_k**, instead of **s_k**.

```
clear
rand(1,1XXX); % XXX=Last three digits of your student ID number, irrelevant to the goal
  of this code, but be sure to include.
Nb=4;
m=rand(1, Nb)>0.5;
coded_m=enc(m);
s=2*coded_m-1;
r=s+randn(1,length(s));
z=(r>0);
m_set(1,:)=[0,0,0,0];... m_set(16,:)=[1,1,1,1];

for k=1:16
  m_k=m_set(k,:);
  coded_m_k=enc(m_k);
s_k=2*coded_m_k-1;
  %D_k(k)=sum((r-s_k).^2 ); %For Soft decision decoding.
  D_k(k)=sum(xor(z,coded_m_k)); %For Hard decision decoding.
end

[T1 T2]=min(D_k);
m_hat=m_set(?,:);
dec_err=sum(m~=m_hat)
```

2.C-1   The square of Euclidean distance between vectors **z** and **coded_m_k** can be calculated as '**D_k(k)=sum((z-coded_m_k).^2)**'. Note that in the second boldfaced line in the m-file above, we used **sum(xor(z,coded_m_k))**, which is the Hamming distance, rather than the Euclidean distance square **sum((z-coded_m_k).^2)**. However, the decoding results for both cases are the identical. Explain why.

2.C-2   Summarize the advantages of using **sum(xor(z,coded_m_k))** over using **sum((z-coded_m_k).^2)**.

2.C-3   Let us compare the performances of hard decision decoding and soft decision decoding [2–5].

(a) Execute the m-file above that implements hard decision decoding repeatedly for at least 20 times and count the number of decoding errors, that is, the cases where **dec_err** is not equal to 0.
    Uncomment the line '**D_k(k)=sum((r-s_k).^2);**' and comment out the line '**D_k(k)=sum(xor(z,coded_m_k));**' to implement soft decision decoding. Then execute the m-file repeatedly for at least 20 times and count the number of decoding errors.

(b) Compare the performances of hard decision decoding and soft decision decoding.

2.C-4   Justify why soft decision decoding outperforms hard decision decoding.

24.3   VITERBI DECODING (TRELLIS-BASED ML DECODING)

[WWW]From the companion website, download **convolutional_code.pdf**. This pdf file consists of three parts:

1. Construction of the state diagram (pp. 2–21).
2. Construction of the Trellis diagram (pp. 22–69).
3. Implementation of the Viterbi decoding [1–6] algorithm (pp. 70–173).

3.A [WWW]In this subsection we construct the state diagram and the trellis diagram of the encoder in Fig. 24.1.

First, carefully review pp. 2–21 of **convolutional_code.pdf** to learn the state diagram, and pp. 22–69 to learn the trellis diagram. Note that the encoder example used in the pdf file is different from the encoder shown in Fig. 24.1.

3.A-1   Fig. 24.4 shows the incomplete state diagram of the encoder in Fig. 24.1. The two bits inside the four circles correspond to the four possible states of [D(2) D(3)] shown in Fig. 24.1. If the input bit $m(k)$ is 0, then the state [D(2) D(3)] will change from the current state to the destination state indicated by a sold arrow; if $m(k)$ is 1, then the state [D(2) D(3)] will change from the current state to the destination state indicated by a dashed arrow. In other words, if the current state, that is, [D(2) D(3)] at $t = k$, is equal to the two bits inside one of the circles, then the next state, that is, [D(2) D(3)] at $t = k + 1$, becomes the two bits inside the destination circle.

Complete the contents inside the two circles marked by '?' in Fig. 24.4.

3.A-2   The two bits specified on each arrow corresponds to $u_1(k + 1)$ and $u_2(k + 1)$, that is, the decoder output when $m(k)$ enters the encoder.

For example, let us assume that the current state [D(2) D(3)] is [0 0] at $t = k$ and the input bit $m(k)$ is 0. Then, according to Fig. 24.1, the next state [D(2) D(3)] at $t = k + 1$ will be [0 0]. Thus the solid arrow points toward the state [0 0] again. Meanwhile,

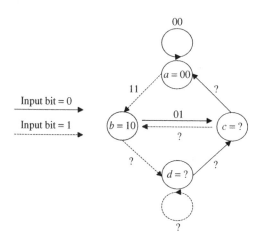

FIGURE 24.4   State diagram of the encoder in Fig. 24.1

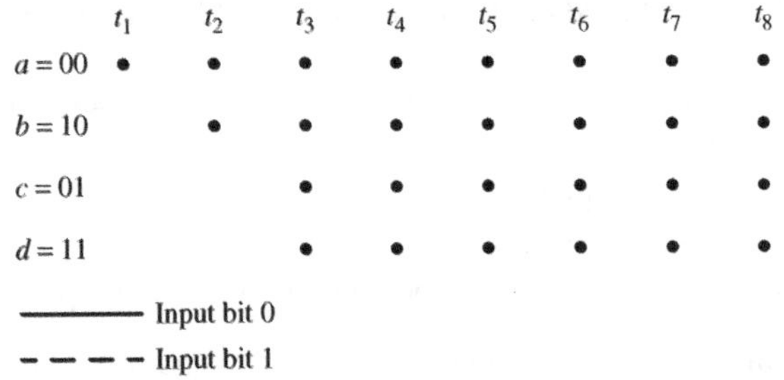

FIGURE 24.5 Trellis diagram of the encoder in Fig. 24.1.

according to Fig. 24.1, the decoder output at $t = k + 1$ is $[u_1(k + 1)$ and $u_2(k + 1)] =$ [0 0]; hence we specify 00 on the arrow. On the contrary, if the input bit $m(k)$ is 1 at the current state $[D(2)\ D(3)]$ of [0 0], then the dashed arrow points toward $[D(2)\ D(3)] = [1\ 0]$ at $t = k + 1$ and the decoder output at $t = k + 1$ is $[u_1(k + 1)$ and $u_2(k + 1)] = [1\ 1]$. Hence we specify 11 on the arrow.

Using this process, complete all the places marked by '?' on the arrows in Fig. 24.4.

3.A-3 Based on the state diagram completed in 3.A-2, complete the trellis diagram in Fig. 24.5. This trellis diagram will be used again in 3.B. Thus it is recommended that the trellis branch be drawn with a pencil; for the decoding algorithm in 3.B, some of the branches can be conveniently erased. The basic rules are as follows.

1. Use a sold line for the state transition branch (simply called "branch" hereafter) if the encoder input bit $m(k)$ is 0 and a dotted line if the encoder input bit $m(k)$ is 1.
2. Specify the corresponding encoder output on the branch, which is called the "branch output."
3. Copy the branch pattern from t_3 to t_4 for the remaining time instants up to t_8.

Capture the completed trellis diagram.

3.B We now use this trellis diagram to manually decode a test input sequence. To this end, generate a hard decision decoder input **z** by using the lines of code below. The vector **coded_m** is the encoder output of the 5-bit message data **m**. The last two bits [0 0] appended at the end of the encoder input are called the "tail bits." The tail bits reset the memory of the encoder after encoding is completed. We will use the vector **z** in the final line as a test input for hard decision decoding in the following problems.

```
>>rand(1,1XXX); %XXX=Last three digits of your student ID number.
>>m=rand(1,5)>0.5;
>>coded_m=enc([m 0 0]);
>>z=coded_m;
>>z([1 2])=not(z([1 2]))
```

3.B-1   Execute the five code lines above and capture the execution results.

3.B-2   We will decode the message **m** from **z** captured in 3.B-1. From the result of the last code line, identify the erroneously detected bits.

3.B-3   [WWW]We go through the decoding process using the trellis diagram created in 3.A-3. The decoding process is explained with animation in the supplemental materials **convolutional_code.pdf** (starting from p. 70). Note again that the encoder used in the pdf file is different from the one in Fig. 24.1 that is considered in this section. The decoding steps are summarized below.

Step1: Select one of the trellis branches between time instants t_1 and t_2. Calculate the distance, that is, the number of different bits between the first two bits of **z** and two bits on the selected branch (branch output). Record the distance on top of the destination (next) state of the selected branch. Do this for all (two in our case) branches between time instants t_1 and t_2.

NOTE: The distance between the branch output and the decoder input is called the "branch metric" [1–6].

Step 2: For all branches between the instants t_2 and t_3, repeat Step 2-1 and Step 2-2 below.

Step 2-1: Calculate the branch metric, that is, the distance between the next two bits of **z** (=[**z(3) z(4)**]) and the selected branch output.

Step 2-2: At the source (previous left-hand) state of the selected branch, there is a branch metric previously recorded in Step 1. Add this previous metric to the current branch metric calculated in Step 2-1 and record their sum on top of the destination (next) state of the selected branch.

NOTE: The connected branch is called a "path" and the sum of the branch metrics is called the "accumulated metric" or path metric [1–6].

Step 3: For all branches between the instants t_3 and t_4, repeat Step 3-1, Step 3-2, and Step 3-3 below.

Step 3-1: Calculate the branch metric, that is, the distance between the next two bits of **z** and the selected branch output.

Step 3-2: At the source state of the selected branch, there is a path metric previously recorded. Add this previous path metric to the current branch metric calculated in Step 3-1 and record their sum on top of the destination state of the selected branch.

Step 3-3: Note that after t_4, two branches merge into one of the four destination states; thus there will be only two path metrics after Step 3-2 for all the branches. Select the branch with a smaller path metric and erase (or mark X on) the unselected branch and its path metric. If the two path metrics are identical, arbitrarily select one of them.

NOTE: The selected path is called a "survivor path" [1–6].

Step 4: Repeat Step 3 for all the remaining time instants.

Step 5: After completing Step 4, select one path with the smallest path metric among the survivor paths. Recall that a solid branch corresponds to an input bit 0 and a dashed branch corresponds to an input bit 1. Using this information, identify the input bit corresponding to each of the branches (seven branches in this example) connected in the selected survivor path. Finally, concatenate the seven input bits to form the decoding result **m**.

(a) Now capture the trellis diagram that shows the manual decoding process and record the first five bits of the seven decoded bits (the last two bits are the tail bits). It is best to highlight the selected path in the trellis diagram with a different color.

(b) Display the value of the original message **m** in the command window. Recall that the first two bits of **z** are erroneously detected. From the result in (a), has the decoding process corrected these errors?

3.C [WWW]In this section we implement trellis-based decoding using a user-defined MATLAB function, **dec( )**, provided below, assuming the encoder in Fig. 24.1. Let **z** denote the hard decision decoder input (code word). Then **dec(z)** outputs the decoding result of z, that is, the message **m** carried in **z**.

In the m-file **dec( )**, comments are provided to explain the eight main variables: **d1, d2, d3, d4, output1, output2, output3,** and **output4**. In the table provided at the bottom of pp. 72–172 of **convolutional_code.pdf**, how these variables are calculated and updated during the decoding process is illustrated in detail.

In the kth iteration of the '**for**' loop, these eight variables are calculated and updated for the transition period from t_k to t_{k+1} in the trellis diagram.

```
% Save the file name as dec.m
function result = dec(z)
d1=0; d2=0; d3=0; d4=0; %Initialize the accumulated metrics of the 4 states
   (00, 10, 01, 11).
output1= []; output2=[]; output3=[]; output4=[];
% In trellis-based decoding, each of the 4 states updates its own survivor path at
   every time instant (t1,t2,t3,...). Since we do not know which one of the 4 survivor
   paths will be selected at the end in Step 5 of 3.B-3, the corresponding encoder
   inputs for all four survivor paths should be kept during decoding. For example, the
   variable output1 is updated to save the encoder input that corresponds to the state
   1 (00) at the current time instant. In other words, the variable output1 memorizes the
   encoder inputs corresponding to the branches forming the survivor path to the
   state 1 of the current time  instant.

for k=1:1:(length(z)/2)
   temp = [z(2*k-1) z(2*k)]; %current 2 bits of the decoder input.

   dis00=sum(xor([0 0],temp));dis10=sum(xor([1 0],temp));
   dis01=sum(xor([0 1],temp));dis11=sum(xor([1 1],temp));
```

%Distances between temp and each of the four patterns. Save these as the
variables as they are used frequently.

```
if k==1
    new_output1 = [output1 0]; new_d1 = d1 + dis00;
    new_output2 = [output2 1]; new_d2 = d2 + dis11;
elseif k==2
    new_output1 = [output1 0]; new_d1=d1+dis00;
    new_output2 = [output1 1]; new_d2=d1+dis11;
    new_output3 = [output2 0]; new_d3=d2+dis01;
    new_output4 = [output2 1]; new_d4=d2+dis10;
else
    %%%%%%%%%%%%%%%%%%%%%%%%%%%%%%%%%%%%%%%%%%%%
    if d1+dis00 < d3+dis11;
        new_output1 = [output1 0]; new_d1=d1+dis00;
    else
        new_output1 = [output3 0]; new_d1=d3+dis11;
    end
    %%%%%%%%%%%%%%%%%%%%%%%%%%%%%%%%%%%%%%%%%%%%
    if d1+dis11 < d3+dis00;
        new_output2 = [output1 1]; new_d2=d1+dis11;
    else
        new_output2 = [output3 1]; new_d2=d3+dis00;
    end
    %%%%%%%%%%%%%%%%%%%%%%%%%%%%%%%%%%%%%%%%%%%%
    if d2+dis01 < d4+dis10
        ???
    else
        ???
    end
    %%%%%%%%%%%%%%%%%%%%%%%%%%%%%%%%%%%%%%%%%%%%
    if ??? < ???
        ???
    else
        ???
    end
    %%%%%%%%%%%%%%%%%%%%%%%%%%%%%%%%%%%%%%%%%%%%
end

% variables update
if k==1
    d1=new_d1; d2=new_d2;
    output1=new_output1; output2=new_output2;
else
    d1=new_d1; d2=new_d2; d3=new_d3; d4=new_d4;
```

```
      output1=new_output1; output2=new_output2;
      output3=new_output3; output4=new_output4;
   end
end

output = [output1; output2; output3; output4];
d=[d1 d2 d3 d4];
ML_state_index=find(d==min(d));
result = output(ML_state_index(1),:);
```

3.C-1   Review the methods to concatenate a new element to a vector in MATLAB. Explain what the command 'a=[b 1]' does if **b** is an arbitrary row vector.

3.C-2   Code lines similar to '**new_output2 = [output1 1]**' and '**new_output1 = [output3 0]**' appear in many places in **dec( )**. Explain what these lines do.

3.C-3   Identify the lines in the m-file that implement each step in 3.B-3. Add a comment to each line to indicate the number for its corresponding step.

3.C-4   Based on the trellis diagram in 3.A-3, complete the six places marked by '???' in the m-file and capture the completed m-file.

3.C-5   For the last two lines '**ML_state_index = find(d==min(d));**' and '**result = output(ML_state_index(1),:);**', explain what the variable on the left-hand side represents and justify how the right-hand side expression is formulated accordingly.

3.C-6   Recall that in 3.B-1, a test decoder input **z** was generated. The m-file **dec.m** was completed in 3.C-4. Execute **dec(z)** in the command window and capture the result. Is the decoding output correct?

3.D   [A]In this section we implement the decoder for other convolutional encoders.

3.D-1   Draw the trellis diagram for the convolutional encoder in Fig. 24.2 up to the time instant t_5. The easiest way would be to determine the differences between the encoders in Figs. 24.1 and 24.2 first and then modify the trellis diagram for the encoder in Fig. 24.1.

3.D-2   [WWW]The code below is the incomplete user-defined function **dec2( )**, which performs trellis-based decoding for the encoder in Fig. 24.2. Complete the m-file and save it as **dec2.m**. Capture the completed m-file.

```
% Save the file as dec2.m
function result = dec2(z)

output1=[]; output2=[]; output3=[]; output4=[];

for k=1:1:(length(z)/3)
   temp = [z(3*k-2) z(3*k-1) z(3*k)];
```

```
    dis000=sum(xor([0 0 0],temp)); dis011=sum(xor([0 1 1],temp));
    dis100=sum(xor([1 0 0],temp)); dis111=sum(xor([1 1 1],temp));
% In the encoder structure in Fig. 24.2, we can see that one more bit is added to the
    encoder output in Fig. 24.1 at each time instance. Therefore, for example, dis10
    in dec.m is modified into dis100 in dec2.m as shown above.

if k==1
    new_output1 = [output1 0]; new_d1 = d1 + dis000;
    new_output2 = [output2 1]; new_d2 = d2 + dis111;
  elseif k==2
    ?
    ...
    ?
  else
    ?
    ...
    ?
  end

  % memory update
  if k==1
    d1=new_d1; d2=new_d2;
    output1=new_output1;   output2=new_output2;
  else
    d1=new_d1; d2=new_d2;  d3=new_d3; d4=new_d4;
    output1=new_output1; output2=new_output2;
    output3=new_output3; output4=new_output4;
  end
end

output = [output1; output2; output3; output4];
d=[d1 d2 d3 d4];
ML_state_index=find(d==min(d));
result = output(ML_state_index(1),:);
```

3.D-3   The two lines **'for k=1:1:(length(z)/3);'** and **'temp = [z(3*k-2) z(3*k-1) z(3*k)];'**
are different from the corresponding lines in dec.m. Explain why these lines should
be modified as they are, or suggest a better way to modify them properly.

3.D-4   Draw the trellis diagram for the encoder in Fig. 24.3 up to t_4.

3.D-5   Based on the trellis diagram in 3.D-4, write a MATLAB function **dec3( )** to
perform trellis-based decoding, save it, and capture it.

3.E   [A]So far, the m-files **dec.m**, **dec2.m**, and **dec3.m** implement the hard decision
decoding; that is, the decoder input is a bit sequence (0s and 1s). In this section

we implement soft decision decoding, with which the decision variables (**r**) for the coded bits that consist of both the received signal and noise are directly passed to the decoder. To implement soft decision decoding, only the part that calculates the distance variable in **dec.m** needs to be modified as below.

```
dis00=sum( ([-1 -1]-temp).^2 );
dis01=sum( ([-1  1]-temp).^2 );
dis10=sum( ([1 -1]-temp).^2 );
dis11=sum( ([1  1]-temp).^2 );
% sum(abs( branch output -temp).^2 ) for complex modulation such as QPSK
```

3.E-1   Justify these modifications.

3.E-2   Explain why the other parts of **dec.m** do not require modifications for soft decision decoding.

3.E-3   Also change the function name into **dec_SD( )** and then save the m-file as **dec_SD.m**.

3.E-4   Modify **dec2.m** properly for soft decision decoding. Also change the function name into **dec2_SD( )** and then save it as **dec2_SD.m**. Capture the revised part only.

3.E-5   Modify **dec3.m** properly for soft decision decoding. Also change the function name into **dec3_SD( )** and then save it as **dec3_SD.m**. Capture the revised part only.

24.4   BER SIMULATION OF CODED SYSTEMS

4.A   [WWW]The m-file below simulates the uncoded BER of binary phase shift keying (BPSK) over an additive white Gaussian noise (AWGN) channel.

```
clear
EbN0dBvector=0:1:9;
DataBitSize=1000;
Eb=1; % because we consider a bipolar signal which takes +1 and −1.
    See the line generating 'ChannelSymbols' below.
for snri=1:length(EbN0dBvector)
  EbN0dB=EbN0dBvector(snri);
  EbN0=10^(EbN0dB/10);
  N0=Eb/EbN0;
  BitErrNum=0;
  TotalBits=0;
  while BitErrNum<100
      DataBits=(rand(1,DataBitSize)>0.5);
      ChannelSymbols=sqrt(Eb)*(2*DataBits−1);
```

```
        r=ChannelSymbols+sqrt(N0/2)*randn(1,length(ChannelSymbols));
        DataBitsHat=r>0;
        BitErrs=sum(DataBits~=DataBitsHat);
        BitErrNum=BitErrNum+BitErrs;
        TotalBits=TotalBits+DataBitSize;
    end
    uncodedBER(snri)=?/?
end
```

4.A-1   Complete the quantities marked by '**?**' in the m-file. Add a comment to every line that has an '=' sign to explain what the variable on the left-hand side represents and justify how the right-hand side expression is properly formulated accordingly. Capture the completed m-file.

4.A-2   Execute the completed m-file above and then execute the following commands to plot the uncoded BER curve. Capture the result.

```
>>figure
>>semilogy(EbN0dBvector,uncodedBER)
>>grid
```

4.B   [WWW]With slight modifications to the m-file in 4.A as shown below, we can simulate the BER of coded systems. Here **enc( )** is the encoding function.

```
clear
DataBitSize=1000;
coderate=1/2;
Eb=1; % because we consider a bipolar signal which takes +1 and −1.
Eb_coded=Eb*coderate; %Energy of channel symbol (coded bit). See the line
    generating 'ChannelSymbols' below.
EbN0dBvector=0:1:8;
for snri=1:length(EbN0dBvector)
    EbN0dB=EbN0dBvector(snri);
    EbN0=10^(EbN0dB/10);
N0=Eb/EbN0;
BitErrNum=0;
TotalBits=0;
while BitErrNum<100
DataBits=(rand(1,DataBitSize)>0.5);
EncodedBits=enc(?);
ChannelSymbols=sqrt(Eb_coded)*(2*EncodedBits-1);
r=ChannelSymbols+sqrt(N0/2)*randn(1,length(ChannelSymbols));
z=r>0; %For hard decision decoding.
```

```
DecodedBits=?(?);
BitErrs=sum(DataBits~=DecodedBits);
BitErrNum=BitErrNum+BitErrs;
TotalBits=TotalBits+DataBitSize;
end
CodedBER(snri)=?/?
end
```

4.B-1   Examine the modified or newly inserted lines.

We define "channel symbol " as the received symbol through the channel. The variable **Eb** denotes the energy per data bit, and the variable **Eb_coded** denotes the energy per channel symbol. The variables **Eb_coded**, **ChannelSymbols**, and **r** are generated sequentially. From these processes we make the following observations:

    (a) The channel symbols correspond to the encoder output.

    (b) For the example we are considering, the channel symbol is BPSK modulated with an energy equal to **Eb_coded**, not **Eb**.

The line '**Eb_coded = Eb*coderate;**' indicates that **Eb_coded** and **Eb** differ only by the factor of **coderate**. Justify the use of this scaling factor.

4.B-2   Complete the places marked by '**?**' in the m-file with an appropriate function name or variable. Execute the completed m-file, and then in the command window, execute the lines of command below. If the simulation takes too long, you can properly decrease the maximum value of elements of **EbN0dBvector** to shorten its length.

Capture the coded BER graph.

```
>>figure
>>semilogy(EbN0dBvector,CodedBER)
>>grid
```

4.B-3   To simulate the coded BER for the encoder in Fig. 24.2, that is, **enc2()**, which lines of the m-file completed in 4.B-2 should be modified and how?

4.B-4   (a) Modify the m-file completed in 4.B-2 to simulate the coded BER of the encoder in Fig. 24.3, that is, **enc3()**. (b) Execute the modified m-file and plot the simulated BER. Capture the simulated BER plot.

4.B-5   (a) To simulate the coded BER with soft decision decoding for the encoder in Fig. 24.1, that is, **enc1()**, which lines of the m-file completed in 4.B-2 should be modified and how? (b) Execute the modified m-file and capture the simulated BER plot.

4.C   Comparison of BER performances of different system configurations.

4.C-1   Uncoded versus coded BER.

(a) Overlay the uncoded BER curve obtained in 4.A-2 on top of the coded BER plots obtained in 4.B-2 using the functions **enc()** and **dec()**. Refer to the note below for techniques to overlay curves from different figures. Capture the resulting BER plot.

NOTE: To copy a curve in a figure and paste it into another figure, first, in the source figure with the curve to be copied, left-click the arrow icon in the menu bar and then right-click the desired curve and choose **Copy** from the pop-up menu. Next, in the destination figure, left-click the arrow icon in the menu bar and then right-click anywhere in the figure and select **Paste** from the pop-up menu to paste the curve copied from the source figure.

(b) Coding gain refers to the SNR difference (typically in dB) to achieve the same BER [1–6]. Based on the overlaid BER curves, measure the coding gain at a BER of 10^{-4}.

4.C-2   [A]Hard decision decoding versus soft decision decoding.

(a) Overlay the coded BER curve with soft decision decoding obtained in 4.B-5 on top of the coded BER curve with hard decision decoding obtained in 4.B-2. Capture the plot that shows the overlaid BER curves.

(b) Based on the overlaid BER curves, measure the SNR gain in dB of soft decision over hard decision at a BER of 10^{-4}.

4.C-3   [A]Coderate = 1/2 versus coderate = 1/3.

(a) In the m-files in 4.B-2 and 4.B-3, modify the line '**EbN0dBvector=0:1:6;**' into '**EbN0dBvector= 0:2:12;**'. Then execute the two modified m-files and plot the simulated BER graphs. Note that the simulation may take a long time depending on the configuration of your PC. Overlay the two simulated coded BER curves in a single figure. Capture the resulting figure.

(b) In the low-SNR region, which code performs better? In the high-SNR region, which code performs better? Justify such BER performance behavior.

4.C-4   [A]Free distance = 3 versus free distance = 5.

(a) Overlay the coded BER curve with a free distance of 3 and code rate of 1/2 obtained in 4.B-4 onto the coded BER curve with a free distance of 5 and also rate of 1/2 obtained in 4.B-2. Capture the resulting figure.

(b) Which code achieves a better BER performance? Why?

REFERENCES

[1] A. J. Viterbi and J. K. Omura, *Principles of Digital Communication and Coding*, New York: McGraw-Hill, 1979.

[2] J. G. Proakis, *Digital Communications*, 5th ed., New York: McGraw-Hill, 2008.

[3] B. Sklar, *Digital Communications: Fundamentals and Applications*, 2nd ed., Boston: Phipe, 2001.

[4] S. Haykin and M. Moher, *Introduction to Analog and Digital Communication*, 2nd ed., Hoboken, NJ: Wiley, 2006.

[5] S. Lin and D. J. Costello, *Error Control Coding*, 2nd ed., Upper Saddle River, NJ: Prentice Hall, 2004.

[6] G. D. Forney, "The Viterbi Algorithm," *Proceedings of the IEEE*, Vol. 61, No. 3, 1973, pp. 268–278.

25

FADING, DIVERSITY, AND COMBINING

- Derive and simulate the bit error rate (BER) in the Rayleigh fading environment and compare it with the BER in the additive white Gaussian noise environment (AWGN).
- Understand the concept of diversity and the diversity combining.
- Compare the performances of various diversity-combining methods.

25.1 RAYLEIGH FADING CHANNEL MODEL AND THE AVERAGE BER

Consider a two-dimensional modulation such as quadrature amplitude modulation (QAM) and MPSK, and denote s as the coordinate of the modulated symbol in the complex plane. Then, the real and imaginary parts of s correspond to the in-phase and quadrature components, respectively. The symbol energy E_s is equal to $E[|s|^2]$.

With this signal model, the receiver's matched filter output in a frequency nonselective fading channel can be expressed as

$$r = hs + n, \qquad (25.1)$$

where h is the signal scaling coefficient due to fading [1–3], which is often called the "fading coefficient," and n is the AWGN term, which follows the complex Gaussian distribution with zero mean and variance $N_0/2$.

1.A For Rayleigh fading, the fading coefficient h follows a complex Gaussian distribution, expressed as $h = z + jy$, where z and y are real-valued independent

Problem-Based Learning in Communication Systems Using MATLAB and Simulink, First Edition.
Kwonhue Choi and Huaping Liu.
© 2016 The Institute of Electrical and Electronics Engineers, Inc. Published 2016 by John Wiley & Sons, Inc.
Companion website: www.wiley.com/go/choi_problembasedlearning

Gaussian variables with zero mean and variance 1/2 [3–5]. The command line below generates a sample of h. Complete the quantities marked by '?'.

```
>>h=?*randn + ?*?*?
```

1.B The fading coefficient can be expressed in polar coordinate form as

$$h = |h|e^{j\angle h}. \tag{25.2}$$

1.B-1 [T]Express the fading magnitude square, that is, $|h|^2$, as a function of z and y.

1.B-2 [T]From the result in 1.B-1 and the fact that z and y are independent Gaussian random variables with zero mean and variance 1/2, determine $E[|h|^2]$.

1.C To create the decision variable D, the receiver derotates the received signal r by using the phase of h as

$$
\begin{aligned}
D &= re^{-j\angle h} \\
&= (hs + n)e^{-j\angle h} \\
&= |h|e^{j\angle h}e^{-j\angle h}s + ne^{-j\angle h} \\
&= |h|s + ne^{-j\angle h}.
\end{aligned}
\tag{25.3}
$$

As a result, the signal term in the decision variable, $|h|s$, has the same phase as that of the original symbol s.

1.C-1 The noise element $ne^{-j\angle h}$ of the decision variable D is still a complex Gaussian random variable, which has the same distribution as n. Therefore, rotating the phase due to fading does not affect the statistical characteristics of the noise in the decision variable D. On the contrary, fading affects the signal term by a scaling factor $|h|$. This scaling factor is the magnitude of a complex Gaussian random variable, which has the Rayleigh distribution [4,5] given as

$$f_{|h|}(x) = \begin{cases} 2x\exp(-x^2) & x \geq 0 \\ 0 & x < 0. \end{cases} \tag{25.4}$$

Execute the following lines of command to plot equation (25.4). Capture the result.

```
>>x=0:0.01:5;
>>f=2*x.*exp(-x.^2);
>>figure
>>plot(x,f)
```

1.C-2 Define $c \overset{\Delta}{=} |h|^2$. Then c is the magnitude square of a complex Gaussian variable and follows the exponential distribution expressed as [4,5]

$$f_c(x) = \exp(-x), \quad x \geq 0. \tag{25.5}$$

Execute the following lines of commands to plot equation (25.5). Capture the result.

```
>>x=0:0.01:5;
>>f_c=exp(-x);
>>figure
>>plot(x,f_c)
```

1.C-3   [T]We can find the expected value of c as $E[c] = \int_0^\infty (?) \times f_c(x)dx$. Determine the quantity marked by '**?**'.

1.C-4   (a) [T]Calculate the average of c using the completed equation in C-3. (b) Is the calculated result equal to the answer in 1.B-2? Also use symbolic math (discussed in Section 1.2 of Chapter 1) to verify the result in (a).

1.C-5   [T]In the absence of fading (i.e., $h = 1$), the energy (mean square) of the signal term in the decision variable D, $E[||h|s|^2]$, equals E_s. For Rayleigh fading for which h follows the distribution given by equation (25.4), calculate the energy of the signal term, $E[||h|s|^2]$, and prove that it is still equal to E_s.

1.C-6   [T]The result in 1.C-5 shows that the average symbol energies of the received signals over a Rayleigh fading channel and a Gaussian channel are same. This is based on the assumption that the power of Rayleigh fading channel coefficients is normalized to unity. Provide a discussion on the relative performances (a conjecture on what you expect to see) of the same signaling scheme over a Rayleigh fading channel and over an AWGN channel.

1.C-7   [A]Prove that the noise term $ne^{-j\angle h}$ in the decision variable D given in equation (25.3) is a complex Gaussian random variable and it has the same distribution as n.

1.D   In a Rayleigh fading channel, $|h|$ in equation (25.3) is a random variable following the distribution in equation (25.4). The instantaneous symbol energy received over the fading channel is $|h|^2 E_s$. Let $c = |h|^2$. Therefore, for BPSK, the instantaneous BER is given as $Q(\sqrt{2cE_b/N_0})$, where $E_b = E$. The average BER can be obtained by taking the expected value of the instantaneous BER over c as

$$BER_{fading} = E_c\left[Q\left(\sqrt{\frac{2cE_b}{N_0}}\right)\right]$$

$$= E_c\left[0.5\left\{1 - erf\left(\sqrt{\frac{cE_b}{N_0}}\right)\right\}\right]. \tag{25.6}$$

1.D-1   [WWW]The following m-file calculates equation (25.6) using symbolic math. Add a comment to each line to explain what that line does. Capture the completed m-file with the comments.

```
clear
syms c EbN0
instantBER=0.5*(1-erf(sqrt(c*EbN0)));
BER_fading=int(instantBER*exp(-c),c,0,inf);
pretty((BER_fading))
```

1.D-2   The closed-form expression for equation (25.6) is given as

$$P_e = \frac{1}{2}\left(1 - \sqrt{\frac{E_b/N_0}{1 + E_b/N_0}}\right).$$

(25.7)

Execute the m-file in 1.D-1 and capture the result. Is the symbolic math result the same as equation (25.7)?

1.D-3   Modify the second line of the m-file in D-1 into **'syms c EbN0 positive'**. Here the argument **positive** is not a symbolic variable but a MATLAB argument to specify that the symbolic variables **c** and **EbN0** are positive (and real, of course). Execute the modified m-file and capture the results. Is the symbolic math result the same as equation (25.7) now?

NOTE: If the type of the symbolic variables is not specified, then symbolic math will try to derive a general solution assuming that the symbolic variables are complex. In such a case, it often fails to find the solution.

25.2   BER SIMULATION IN THE RAYLEIGH FADING ENVIRONMENT

2.A   [WWW]The following m-file simulates the BER of BPSK over a Rayleigh fading channel.

```
clear
EbN0dB_vector=0:2:20;

Eb=1;

for snr_i=1:length(EbN0dB_vector)
   EbN0dB=EbN0dB_vector(snr_i);
   EbN0=10.^(EbN0dB/10);
   N0=Eb/EbN0;
   sym_cnt=0;
   err_cnt=0;
   while err_cnt<500
      s=sqrt(Eb)*sign(rand-0.5);
      h=sqrt(1/2)*(randn+j*randn);
```

```
    n=sqrt(N0/2)*(randn+j*randn);

    r=?*s + ?;

    D=r*exp(-j*angle(h));
    s_hat=sign(D);
    if ?
        err_cnt=err_cnt+1;
    end
    sym_cnt=sym_cnt+1;
  end
  BER(snr_i)=err_cnt/sym_cnt
end
figure
semilogy(EbN0dB_vector, BER)
xlabel('E_b/N_0 [dB]')
ylabel('BER')
grid
```

2.A-1 (a) Explain why the three parts in bold in the m-file are set as they are. (b) Complete the places marked by '**?**', and add a comment to each line of the m-file. For lines that involve the operator '=', use the comment to explain what the variable on the left-hand side represents and justify how the right-hand side expression is properly formulated accordingly. Capture the completed m-file.

2.A-2 Explain why the line '**h=sqrt(1/2)*(randn+j*randn);**' should not be placed before the '**while**' loop.

2.A-3 If the line '**h=sqrt(1/2)*(randn+j*randn);**' is placed before the '**while**' loop, what will happen to the resulting BER.

2.B Execute the completed m-file in 2.A and capture the resulting BER graph.

2.C We should find that the BER values in 2.B are all equal to 1. This is because the line '**s_hat=sign(D);**' is incorrect. It should be corrected as '**s_hat = sign(real(D));**'. Explain why it is necessary to take the real part of the decision variable **D** although the signal term of **D** (**sqrt(Eb)** or **-sqrt(Eb)**) is already real-valued.

2.D Modify the line '**s_hat=sign(D);**' into '**s_hat=sign(real(D));**' and execute the m-file. Capture the BER result.

2.E The theoretical BER of BPSK signaling over a Rayleigh fading channel was given in equation (25.7). Execute the following commands in the command window to overlay the theoretical BER on top of the simulated BER curve obtained in 2.D. Capture the resulting figure.

```
>>EbN0_vector=10.^(EbN0dB_vector/10);
>>BER_theory=0.5*(1-sqrt((EbN0_vector)./(1+EbN0_vector)));
>>hold on;
>>semilogy(EbN0dB_vector, BER_theory,'r')
>>legend('Rayleigh fading, Simulation', 'Rayleigh fading, Theory')
```

2.F   Do the simulated and theoretical BER values match each other?

2.G   In this section we compare the BER performances of BPSK over Rayleigh fading and Gaussian channels.

2.G-1   Rewrite the BER expression of BPSK in an AWGN channel as a function of E_b/N_0.

2.G-2   Execute the commands below to overlay the BPSK BER curve in an AWGN channel on top of the BER curve obtained in 2.D or 2.E. Capture the result.

```
>>BER_AWGN=0.5*erfc(sqrt(EbN0_vector));
>>hold on;
>>semilogy(EbN0dB_vector, BER_AWGN,'g')
>>legend('Rayleigh fading, Sim', 'Rayleigh fading, Theory', 'AWGN')
>>axis([0 20 1e-6 1])
```

2.H   Summarize the characteristics of the BER curves of BPSK over Rayleigh fading and Gaussian channels in terms of the changing rate of BER and performance gap as E_b/N_0 increases and decreases.

2.I   Are the observations made in 2.H from simulation consistent with what was discussed in 1.C-6?

2.J   [A]Next we modify the m-file in 2.D to simulate the BER performance of QPSK over a Rayleigh fading channel. The modified m-file is given below.

2.J-1   The parts in bold are the modified parts from the previous m-file for BPSK BER simulation. Add a comment to each of these lines to justify the modifications made. Capture the completed m-file.

```
clear
EbN0dB_vector=0:2:20;
Eb=1;
for snr_i=1:length(EbN0dB_vector)
  EbN0dB=EbN0dB_vector(snr_i);
  EbN0=10.^(EbN0dB/10);
  N0=Eb/EbN0;
  sym_cnt=0;
```

```
err_cnt=0;
while err_cnt<500
  s=sqrt(Eb)*sign(rand-0.5)+j*sqrt(Eb)*sign(rand-0.5);
  h=sqrt(1/2)*(randn+j*randn);
  n=sqrt(N0/2)*(randn+j*randn);
  r=h*s+n;
  D=r*exp(-j*angle(h));
  s_hat=sign(real(D));
  s_hat2=sign(imag(D));
  if sign(real(s)) ~= s_hat
    err_cnt=err_cnt+1;
  end
  if sign(imag(s)) ~= s_hat2
    err_cnt=err_cnt+1;
  end
  sym_cnt=sym_cnt+2;
end
BER(snr_i)=err_cnt/sym_cnt
end
figure
semilogy(EbN0dB_vector, BER)
xlabel('E_b/N_0 [dB]')
ylabel('BER')
grid
```

2.J-2   Execute the m-file in 2.J-1 and capture the simulated BER plot.

2.J-3   Compare the simulated BER curve of QPSK with the BER curve of BPSK obtained in 2.E. They should be identical (ignore the small simulation errors). Justify why they should be the same.

25.3   DIVERSITY

In 1.D we defined the variable c, which determines the instantaneous symbol energy in a Rayleigh fading channel. From the distribution of c, we can easily explain why the BER performance of BPSK in a Rayleigh fading channel is significantly worse than that in an AWGN channel (no fading). The instantaneous symbol energy cE_b could change from 0 to infinity due to fading, and the instantaneous BER given as $Q(\sqrt{2cE_b/N_0})$ almost exponentially decreases as c increases. If c becomes larger than 1, then the BER is lower than the BER over an AWGN channel. However, the error rate will be dominated by the bit errors when c is smaller than 1. For a mathematical proof, refer to Jensen's inequality [4, 5], which is covered in most of the existing textbooks.

If there is a method that will drastically reduce the probability of encountering a small instantaneous symbol energy, then the BER performance will improve significantly. One such method is the diversity technique [5]. Diversity here refers to the mechanism that multiple independently faded copies of the same signal are available for use in the detection process. Common diversity methods include spatial diversity (refer to Chapter 28), temporal/time diversity, frequency diversity, and multipath diversity. Spatial diversity could be achieved by transmitting the same data through multiple transmit antennas or receiving the same data through multiple receive antennas. For such schemes to be effective, the transmit antennas or the receive antennas must operate independently or at least have sufficiently low correlations. Multipath diversity is unique [5]: if multiple received signal paths are resolvable, then this case may be considered a form of time diversity; however, the resolvable paths are closely related to the fact that the channel is frequency selective, that is, the different frequency components of the desired signal are faded differently (refer to Chapter 27).

In this section we consider the scenario that L independently received copies, $r(1), r(2), ..., r(L)$, all carrying the same transmitted signal s, are available for use in the detection process. These copies could be exploited to achieve a maximum diversity order of L. The received signal copies are expressed as

$$r(1) = h(1)s + n(1), \; r(2) = h(2)s + n(2), ..., \; r(L) = h(L)s + n(L), \quad (25.8)$$

where $h(1), h(2), ..., h(L)$ are assumed to be independent and identically distributed (i.i.d.), all having the same distribution as h created in 1.A; the noise terms $n(1), n(2), ..., n(L)$ are also i.i.d. with the same distribution as n in equation (25.1), and the transmitted signal s is a BPSK signal, taking on the values of $\sqrt{E_s/L}$ or $-\sqrt{E_s/L}$ with equal probability.

3.A [T]Assuming a noiseless and the nonfading condition ($h(1) = ... = h(L) = 1$ and $n(1) = ... = n(L) = 0$), prove that the total received symbol energy from all L branches, that is, $|r(1)|^2 + |r(2)|^2 + ... + |r(L)|^2$, equals E_s.

25.4 COMBINING METHODS

This section investigates techniques for combining the L branches of the received signals to form the decision variable [5,6]. Three commonly used combining methods are as follows:

1. Selection diversity combining (SDC)
2. Equal gain combining (EGC)
3. Maximum ratio combining (MRC)

The fading coefficients $h(1), h(2), ..., h(L)$ are assumed to be known in the receiver. In practice, most communications channels can be classified as "slow" fading channels, for which the fading coefficients will be nearly constant over many symbol periods. This allows the receiver to estimate the fading coefficients.

4.A   Selection diversity combining.

In SDC, the branch with the largest fading coefficient is selected for detection and the rest are not used. The decision variable D is generated as

$$\text{Step 1. Set } k_{\text{best}} = \arg\max_k |h(k)|,$$
$$\text{Step 2. Set } D = r(k_{\text{best}})e^{-j\angle h(k_{\text{best}})}. \tag{25.9}$$

4.A-1   [WWW]The following m-file simulates the BER of BPSK over a Rayleigh fading channel with three SDC branches, that is, $L = 3$. Complete the places marked by '?'. Add a comment to each of the lines in bold to explain what the line does. Especially for the lines with '=', explain what the variable on the left-hand side represents and justify how the right-hand side expression is properly formulated accordingly.

Capture the completed m-file.

```
clear
EbN0dB_vector=0:3:15;
Eb=1;
L=3;
for snr_i=1:length(EbN0dB_vector)
  EbN0dB=EbN0dB_vector(snr_i);
  EbN0=10.^(EbN0dB/10);
  N0=Eb/EbN0;
  sym_cnt=0;
  err_cnt=0;
  while err_cnt<100 % If you increase err_cnt (currently 100), the accuracy increases
and time also increase.
    b=sign(rand-0.5); %BPSK symbol {1,-1}
    s=sqrt(Eb/L)*b;
    for k=1:L
      h(k)=sqrt(1/2)*(randn+j*randn);
      n(k)=sqrt(N0/2)*(randn+j*randn);
      r(k)=?*s+?;
    end

    [T1 T2]=max(?); % Refer to (25.9). To see how to use max( ), execute '>>help
max' in the command window.
    D=r(?)*exp(-j*angle(h(?))); %Refer to (25.9).

    b_hat=sign(real(D));
    if b_hat~=b;
      err_cnt=err_cnt+1;
    end
    sym_cnt=sym_cnt+1;
```

```
    end
      BER(snr_i)=err_cnt/sym_cnt
  end
  figure
  semilogy(EbN0dB_vector, BER)
  xlabel('E_b/N_0 [dB]')
  ylabel('BER')
  grid
```

4.A-2   Execute the completed m-file and capture the simulated BER.

4.A-3   Execute the m-file separately for each of two other cases: $L = 1$ and 5. Then plot the three BER curves in a single figure. Methods to overlay curves from different figures in a single plot were discussed in Section 4.C-1 of Chapter 24. Finally, execute **legend('L=1','L=3','L=5')** in the command window and capture the resulting plot. Save this figure in .fig format and name it **Ch25_4A_3.fig**, as it will be needed later in this chapter.

4.A-4   (a) Analyze how the slope of the BER curves changes as the diversity order L increases, and intuitively explain the reason that causes such characteristics. (b) Explain why the BER becomes slightly worse in the low-SNR region as the diversity order L increases.

4.B   Equal gain combining.

In EGC, $r(1), r(2), ..., r(L)$ are combined with the same weight regardless of the magnitudes of the L instantaneous fading coefficients. For coherent combining, the phase rotation due to fading at each branch is compensated first and the decision variable D is written as

$$D = \sum_{k=1}^{L} r(k)e^{-j\angle h(k)}. \qquad (25.10)$$

4.B-1   In the m-file completed in 4.A-1, modify the right-hand side of the line '**D=r(?)*exp(-j*angle(h(?)));**' into '**D=sum(r.*exp(-j*angle(h)));**' and complete this line properly to implement the right-hand side of equation (25.10). Capture the modified m-file.

4.B-2   In addition, modify the line '**EbN0dB_vector=0:3:15**' into '**EbN0dB_vector=0:3:12**'. Execute the modified m-file for each of the following cases: $L = 1$, $L = 3$, and $L = 5$. Then overlay the three BER curves in a single figure. After this, execute **legend('L=1', 'L=3', 'L=5')** in the command window and capture the figure. Save this figure in .fig format and name it **Ch25_4B_2.fig**, as it will be needed later in this chapter.

4.B-3   Analyze how the BER and the slope of the BER curves change as the diversity order L changes.

4.B-4 Compare the BER results with EGC in 4.B-2 and with SDC in 4.A-3. This can be done best by overlaying all six curves in a single figure. Focus on the relative BER values and the slopes of the BER curves of the same diversity order obtained by using the two combining techniques.

4.C Maximum ratio combining.

Maximum ratio combining is similar to EGC in the sense that all branches are combined in both schemes. Unlike EGC, MRC employs different combining weights that are proportional to the fading magnitude of each branch. The decision variable is expressed as

$$D = \sum_{k=1}^{L} |h(k)||r(k)e^{-j\angle h(k)}. \tag{25.11}$$

4.C-1 [T]Show that equation (25.11) can be simplified as

$$D = \sum_{k=1}^{L} h^*(k)r(k). \tag{25.12}$$

4.C-2 In the m-file completed in 4.A-1, modify the right-hand side of the line **'D=r(?)*exp(-j*angle(h(?)));'** into **'D=sum(conj(?).*?);'**, and complete this line properly to implement the right-hand side of equation (25.12). Capture the completed line.

4.C-3 In addition, modify the line **'EbN0dB_vector=0:3:15'** into **'EbN0dB_vector=0:3:12'** as done in 4.B-2. Execute the modified m-file for each of the following cases: $L = 1$, $L = 3$, and $L = 5$. Then overlay the three BER curves in a single figure. After this, execute **legend('L=1', 'L=3', 'L=5')** in the command window and capture the resulting figure. Save this figure in .fig format and name it **Ch25_4C_3.fig**, which will be needed later in this chapter.

4.C-4 Analyze how the BER and the slope of the BER curves change as the diversity order L changes.

4.C-5 Compare the BER results with MRC in 4.C-3 and with EGC in 4.B-2. As in 4.B-4, this can be done best by overlaying all six curves in a single figure. Focus on the relative BER values and the slopes of the BER curves of the same diversity order obtained by using the two combining techniques.

4.C-6 [A,T]Prove that MRC maximizes the SNR among all three combining methods using the Cauchy Schwarz inequality [7].

4.D Now we compare the performances all three combining methods.

4.D-1 Open the figure files (.fig files) saved in 4.A-3, 4.B-2, and 4.C-3 if you have closed these figures. Overlay all of the nine BER curves in one figure. Change the line color of the BER curves for SDC to blue, for EGC to green, and for MRC to red. Add a legend for all nine curves. Then capture the completed figure.

(a) For the same diversity order L, which combining scheme performs the best?

(b) For $L = 5$, analyze the BER gaps among the three schemes as SNR changes and summarize the observations.

4.D-2   Based on the decision variables in equations (25.9), (25.10), and (25.12), which combining method has the lowest implementation complexity and which one is most difficult to implement? Why?

4.E   In this problem we investigate the diversity gain.

4.E-1   Summarize the common trend in the change of BER values with the three methods as L increases and explain the reason that causes this trend.

4.E-2   Revisit the m-file completed in 4.C for MRC. Set **L=5** and modify the '**for**' loop inside the '**while**' loop as shown below. The distribution of **h(k)** after this modification remains the same as that in the original version. Explain why.

```
hk=sqrt(1/2)*(randn+j*randn);
for k=1:L
   h(k)=hk;
   n(k)=sqrt(N0/2)*(randn+j*randn);
   r(k)=?*s+?;
end
```

4.E-3   Execute the modified m-file and capture the BER graph.

4.E-4   (a) Compare the BER curve in 4.E-3 with the BER curve obtained from the original m-file. (b) Observe that these two BER curves are significantly different. Justify it; that is, explain why the modified m-file cannot achieve any diversity gain although the distribution of **h(k)** remains the same as before.

4.E-5   Based on the results in 4.E-3 and 4.E-4, determine the conditions on the statistical properties of the fading coefficients **h(1)**, **h(2)**, ..., **h(L)**, so that the diversity gain is maximized and the BER is minimized accordingly.

REFERENCES

[1] T. S. Rappaport, *Wireless Communications: Principles and Practice*, 2nd ed., Upper Saddle River, NJ: Prentice Hall, 2002.

[2] D. Tse and P. Viswanath, *Fundamentals of Wireless Communication*, Cambridge, UK: Cambridge University Press, 2005.

[3] B. Sklar, "Rayleigh Fading Channels in Mobile Digital Communication Systems Part I: Characterization," *IEEE Communications Magazine*, Vol. 35, No. 7, 1997, pp. 90–100.

[4] A. Papoulis, *Probability, Random Variables, and Stochastic Processes*, New York: McGraw-Hill, 1965.

[5] J. G. Proakis, *Digital Communications*, 5th ed., New York: McGraw-Hill, 2008.

[6] D. G. Brennan, "Linear Diversity Combining Techniques," *Proceedings IRE.*, Vol. 47, 1959, pp. 1075–1102.

[7] G. Strang, *Linear Algebra and Its Applications*, 4th ed., Belmont, CA: Brooks Cole, 2005.

26

ORTHOGONAL FREQUENCY DIVISION MULTIPLEXING IN AWGN CHANNELS

- Generate orthogonal frequency division multiplexing (OFDM) signals.
- Implement the demodulation process of OFDM signals.
- Simulate the bit error rate (BER) performance of OFDM signaling in AWGN environments.

26.1 ORTHOGONAL COMPLEX SINUSOID

1.A [T]Suppose that two complex signals $x(t)$ and $y(t)$ are orthogonal over the range $\alpha \leq t \leq \beta$. Express this proposition in an equation.

1.B [T]Consider the following two complex sinusoids $a(t)$ and $b(t)$:

$$a(t) = Ae^{j(2\pi ft + \theta_a)},$$
$$b(t) = Be^{j\{2\pi(f+\Delta f)t + \theta_b\}}. \tag{26.1}$$

Show that $\Delta f = n/T$ with any nonzero integer n is a sufficient and necessary condition for $a(t)$ and $b(t)$ to be orthogonal over the interval $t_0 \leq t \leq t_0 + T$, regardless of the amplitudes and phases of the sinusoids.

1.C. [T]Given the orthogonality condition in 1.B.

(a) Show that the minimum frequency separation between two complex orthogonal sinusoids over a time period of T is $1/T$.

(b) Discuss how the minimum frequency separation will change as T increases.

Problem-Based Learning in Communication Systems Using MATLAB and Simulink, First Edition.
Kwonhue Choi and Huaping Liu.
© 2016 The Institute of Electrical and Electronics Engineers, Inc. Published 2016 by John Wiley & Sons, Inc.
Companion website: www.wiley.com/go/choi_problembasedlearning

1.D. [T]Let $\Delta f = 1/T$ in equation (26.1). Show that $a(t - \tau_a)$ and $b(t - \tau_b)$ are still orthogonal over the time period $x \le t \le x + T$ for any values of τ_a and τ_b.

26.2 GENERATION OF ORTHOGONAL FREQUENCY DIVISION MULTIPLEXING SIGNALS

2.A. In orthogonal frequency division multiplexing (OFDM) systems [1, 2], N_c complex sinusoids, also called subcarriers, are used to modulate up to N_c data symbols for transmission in parallel over an ODFM symbol duration T. The frequencies of any two adjacent subcarriers are separated by the minimum separation required to maintain orthogonality between them. The OFDM signal over one OFDM symbol duration T is expressed as

$$x(t) = \sum_{k=0}^{N_c-1} s(k) \frac{e^{j2\pi k f_\Delta t}}{\sqrt{T}}, \quad 0 \le t \le T, \tag{26.2}$$

where $s(k)$ denotes the kth data symbol and f_Δ is the frequency separation between two adjacent subcarriers.

2.A-1 [T]To ensure that the data symbols transmitted in parallel can be demodulated free of interference at the receiver, the complex sinusoids must be mutually orthogonal. Show that to maintain orthogonality among the subcarriers, the minimum subcarrier frequency spacing f_Δ equals $1/T$. Also derive the total bandwidth (minimum) of the OFDM signal $x(t)$.

2.B System parameter setting.
     Consider an OFDM system with the following system parameters:

- Modulation: quadrature phase shift keying (QPSK)
- OFDM symbol duration: $T = 10^{-4}$ seconds
- Channel: AWGN (Extension to multipath fading channels will be discussed in the next chapter.)
- Number of subcarriers: $N_c = 16$

2.B-1 [T]Determine the minimum subcarrier frequency spacing f_Δ to maintain mutual orthogonality among the subcarriers.

2.B-2 [T]Substitute the answer to 2.B-1 into equation (26.2) and show that the highest frequency of all subcarriers is $(N_c - 1)/T$.

2.B-3. [T]We want to generate the sampled version of the waveform expressed in equation (26.2) in MATLAB. Let **t_step** denote the sample interval variable; the sampling frequency is thus **1/t_step**.

Using the Nyquist sampling criterion, determine the maximum value of **t_step** if the bandwidth of the OFDM signal $x(t)$ in equation (26.2) is N_c/T (The bandwidth of OFDM signal will be covered in Section 3 of this chapter).

2.C   Generate the sampled waveform of the OFDM signal in equation (26.2) as a MATLAB vector via the following steps.

2.C-1   Set the maximum sample interval to **t_step** $= T/N_c$. Then show that the number of samples per OFDM symbol is equal to the number of subcarriers.

2.C-2   The following lines of commands illustrate the process to create the sampled vector of the third subcarrier (which corresponds to index $k = 2$ in equation (26.2), since k starts from 0), that is, $\frac{1}{\sqrt{T}}e^{j2\pi k f_s t}$, $k = 2$. Complete the quantities marked by '?'.

```
>>k=2;
>>Nc=16;T=10e-5;
>>f_delta=1/?;
>>t_step=T/Nc;
>>t_vector=0:t_step:(T-t_step) %or  t_vector= t_step*(0:Nc-1)
>>sub_carrier=1/sqrt(T)*exp(j*2*pi*?*?*t_vector);
```

2.C-3   Execute the commands in 2.C-2. Then execute the commands below and capture the result. Assess whether the third subcarrier is generated correctly and justify your assessment. This can be done by measuring the frequency and phases of the real and imaginary parts from the plots and comparing them with $\frac{1}{\sqrt{T}}e^{j2\pi k f_s t}$, $k = 2$.

```
>>figure
>>plot(t_vector, real(sub_carrier))
>>hold on
>>plot(t_vector, imag(sub_carrier),'r')
>>hold off
```

2.C-4   [WWW]The following m-file generates the sampled version of the OFDM signal $x(t)$ in equation (26.2) as a MATLAB vector **xt** based on the system parameters given at the beginning of 2.B. Complete the quantity marked by '?' in the third to the last and capture the completed line.

```
clear
Nc=16;T=10e-5;
f_delta=1/T;
t_step=T/Nc;
t_vector=0:t_step:(T-t_step) % or t_vector=t_step*(0:Nc-1)
```

```
for k=0:(Nc-1)
    k_th_subcarrier=1/sqrt(T)*exp(j*2*pi*k*f_delta*t_vector);
    subcarrier_matrix(k+1,:)=k_th_subcarrier;
    %Save subcarriers as the rows in subcarrier_matrix in order to use them for OFDM
modulation and demodulation.
end

s_vector=sign(rand(1,Nc)-0.5)+j*sign(rand(1,Nc)-0.5);
%Nc number of QPSK symbols

xt=zeros(1,length(t_vector));
for k=0:(Nc-1)
    s_k=s_vector(k+1);
    k_th_subcarrier=subcarrier_matrix(k+1,:);
    xt=xt+ ?*k_th_subcarrier ; %Refer to (26.2).
end
xt
```

2.C-5   Execute m-file and capture the result.

2.C-6   Execute the following two lines and capture the results. The first command line computes the inverse fast Fourier transform (IFFT) of the symbol vector. Complete the equation $x(t) = ? \times$ IFFT of $[s(1), s(2), ..., s(N_c)]$.

```
>>ifft(s_vector)
>>sqrt(T)/Nc*xt
```

2.C-7   [WWW]From the results obtained in 2.C-6, we can significantly simplify the m-file completed in 2.C-4 as follows. Complete the last line and capture the m-file.

```
clear
Nc=16;T=10e-5;
f_delta=1/T;
s_vector=sign(rand(1,Nc)-0.5)+j*sign(rand(1,Nc)-0.5);
xt=?*ifft(?)
```

2.C-8   From the results obtained in 2.C-7, OFDM signal generation is equivalent to performing how many points of IFFT of the symbol stream with a proper scaling factor?

2.C-9   [A,T]Verify the conclusion that the IFFT of the symbol vector $[s(1), s(2), ..., s(N_c)]$ equals the OFDM signal scaled by $\sqrt{T}/N_c$, that is, $\frac{\sqrt{T}}{N_c}x(t)$, through the following problems.

(a) Write the expression for the nth sample of the sampled OFDM waveform **xt** (i.e., **xt(n)**). The steps implemented in the m-file in 2.C-4 to generate **xt** would be a good starting point.

(b) Execute **help fft** in the MATLAB command window and review the mathematical expression of the IFFT output. Show that the expression obtained in (a) scaled by $\sqrt{T/N_c}$ is equal to the nth output of the IFFT of $[s(1), s(2), \ldots, s(N_c)]$.

26.3 BANDWIDTH EFFICIENCY OF OFDM SIGNALS

3.A First, we revisit the single-carrier modulation methods. Let the symbol rate be R_s.

3.A-1 [T]Express the symbol duration as a function of R_s.

3.A-2 [T]Suppose that the raised cosine pulse with roll-off factor of α is used for pulse shaping. (a) Sketch the spectrum of the passband signal. (b) Using the sketch in (a), show that the passband null-to-null bandwidth is equal to $R_s(1 + \alpha)$.

3.B Consider the OFDM signaling with N_c subcarriers. For a fair comparison of its bandwidth with that of the single-carrier system, we also let the symbol rate be R_s here.

3.B-1 [T]The OFDM symbol duration T defined in equation (26.2) equals N_c times of the symbol duration for the single-carrier system. Justify this relationship using the fact that during each OFDM symbol interval, multiple data symbols are transmitted in parallel.

3.B-2 [T]Based on the results in 3.B-1 and 3.A-1, show that $T = N_c/R_s$.

3.B-3 [T]Show that the subcarrier frequency spacing f_Δ is equal to R_s/N_c using the relationship between f_Δ and T in 2.A-1 and the relationship established in 3.B-2.

3.B-4 [T]The bandwidth of the OFDM signal equals the separation between the highest frequency and lowest frequency of the signal spectrum. Show that it can be well approximated as $(N_c - 1)f_\Delta = \frac{N_c-1}{N_c}R_s$.

3.B-5 [T](a) For a sufficiently large N_c, which is the case for most practical OFDM systems, show that the OFDM signal bandwidth is further approximated as R_s. (b) Determine the ratio of the OFDM signal bandwidth (R_s) to the single-carrier system bandwidth given in 3.A-3. (c) Based on the result in (b), compare the bandwidth efficiency of OFDM and single-carrier systems.

26.4   DEMODULATION OF OFDM SIGNALS

4.A   [T]The symbol $s(n)$ modulated on the nth subcarrier as expressed by equation (26.2) can be extracted out as

$$s(n) = \int_0^T x(t)\frac{e^{-j2\pi n f_\Delta t}}{\sqrt{T}}dt. \tag{26.3}$$

Substitute equation (26.2) into equation (26.3) and prove equation (26.3).

4.B   In this section we implement the right-hand side of equation (26.3) through numerical integration in MATLAB and verify the result. Numerical integration was discussed in Section 2.1 of Chapter 2.

In the m-file in 2.C-4, the sampled vector of the nth subcarrier $\frac{1}{\sqrt{T}}e^{-j2\pi n f_\Delta t}$ is stored in the $(n+1)$th row of the matrix **subcarrier_matrix**, that is, **subcarrier_matrix(n+1,:)**. The three lines of code below calculate the right-hand side of equation (26.3) via numerical integration. The result is then compared with $s(n)$ (**s_vector(n+1)** in MATLAB).

Execute the three code lines below for several values of n (integer) from 0 to N_c-1 and capture the results.

NOTE: If the workspace has been cleared after executing the m-file in 2.C-4, then execute this m-file before executing the three lines. Is the OFDM demodulation scheme expressed in equation (26.3) verified for all values of n simulated?

```
>>n=? ; %Select an arbitrary integer from 0 to Nc-1
>>t_step*sum(xt.*conj(subcarrier_matrix(n+1,:)))
>>s_vector(n+1)
```

26.5   BER SIMULATION OF OFDM SYSTEMS

5.A   [WWW]In a practical OFDM communications system, data are transmitted in frames. Each frame typically consists of many OFDM symbols. The m-file below generates one OFDM frame. The variable **Nf** denotes the number of OFDM symbols per frame.

For the lines in bold, explain what the variable on the left-hand side represents and justify how the right-hand side expression is properly formulated accordingly.

```
clear
Nf=10;
Nc=16;T=10e-5;
f_delta=1/T;
t_step=T/Nc;
```

```
t_vector=0:t_step:(T-t_step);
for k=0:(Nc-1)
    subcarrier=1/sqrt(T)*exp(j*2*pi*k*f_delta*t_vector);
    subcarrier_matrix(k+1,:)=subcarrier;
end
xt_frame=[];
for m=1:Nf
s_vector=sign(rand(1,Nc)-0.5)+j*sign(rand(1,Nc)-0.5);

xt=zeros(1,length(t_vector));
for k=0:(Nc-1)
    s_k=s_vector(k+1);
    xt=xt+s_k*subcarrier_matrix(k+1,:);
end
xt_frame=[xt_frame xt];
end
```

5.B   In this section we simulate the BER of OFDM transmission over an AWGN channel. Here are the simulation steps:

Step 1: Create OFDM symbols.
Step 2: Concatenate OFDM symbols to form an OFDM frame.
Step 3: Create the sampled noise vector and add it to the transmitted signal to create the received signal.
Step 4: Divide the received OFDM frame into OFDM symbols.
Step 5: Demodulate the data symbols contained in each received OFDM symbol.
Step 6: Compare the demodulated and transmitted data symbols to calculate the error rate.

5.B-1   [WWW]The m-file below implements the steps above and simulates the BER. Identify the line(s) that corresponds to each step above. Add a comment to the line(s) to indicate the corresponding step it implements. Capture the m-file with comments added.

```
clear
Nf=10;
Nc=16;T=10e-5;
f_delta=1/T;
t_step=T/Nc;
t_vector=0:t_step:(T-t_step); %=t_step*(0:Nc-1)
Ns=length(t_vector); %Number of samples in one OFDM symbol duration T
Eb=1;
EbN0dBvector=0:3:9;
for k=0:(Nc-1)
```

```
      k_th_subcarrier=1/sqrt(T)*exp(j*2*pi*k*f_delta*t_vector);
      subcarrier_matrix(k+1,:)=k_th_subcarrier;
   end
 for snr_i=1:length(EbN0dBvector)
   EbN0dB=EbN0dBvector(snr_i);
   EbN0=10^(EbN0dB/10);
   N0=Eb/EbN0;
   vn=N0/(2*t_step); % % Refer to Problem 1 of Exercise 21

   bitcnt=0; errcnt=0;
   while errcnt<100

%%%%%%%%%%%%%% Transmitter %%%%%%%%%%%%%%%%%%%
      OFDM_frame=[];
      for m=1:Nf
        data_symbols_in_OFDMsymbol=sign(rand(1,Nc)-0.5)+j*sign(rand(1,Nc)-0.5);
        data_symbols_in_OFDMframe(m,:)=data_symbols_in_OFDMsymbol;
        xt=zeros(1,Ns);
        for k=0:(Nc-1)
          s_k=data_symbols_in_OFDMsymbol(k+1);
          xt=xt+s_k*subcarrier_matrix(k+1,:);
        end
        OFDM_frame=[OFDM_frame xt];
      end

   %%%%%%%%%% AWGN Channel and received signal generation %%%%%%%%
   noise=sqrt(vn)*(randn(1,length(OFDM_frame))+j*randn(1,length(OFDM_frame)));
   rt_frame=OFDM_frame+noise;

   %%%%%%%%%%%%%%%%% Receiver %%%%%%%%%%%%%%%%%%%%%%%%%%%%%
      for m=1:Nf
        mth_OFDMsymbol_in_rt=rt_frame((m-1)*Ns+(1:Ns));
        for k=0:(Nc-1)
          k_th_subcarrier =subcarrier_matrix(k+1,:);
          D=t_step*sum( mth_OFDMsymbol_in_rt.*conj(k_th_subcarrier));
          estimated_data_symbols_in_OFDMframe(m,k+1)=sign(real(D))+j*sign
(imag(D));
        end
      end

   Ierrs=sum(sum(real(data_symbols_in_OFDMframe)~=real(estimated_
   data_symbols_in_OFDMframe)));
   Qerrs=sum(sum(imag(data_symbols_in_OFDMframe)~=imag(estimated_
   data_symbols_in_OFDMframe)));
   errcnt=errcnt+(Ierrs+Qerrs);
```

```
      bitcnt=bitcnt+Nc*Nf*2;
   end

   BER(snr_i)=errcnt/bitcnt
   BERtheory(snr_i)=0.5 *erfc(sqrt(EbN0));
 end
 figure
 semilogy(EbN0dBvector, BER;'b')
 hold on
 semilogy(EbN0dBvector, BERtheory;'r')
 grid
 legend('Simulated, OFDM;'Theoretical, single-carrier QPSK(or BPSK)')
```

5.B-2   For each of the lines in bold, explain what the variable on the left-hand side represents and justify how the right-hand side expression is formulated accordingly.

5.B-3   Execute the m-file and capture the simulated BER graph.

5.B-4   Change the values of **Nf**, **Nc**, and **T** to a different set of values. (a) Execute the m-file again with this set of new values. Capture the BER graph and record the parameter values on the graph. (b) Check whether or not the simulated BER matches the theoretical BER of single-carrier QPSK (or BPSK), regardless of the parameter setting.

5.B-5   Modify the m-file so that each subcarrier carries a BPSK symbol, instead of a QPSK symbol. Document all the modified lines and add a short explanation to each of these lines.

5.B-6   Execute the modified m-file for BPSK modulation. Capture the simulated BER graph.

5.B-7   Based on the simulation results in 5.B-3 and 5.B-6, compare the BER performances of the single-carrier system and the OFDM system in an AWGN channel. Explain why their BER performances should be or should not be the same.

REFERENCES

[1] A. R. S. Bahai and B. R. Saltzberg, *Multi-Carrier Digital Communications: Theory and Applications of OFDM*, Alphen aan den Rijn, Netherlands: Kluwer, 1999.

[2] R. Prasad, *OFDM for Wireless Communications Systems*, London: Artech House Publishers, 2004.

27

ORTHOGONAL FREQUENCY DIVISION MULTIPLEXING OVER MULTIPATH FADING CHANNELS

- Create the impulse response of a multipath channel and analyze the relationship between the power-multipath magnitude profile and channel response.
- Generate orthogonal frequency division multiplexing (OFDM) signals with cyclic prefix (CP) added and demodulate them in multipath fading environments.
- Analyze the bit error rate (BER) performance of OFDM systems over multipath fading channels.

27.1 MULTIPATH FADING CHANNELS

1.A [WWW]The user-defined MATLAB function below, **ht_mp_ch()**, outputs the sampled version of the impulse response $h_{mp}(t)$ of the multipath fading channel [1–3] as a vector. The last line is for normalizing the total energy of $h_{mp}(t)$ to unity.

The input variables are as follows.

- **L**: Number of multipath components.
- **max_delay**: The maximum excess delay of the channel in seconds. The delay of the first arrival path is assumed to be zero seconds; the arrival times of the remaining **L**-1 multipath components are assumed to be uniformly distributed over [0 **max_delay**].

Problem-Based Learning in Communication Systems Using MATLAB and Simulink, First Edition.
Kwonhue Choi and Huaping Liu.
© 2016 The Institute of Electrical and Electronics Engineers, Inc. Published 2016 by John Wiley & Sons, Inc.
Companion website: www.wiley.com/go/choi_problembasedlearning

- **decay_base**: The base of the exponentially decaying profile. The multipath amplitude at t = **max_delay** decays to the value of (**decay_base** × the amplitude of the first arrival path).
- **t_step**: Sample interval of the output vector **impulse_response**, that is, the sampled version of the multipath channel impulse response.

```
function impulse_response=ht_mp_ch(max_delay,L,decay_base,t_step)

t_vector=0:t_step:max_delay;
mp_tmp=0*(t_vector);

path_delays=[0 sort(rand(1,L-1)*max_delay)];
impulse_positions=floor(path_delays/t_step);
mp_tmp(impulse_positions+1)=exp(j*2*pi*rand(1,L));
mp_tmp=mp_tmp.*(decay_base.^(t_vector/max_delay));
impulse_response=mp_tmp/sqrt(sum(abs(mp_tmp).^2));
```

1.A-1   Create and save the m-file above as **ht_mp_ch.m**. We simulate the system assuming with **L**=10 paths, the channel maximum excess delay = 1 μs, **decay_base** = 1/8, and the sample interval of the channel impulse response = 0.01 μs. Complete the four quantities marked by '**?**' in the lines of code below.

```
>>figure
>>A=?;B=?;C=?,D=?;
>>ht=ht_mp_ch(A,B,C,D);
>>stem(0:D:A, abs(ht),'');
```

1.A-2   Execute the last two lines in 1.A-1 repeatedly at least 10 times and check the resulting multipath amplitudes. Capture two of the realizations and examine them. Are these results what you expect to see?

1.B [WWW]The m-file below plots the multipath magnitude profile $|h_{mp}(t)|$ and its magnitude spectrum denoted by $|H_{MP}(f)|$.

```
clear
L=5; max_delay=1e-6; decay_base=1e-6; t_step=1e-8;
ht=ht_mp_ch(max_delay,L,decay_base,t_step);

ht4plot=[ht zeros(1,1024-length(ht))]

figure(1)
subplot(2,1,1);
```

```
stem((0:1023)*t_step,abs(ht4plot), '');
title('Multipath magnitude profile(=|h_{mp}(t)|')');xlabel('time [sec]');

axis([-100*t_step 1023*t_step 0 1]);grid on
subplot(2,1,2);
plot((0:1023)/1023/t_step,abs(fft(ht4plot)));
title('Channel response in the frequency domain (=|H_{MP}(f)|')');xlabel('frequency [Hz]');
axis([0 1/t_step 0 max(abs(fft(ht4plot))+0.2)]);grid on
```

1.B-1   Execute the m-file repeatedly at least five times for each of the following values of **max_delay**: 1e-6, 1e-7, and 1e-8. Capture one sample figure for each case.

1.B-2   (a) Describe the shape that the multipath magnitude profile $|h_{mp}(t)|$ converges to as max_delay approaches 0.

(b) Describe the shape that the magnitude spectrum $|H_{MP}(f)|$ converges to as **max_delay** approaches 0.

(c) Justify the answer in (b) on the basis of the relationship between $|h_{mp}(t)|$ and $|H_{MP}(f)|$.

1.B-3   Set max_delay to 1e-6 and execute the m-file for each of the following cases: **decay_base** = 1, 1e-2, 1e-4, and 1e-6. Capture one sample figure for each case and describe the changes in the shapes of $|h_{mp}(t)|$ and $|H_{MP}(f)|$ as **decay_base** decreases. Also justify the changes.

1.B-4   If the magnitude spectrum of the channel response, that is, $|H_{MP}(f)|$, varies appreciably within the signal bandwidth, the channel is classified as being "frequency selective" to the signal. If $|H_{MP}(f)|$ is fairly flat in the signal bandwidth, then the channel is "frequency nonselective" or "frequency flat" to the signal.

From the observations made in 1.B-2 and 1.B-3, summarize the conditions on **decay_base** and **max_delay** for the channel to be frequency selective to a channel with a certain bandwidth.

1.B-5   [T]The multipath channel can be modeled as a linear time invariant filter with impulse response $h_{mp}(t)$. Express the channel output $y(t)$ in terms of an input signal $x(t)$ and $h_{mp}(t)$.

1.B-6   [T]Let $X(f)$, $Y(f)$, and $H_{MP}(f)$ denote the Fourier transforms of $x(t)$, $y(t)$, and $h_{mp}(t)$, respectively. From the time domain relationship obtained in 1.B-5, prove the corresponding frequency domain relationship $Y(f) = H_{MP}(f)X(f)$.

1.B-7   According to the frequency domain relationship equation $Y(f) = H_{MP}(f) \times X(f)$, the flatter the channel frequency response is over the signal band, the less the transmitted signal is distorted. In the extreme case that the channel is frequency flat; that is, $H_{MP}(f)$ equals a constant, the transmitted signal simply undergoes a constant scaling. From these observations, answer the following questions:

(a) For a given bandwidth of the transmitted signal, determine the conditions on the multipath channel parameters **decay_base** and **max_delay** that result in less distortion to the transmitted signal spectrum.

(b) For a given channel response $H_{MP}(f)$, determine the conditions on the bandwidth of $X(f)$ that result in less distortion to the received spectrum $Y(f)$.

(c) From the conditions determined in (b), we should be able to conclude that the transmitted signal will be distorted less by the channel if the period of the transmitted data symbols is larger. Provide a proper justification of this conclusion.

1.B-8   Recall that in 3.B-1 of Chapter 26, we concluded that the OFDM symbol duration is N_c times of the single-carrier symbol duration, where N_c is the number of subcarriers. From this and the conclusion made in 1.B-7(c), compare the robustness of OFDM and single-carrier modulations (and assume the same data rate) against frequency selective fading.

27.2   GUARD INTERVAL, CP, AND CHANNEL ESTIMATION

2.A   In a multipath fading channel, the received signal is a sum of the scaled and delayed copies of the same transmitted signal. If the maximum excess delay relative to the symbol duration is small (e.g., 10–20% of the symbol duration), then only a small portion of the delayed versions of the preceding symbol will overlap with a current symbol. Even such a small overlap will cause undesirable effects. In a single-carrier system, this causes ISI between adjacent symbols. In an OFDM system, in addition to ISI (inter-OFDM-symbol interference), since the starting and ending boundaries of the delayed versions of the same OFDM symbol are not aligned up, the subcarriers are no longer mutually orthogonal, causing intercarrier interference (ICI) [4, 5].

Here we focus on OFDM systems. In order to minimize the impacts of ICI and ISI, a guard interval with a minimum length equal to the channel excess delay can be left in front of each OFDM symbol. In addition, the last portion of each OFDM symbol of the same length as the guard interval is copied and inserted in the guard interval. This portion of the signal is called the "cyclic prefix." In the receiver, the CP for the first arrival path is removed before OFDM demodulation. Since the OFDM symbol boundaries of the delayed versions lie within the guard interval of the following OFDM symbol, ISI is removed. In addition, the insertion of the CP will still maintain the orthogonality among the subcarriers despite the delayed versions, and this allows ICI to be eliminated after OFDM demodulation.

2.A-1   [WWW]Copy the m-file in 5.B-1 of Chapter 26, give it a new file name, and save it. Insert the lines in bold shown below to the m-file. In the modified m-file, the guard interval is set to 1/4 of the OFDM symbol duration **T**. The last 25% of each OFDM symbol is copied and inserted in the guard interval before the OFDM symbol.

For each of the lines in bold, explain what the variable on the left-hand side represents and justify how the right-hand side expression is formulated accordingly.

```
clear
...
Ns=length(t_vector); %Number of samples per OFDM symbol(T seconds) before
    inserting CP.
GI=1/4;
Ns_in_GI=ceil(Ns*GI);
Ns_total=Ns+Ns_in_GI;
Eb=1;
...
for snr_i=1:length(EbN0dBvector)
  ...
  while errcnt<100
    ...
    for m=1:Nf
      ...
      for k=0:(Nc-1)
        ...
      end
      xt_tail=xt((Ns-Ns_in_GI+1):Ns);
      xt=[xt_tail xt];
      OFDM_frame=[OFDM_frame xt];
    end
    ...
    for m=1:Nf
      first_index_of_mth_OFDM symbol=(m-1)*Ns_total+ Ns_in_GI + 1 ;
      mth_OFDM symbol_in_rt=rt_frame( first_index_of_mth_OFDM symbol+
(0:Ns-1) );
      for k=0:(Nc-1)
        ...
      end
    end
    ...
  end
  ...
end
...
```

2.A-2 [WWW]In order to implement the multipath fading channel, modify the m-file in 2.A-1 further by adding the lines in bold in the code below. In the modified m-file, the received signal **rt_frame** is the OFDM frame that has passed through the multipath channel. Explain why the **line 'ht = ht_mp_ch(max_delay, L, decay_base, t_step);'** should be placed inside, not outside of the '**while**' loop.

```
clear
Nf=10;
L=5; max_delay=1.25e-5; decay_base=1;
Nc=16; T=8*max_delay;
t_step=(T/Nc)/16;
f_delta=1/T;
t_vector=0:t_step:(T-t_step); %=t_step*(0:Nc-1)
Ns=length(t_vector); %Number of samples per OFDM symbol(T seconds) before
    inserting CP.
GI=1/4;
Ns_in_GI=ceil(Ns*GI);
Ns_total=Ns+Ns_in_GI;
Eb=1;
EbN0dBvector=0:3:9;

...
for snr_i=1:length(EbN0dBvector)
  ...
  while errcnt<500
    ...
    for m=1:Nf
       ...
    end
    ht=ht_mp_ch(max_delay,L,decay_base,t_step);
    OFDM_frame_after_ht=conv(OFDM_frame,ht);
    frame_length=length(OFDM_frame_after_ht);
    noise=sqrt(vn)*(randn(1,frame_length)+j*randn(1,frame_length));
    rt_frame=OFDM_frame_after_ht+noise;

    for m=1:Nf
        ...
      end
    end
    ...
  end
  ...
end
...
```

2.A-3   In the added line 'OFDM_frame_after_ht=conv(OFDM_frame,ht);', explain what the variable **OFDM_frame_after_ht** represents and justify how the right-hand side expression is properly formulated accordingly.

2.A-4   According to the modified m-file in 2.A-2, do the **Nf** OFDM symbols in one frame go through the same multipath channel or each of the **Nf** OFDM symbols goes through a different multipath channel? In other words, does the impulse response of the channel change within one frame? Justify your answer.

2.A-5   Execute the modified m-file in 2.A-2 and capture the result of the vector **BER** displayed in the command window. Check to see whether the simulated BERs are around 0.5, regardless of the signal-to-noise ratio (SNR) values.

2.A-6   To simulate the noiseless case, modify the line '**rt_frame = OFDM_frame_ after_ht+noise;**' into '**rt_frame = OFDM_frame_after_ht;**'. Execute the modified m-file and check the BER values. The BERs should still be around 0.5. Give the reason why.

2.B   Recall that in the AWGN channel environment, the decision variable D_k (= **D(k)** in the m-file) of the k-th subcarrier is modeled as

$$D_k = s_k + n_k, \quad k = 0, 1, 2, \ldots, N_c - 1, \tag{27.1}$$

where s_k is the data symbol on the kth subcarrier and n_k is the noise term.

In the frequency selective multipath fading channel, we saw in Section 27.1 that the channel frequency response $H_{MP}(f)$ is not a constant over the signal band. Hence the fading gains for different subcarriers are in general not identical. The decision variable D_k of the kth subcarrier can be expressed as

$$D_k = F_k s_k + n_k, \quad k = 0, 1, 2, \ldots, N_c - 1, \tag{27.2}$$

where F_k denotes the fading gain of the kth subcarrier at the frequency f_k and is obtained as $F_k = H_{MP}(f_k)$.

2.B-1   Justify equation (27.2).

2.B-2   [WWW]In the OFDM receiver, the fading gain F_k for each subcarrier should be estimated and applied to form the proper decision variable $D_{k, \text{compensated}}$ as

$$\begin{aligned} D_{k,\text{compensated}} &= D_k/F_k \\ &= s_k + n_k/F_k. \end{aligned} \tag{27.3}$$

The process to estimate the fading gain F_k is called "channel estimation," and the process expressed in equation (27.3) is called "channel compensation" (or channel inversion) [5].

Commonly used OFDM channel estimation uses a pilot approach; known data symbols called "pilot symbols" are transmitted. Within each frame there exist many OFDM symbols, and the pilots are transmitted in some or all OFDM symbols, depending on how fast the channel changes in time. Let p_k denote the pilot symbol on the kth subcarrier and $D_{k,\text{pilot}}$ denote the corresponding decision variable. The estimated fading gain of the kth subcarrier denoted by $\hat{F}_k$ is obtained as

$$\hat{F}_k = D_{k,\text{pilot}}/p_k = \left(F_k p_k + n_k\right)/p_k = F_k + n_k/p_k. \tag{27.4}$$

The estimation error term n_k/p_k can be reduced by transmitting the pilot multiple times or by boosting up the pilot symbol energy.

Next we simulate the BER of OFDM systems over multipath fading channels assuming that perfect channel estimates are available. To this end, in the m-file, the noise term is not added for pilot symbols, which allows us to obtain perfect channel estimates. In practice, this approximates well the case that the pilots are received with a very high SNR. Also the pilot symbols are set to 1 for all subcarriers for simplicity.

The lines in bold in the m-file below are added for estimating the channel gain F_k, $k = 0, 1, \ldots, N_c$. Note that in the m-file, the variable **F(k+1)** corresponds to F_k because the vector index starts from 1 in MATLAB but OFDM subcarrier index starts from 0. The channel impulse response **ht** is generated randomly. The complex fading gains F_k, $k = 0, 1, \ldots, N_c$ are calculated by using equation (27.4).

Add a comment for each of the lines in bold to explain what the variable on the left-hand side represents and justify how the right-hand side expression is properly formulated accordingly.

```
clear
L=5; max_delay=1.25e-5; decay_base=1;
Nc=16; T=8*max_delay;
t_step=(T/Nc)/16;
f_delta=1/T;
t_vector=0:t_step:(T-t_step); %=t_step*(0:Nc-1)
Ns=length(t_vector); %Number of samples per OFDM symbol(T seconds) before insert-
ing CP
GI=1/4;
Ns_in_GI=ceil(Ns*GI);
Ns_total=Ns+Ns_in_GI;
for k=0:(Nc-1)
    subcarrier=1/sqrt(T)*exp(j*2*pi*k*f_delta*t_vector);
    subcarrier_matrix(k+1,:)=subcarrier;
end

pilot_sk=ones(1,Nc);
xt=zeros(1,Ns);
for k=0:(Nc-1)
    s_k=pilot_sk(k+1);
    xt=xt+s_k*subcarrier_matrix(k+1,:);
end
xt_tail=xt((Ns-Ns_in_GI+1):Ns);
```

```
pilot_OFDM symbol=[xt_tail  xt];

ht=ht_mp_ch(max_delay,L,decay_base,t_step);

rt_pilot=conv(pilot_OFDM symbol,ht);
for k=0:(Nc-1)
  D=t_step*sum(rt_pilot(Ns_in_GI+(1:Ns)).*conj(subcarrier_matrix(k+1,:)));
  F(k+1)=D/pilot_sk(k+1);
end
stem(abs(F));grid on
```

2.B-3   Modify the line '**T=8*max_delay**' into '**T=8*1.25e-5**' and execute the m-file for each of the following values of **max_delay**: 1.25e-5, 1.25e-6, and 1.25e-7.

(a) Capture the results for each case.
(b) Describe how the fading gains change as **max_delay** decreases. Justify the results.

2.B-4   Modify the m-file to set '**T=8*1.25e-5**', '**max_delay=1.25e-5**', and '**L=1**'. Execute the modified m-file and capture the result. Justify the result.

27.3   BER SIMULATION OF OFDM SYSTEMS OVER MULTIPATH FADING CHANNELS

[WWW]The m-file below implements the whole OFDM system and simulates the BER in the multipath fading channel.

```
clear
Nf=10;
L=5; max_delay=1.25e-5; decay_base=1;
Nc=16; T=8*max_delay;
t_step=(T/Nc)/16;
f_delta=1/T;
t_vector=0:t_step:(T-t_step); %=t_step*(0:Nc-1)
Ns=length(t_vector); %Number of samples per OFDM symbol(T seconds) before
     inserting Cyclic Prefix
GI=1/4;
Ns_in_GI=ceil(Ns*GI);
Ns_total=Ns+Ns_in_GI;
Eb=1;
EbN0dBvector=0:3:18;
```

```
for k=0:(Nc-1)
   subcarrier=1/sqrt(T)*exp(j*2*pi*k*f_delta*t_vector);
   subcarrier_matrix(k+1,:)=subcarrier;
end

pilot_sk=ones(1,Nc);
xt=zeros(1,Ns);
for k=0:(Nc-1)
  s_k=pilot_sk(k+1);
  xt=xt+s_k*subcarrier_matrix(k+1,:);
end
xt_tail=xt((Ns-Ns_in_GI+1):Ns);
pilot_OFDM symbol=[xt_tail xt];

for snr_i=1:length(EbN0dBvector)
   EbN0dB=EbN0dBvector(snr_i);
   EbN0=10^(EbN0dB/10);
   N0=Eb/EbN0;
   vn=N0/(2*t_step); % Refer to Section 1.C and 1.D of Chapter 21.

   bitcnt=0; errcnt=0;
   while errcnt<1000
%Reduce 1000 to a smaller value if the simulation takes long time. The simulation
    error will increase instead.

     OFDM_frame=[];
     for m=1:Nf
        datasymbols_in_OFDM symbol=sign(rand(1,Nc)-0.5)+j*sign(rand(1,Nc)-0.5);
        datasymbols_in_OFDMframe(m,:)=data symbols_in_OFDM symbol;
        xt=zeros(1,Ns);
        for k=0:(Nc-1)
           s_k=datasymbols_in_OFDM symbol(k+1);
           xt=xt+s_k*subcarrier_matrix(k+1,:);
        end
        xt_tail=xt((Ns-Ns_in_GI+1):Ns);
        xt=[xt_tail xt];
        OFDM_frame=[OFDM_frame xt];
     end

     ht=ht_mp_ch(max_delay,L,decay_base,t_step);
     OFDM_frame_after_ht=conv(OFDM_frame,ht);
     frame_sample_length=length(OFDM_frame_after_ht);
    noise=sqrt(vn)*(randn(1,frame_sample_length)+j*randn(1,frame_sample_length));
     rt_frame=OFDM_frame_after_ht+noise;
```

```
    rt_pilot=conv(pilot_OFDM symbol,ht);
    for k=0:(Nc-1)

        D=t_step*sum(rt_pilot(Ns_in_GI+(1:Ns)).*conj(subcarrier_matrix(k+1,:))/
        sqrt(T));
        F(k+1)=D/pilot_sk(k+1);
    end

    for m=1:Nf
        first_index_of_mth_OFDMsymbol=(m-1)*Ns_total+ Ns_in_GI + 1 ;
        mth_OFDM symbol_in_rt=rt_frame( first_index_of_mth_OFDMsymbol+
        (0:Ns-1) );
        for k=0:(Nc-1)
            D=t_step*sum( mth_OFDMsymbol_in_rt.*conj(subcarrier_matrix(k+1,:))
            /sqrt(T));
            Dc=D/F(k+1); %Dc means the channel compensated decision
            variable in (27.3).
            estimated_data symbols_in_OFDMframe(m,k+1)=sign(real(Dc))+j*sign
            (imag(Dc));
        end
    end

    Ierrs=sum(sum(real(data symbols_in_OFDMframe)~=real(estimated_data
        symbols_in_OFDMframe)));
    Qerrs=sum(sum(imag(data symbols_in_OFDMframe)~=imag(estimated_data
        symbols_in_OFDMframe)));
    errcnt=errcnt+(Ierrs+Qerrs);
    bitcnt=bitcnt+Nc*Nf*2;
    end
    BER(snr_i)=errcnt/bitcnt
    %BERtheory(snr_i)=qfunc(sqrt(2*EbN0))
    BERtheory(snr_i)=1/2 - EbN0^(1/2)/(2*(EbN0 + 1)^(1/2))
end
figure
semilogy(EbN0dBvector, BER,'b')
hold on
semilogy(EbN0dBvector, BERtheory,'r')
grid
legend('BER simulation','BER theory (Rayleigh fading)')
```

3.A   The three parts in bold implement the following functions:

- The first part: creates the pilot symbols.
- The second part: estimates the fading channel coefficients for all **Nc** subcarriers.
- The third part: implements equation (27.3) to form the decision variable.

These functions should be implemented in the proper locations in the code. Discuss why it is appropriate to implement the three parts at their current locations as shown in the m-file.

3.B   The effect of the number of paths **L** on the BER performance.

3.B-1   Execute the m-file in 3.A for each of the following cases: **L** = 2, 3, 5, and 7. Overlay the simulated BER curves for all cases in one figure and capture the plot (the process to copy curves displayed in different figures generated in MATLAB and overlay them in one figure was discussed in 4.C-1 of Chapter 24)

3.B-2   Compare the simulated BER values with the theoretical BER of coherent QPSK over Rayleigh fading channels. We should see that as **L** increases, the simulated BER approaches the theoretical BER of coherent quadrature phase shift keying (QPSK) over Rayleigh fading channels. Determine the distribution of the fading gain for each subcarrier by invoking the central limit theorem (discussed in Section 15.3 of Chapter 15) and then explain the result above.

3.C   The effect of the guard interval length **GI** on the BER performance.

3.C-1   Set **L** = 5 again and execute the m-file for each of the following cases: **GI** = 1/2, 1/4, 1/8, 1/32, 1/64, and 0. Overlay the simulated BER curves in one figure and capture the plot.

3.C-2   The results in 3.C-1 should show the following BER characteristics.

- The BER results are almost the same for **GI** = 1/2, 1/4, and 1/8.
- The BER increases significantly as **GI** decreases to 1/32, 1/64, and 0.

Justify these simulation results (notice the value of **max_delay** set in the m-file).

3.C-3   Set **GI** = 1/64. Modify the line 'rt_frame = **OFDM_frame_after_ht** + noise;' into 'rt_frame = **OFDM_frame_after_ht** + 0*noise;' and then execute the modified m-file.

  (a) Capture the BER result. Note that although without further modifications to the m-file the BER figures will show the x axis range to be from 0 to 18 dB, the effective SNR is actually infinity because noise is not added to the received signal.
  (b) The result should show that the BER value is not 0 despite the infinite SNR values. Explain why.

3.D   The effect of the multipath decaying factor, **decay_base**, on the BER performance.

3.D-1   Set **L** = 5 and **GI** = 1/4. Restore the line 'rt_frame = **OFDM_frame_after_ht** + 0*noise;' back to the original 'rt_frame = **OFDM_frame_after_ht**+noise;'. Then execute the m-file for each of the following cases: **decay_base**=1, 1e-1, and 1e-6. Overlay the three simulated BER curves in the single graph and capture the plot.

3.D-2   The results in 3.D-1 should show that the BER values become significantly lower than the theoretical BER in the Rayleigh fading channel as **decay_base** decreases.

(a) Provide a theoretical justification of this result.

(b) As **decay_base** approaches 0, the BER values converge to the BER performance in which type of channel?

3.E   The effect of the number of subcarriers, **Nc**, on the BER performance.

3.E-1   Set **L** = 5, **GI** = 1/4, and **decay_base** = 1. Execute the m-file for each of the following cases: **Nc** = 4, 16, and 64. Overlay the three simulated BER curves in a single graph and capture the plot. Are the simulated BERs approximately the same regardless of the value of **Nc**?

3.E-2   Set **GI** = 1/16 and execute the m-file for each of the following cases: **Nc** = 4, 16, and 64. Overlay the three simulated BER curves in one figure and capture the plot. Does the BER increase or decrease as **Nc** increases and why?

3.E-3   Note that the fourth line of the m-file sets two parameters, **Nc** = 16 and **T=8*max_delay**. If **Nc** is set larger than 16, will the bandwidth of the OFDM signal increase or decrease? Why?

3.E-4   In order to simulate the performance of the system with different values of **Nc** while the signal bandwidth is kept the same as the case with **Nc** = 16, the OFDM symbol duration **T** should be scaled by **Nc/16**. This can be done by changing the line 'T=8*max_delay;' into 'T=(Nc/16)*8*max_delay;'. Justify this modification to maintain the same bandwidth for different values of Nc.

3.E-5   Modify the line 'T=8*max_delay;' into 'T=(Nc/16)*8*max_delay;' and repeat 3.E-2. Overlay the three simulated BER curves in one figure and capture the plot. Does the BER increase or decrease as **Nc** increases and why?

3.E-6   Judged from the results in 3.E-5, for a given total OFDM bandwidth, will an increase to **Nc** help mitigate the effects of ICI and ISI caused by multipath fading?

REFERENCES

[1] T. S. Rappaport, *Wireless Communications: Principles and Practice*, 2nd ed., Upper Saddle River, NJ: Prentice Hall, 2002.

[2] D. Tse and P. Viswanath, *Fundamentals of Wireless Communication*, Cambridge, UK: Cambridge University Press, 2005.

[3] J. G. Proakis, *Digital Communications*, 5th ed., New York, NY: McGraw-Hill, 2008.

[4] A. R. S. Bahai and B. R. Saltzberg, *Multi-Carrier Digital Communications: Theory and Applications of OFDM*, Alphen aan den Rijn, Netherlands: Kluwer, 1999.

[5] R. Prasad, *OFDM for Wireless Communications Systems*, London: Artech House, 2004.

28

MIMO SYSTEM—PART I: SPACE TIME CODE

- Investigate the structure of Alamouti code, one of the most commonly used space time codes.
- Implement Alamouti code and verify its performance through the simulation.
- Analyze the rates and the diversity orders of various space time block codes.

28.1 SYSTEM MODEL

1.A [WWW]Received signal model for the single transmit antenna system.

The m-file below simulates the received signal in a slow and frequency flat Rayleigh fading channel. For the last five lines in bold, explain what the variable on the left-hand side represents and justify how the right-hand side expression is formulated accordingly.

```
clear
EsN0= ; %Es/N0; set it to a desired value
s= ;
% Create the complex symbols according to the modulation method, for example
% in case of QPSK, s=sign(rand-0.5)+j*sign(rand-0.5), and in case of BPSK,
s=sign(rand-0.5);
```

Problem-Based Learning in Communication Systems Using MATLAB and Simulink, First Edition.
Kwonhue Choi and Huaping Liu.
© 2016 The Institute of Electrical and Electronics Engineers, Inc. Published 2016 by John Wiley & Sons, Inc.
Companion website: www.wiley.com/go/choi_problembasedlearning

```
Es=abs(s)^2;
N0=Es/EsN0;
h=sqrt(1/2)*(randn+j*randn);
n=sqrt(N0/2)*(randn+j*randn);
r=h*s+n;
```

NOTE: If a BER simulation includes creating the received waveform and calculating the decision variable from the received waveform as done in Chapters 21, 22, 23, 26, and 27, then we call it "waveform level BER simulation." On the contrary, if a BER simulation starts with the decision variable model and skips the process of generating the received signal waveform as done in Chapters 24, 25, and this chapter, then we call it "decision variable level simulation."

1.B [T]System with two transmit antennas.

The complex Gaussian variable **h** in the m-file in 1.A is the fading gain. As discussed in Section 25.1 of Chapter 25, **abs(h)** has the Rayleigh distribution. Also the multiplicative distortion—the desired signal is multiplied by **h** at the receiver—as implemented in the last line of the m-file above degrades the BER performance significantly. The discussion in Chapter 25 has shown that spatial diversity is a very effective way to mitigate the effects of fading. However, implementing multiple antennas at both sides of a communications link is not necessarily feasible for many applications. If one side of the link must be simple and have only one antenna (e.g., the mobile side), but the other side can have multiple antennas, how to realize spatial diversity at the one-antenna terminal?

Let us consider a system where the transmitter has two antennas, one named "antenna A" and the other "antenna B," and the receiver has only one antenna. The goal is to design a scheme to provide spatial diversity at the receiver. Let **hA** and **hB** be two independent complex Gaussian random variables that denote the Rayleigh fading coefficients from antenna A and antenna B to the receive antenna, respectively.

There are a number of ways to use the two transmit antennas. Next we consider two simple methods.

1.B-1   The first scheme is described in Table 28.1, where T1, T2, … denote the transmission time slots and **s1, s2, s3**, … are the data symbols to be transmitted. The transmission delay from the transmitter to the receiver is neglected without affecting the description of the transmission and reception processes.

TABLE 28.1   Space Time Symbol Mapping Method I.

Space	Time				
	T1	T2	T3	T4	…
Antenna A	s1	s2	s3	s4	…
Antenna B	s1	s2	s3	s4	…

At T1, the received signal is the sum of **hA*s1** and **hB*s1**.

(a) The m-file below simulates the received signal for the method described in Table 28.1. Justify the use of the scaling factor 2 in the fourth line '**Es=2*abs(s1)^2;**' that calculates the transmitted energy per symbol.

```
clear
EsN0= ; %Es/N0, set to a desired value
s1= ;   % Create any symbol randomly according to the modulation method
Es=2*abs(s1)^2; N0=Es/EsN0;

hA=sqrt(1/2)*(randn+j*randn);
hB=sqrt(1/2)*(randn+j*randn);
n=sqrt(N0/2)*(randn+j*randn);
sA=s1; % antenna A transmit signal
sB=s1; % antenna B transmit signal
r=hA*sA+hB*sB+n;
```

(b) The BER performance cannot be improved by using the method in Table 28.1 compared with the single antenna system discussed in 1.A. Prove that the diversity order of this scheme is still 1 even though both transmit antennas are used in the transmission process.

HINT: You may rewrite the received signal as '**r=(hA+hB)/sqrt(2)*s +n**', and then determine the distribution of the effective fading coefficient '**(hA+hB)/sqrt(2)**'.

1.B-2   The diversity gain can be realized only if multiple independent observations of the same transmitted signals are available at the receiver, like the receive diversity case described by equation (25.8) in Chapter 25. The second scheme is described in Table 28.2.

(a) The code fragment below generates the received signal for the method described in Table 28.2. Since the same transmitted data symbol is received independently twice over two adjacent time slots, one from antenna A and one from antenna B, these two observations of the same data symbol should be combined before detection. Complete the last line of the code fragment to form the decision variable by employing MRC (which was discussed in Chapter 25).

TABLE 28.2   Space Time Symbol Mapping Method II.

Space	T1	T2	T3	T4	
			Time		
Antenna A	s1	No signal	s2	No signal	...
Antenna B	No signal	s1	No signal	s2	...

```
.
.
.
hA=sqrt(1/2)*(randn+j*randn);
hB=sqrt(1/2)*(randn+j*randn);
n1=sqrt(N0/2)*(randn+j*randn); %received noise  at T1
n2=sqrt(N0/2)*(randn+j*randn); %received noise  at T2
r1=hA*s1+n1; % received signal at T1
r2=hB*s1+n2; % received signal at T2
z=conj(?)*r1+conj(?)*r2; % decision variable after MRC.
```

(b) Substitute the expressions of **r1** and **r2** into the expression of **z** and show that the signal scaling factor in the decision variable **z** is **abs(hA)^2 +abs(hB)^2**.

(c) Also show that the noise term in **z** is **conj(hA)*n1+conj(hB)*n2** and that it has a zero mean and instantaneous variance equal to **(abs(hA)^2+abs(hB)^2)*N0**.

(d) From (b) and (c), show that the instantaneous E_b/N_0 with the decision variable **z** obtained from MRC is **(abs(hA)^2+abs(hB)^2)*EbN0**.

(e) From the instantaneous E_b/N_0, show that the method in Table 28.2 achieves a diversity order of 2.

(f) However, the method in Table 28.2 has a major disadvantage compared with the single transmit antenna system discussed in 1.A in terms of bandwidth efficiency. Explain why.

28.2 ALAMOUTI CODE

Alamouti code [1] is the simplest and most widely used space time block code (STBC) [1–5]. This method transmits the data symbol pairs as shown in Table 28.3. For proper operation of Alamouti code, the fading coefficients **hA** and **hB** of the two transmission antennas should remain the same over at least two consecutive time slots.

2.A   Let **s1** and **n1** denote, respectively, the signal and the noise terms received at the first time slot T1, and let **s2** and **n2** denote, respectively, the signal and the noise terms received at the second time slots T2. Then, **r1** and **r2** can be created as shown below.

TABLE 28.3   Space Time Symbol Mapping of Alamouti Code.

	Time						
Space	T1	T2	T3	T4	T5	T6	...
Antenna A	s1	conj(s2)	s3	conj(s4)	s5	conj(s6)	...
Antenna B	s2	-conj(s1)	s4	-conj(s3)	s6	-conj(s5)	...

```
s1= ;
s2= ;
.

.

.
hA=sqrt(1/2)*(randn+j*randn);
hB=sqrt(1/2)*(randn+j*randn);
n1=sqrt(N0/2)*(randn+j*randn); %Received noise at T1.
n2=sqrt(N0/2)*(randn+j*randn); %Received noise at T2.
r1=hA*s1+hB*s2+n1; % received signal at T1
r2= ? ; % received signal at T2
```

Complete the last line to create **r2** and capture the competed line.

2.B [A]The receiver is assumed to have perfect channel state information (CSI) (the channel gains **hA** and **hB**). The data symbol pair **s1** and **s2** are estimated by using **r1**, **r2, hA**, and **hB**. Next we decode Alamouti code using MLD on the basis of exhaustive search implemented through the following steps:

Step 1. Create a data symbol pair (**s1, s2**) and the fading gains (**hA, hB**) from the two transmit antennas to the receive antenna.

Step 2. Perform Alamouti coding on the data symbol pair (**s1, s2**) and create the received signals **r1** and **r2**.

Step 3. Prepare all possible data symbol pairs (candidate pairs) of (**s1, s2**) for exhaustive MLD. For example, for binary phase shift keying (BPSK), the pairs are $\{(1,1), (1,-1), (-1,1), (-1,1)\}$.

Step 4. For each candidate pair, perform Step 4-1 and Step 4-2.

Step 4-1. Perform Alamouti coding on the current candidate data symbol pair and construct the virtual received signal without noise.

Step 4-2. Calculate the Euclidean distance between the received signal (**r1, r2**) in Step 2 and the virtual received signal constructed in Step 4-1.

Step 5. Among all candidate data symbol pairs, select the candidate pair that has the smallest Euclidean distance with the estimated data symbol pair.

2.B-1   Review the MLD concept and justify why Step 4 and Step 5 above implement the MLD process.

2.B-2   [WWW]The m-file below simulates the BER of a BPSK system with Alamouti STBC.

(a) Identify the code lines that correspond to each of the steps described above.
   Complete the places marked by '?'. For each of the lines in bold, add a comment to explain what the variable on the left-hand side represents and justify how the right-hand side expression is properly formulated accordingly. Capture the completed m-file.

```
clear
EbN0dB_vector=[0 3 6 9 12 15]; % Set EbN0 values in dB.
Eb=2; %Total bit energy for two time slots T1 and T2. Refer to Table 28.3.

for snri=1:length(EbN0dB_vector)
    EbN0dB=EbN0dB_vector(snri);
    EbN0=10^(EbN0dB/10);
    N0=Eb/EbN0;
    errcnt=0; symcnt=0;
    while errcnt<500 %Decrease errcnt limit(=500) if the simulation takes long time but
the result will be less accurate.

        s1=sign(rand-0.5);
        s2=sign(rand-0.5);

        %%%Creation of fading coefficients and noise %%%%%%%%%%%%%%
        hA=sqrt(1/2)*(randn+j*randn);
        hB=sqrt(1/2)*(randn+j*randn);
        n1=sqrt(N0/2)*(randn+j*randn);   %Received noise at T1.
        n2=sqrt(N0/2)*(randn+j*randn);   %Received noise at T2.
        %%%%%%%%%%%%%%%%%%%%%%%%%%%%%%%%%%%%%%%%%%%%

        %%%%Generating Alamouti coded received signal%%%%%%%%%%%%%
        r1=hA*s1+hB*s2+n1;  %received signal at T1.
        r2=?;          %received signal at T2. Refer to Table 28.3
        %%%%%%%%%%%%%%%%%%%%%%%%%%%%%%%%%%%%%%%%%%%%%%%%

        %%%%%%ML decoding %%%%%%%%%%%%%%%%%%%%%%%%%%%%%%%%%%%
        s1s2pairCandidate=[[1,1];[1,-1];[-1,1];[-1,-1]];
        N_set=4; %Number of candidates
        for k_set=1:N_set
            s1candidate=s1s2pairCandidate(k_set,1);
            s2candidate=s1s2pairCandidate(k_set,?);
            r1candidate=hA*s1candidate+hB*s2candidate;
        %Create a virtual r1 assuming that (s1, s2)=(s1candidate, s2candidate) and there
is no noise.
            r2candidate=? ;

            distance(k_set)=sum(abs([r1, r2]-[r1candidate,r2candidate]).^2);
        end
        [A B]= min( ? );
        s1_hat=s1s2pairCandidate(B,1);
        s2_hat=s1s2pairCandidate(B,2);
        %%%%%%%%%%%%%%%%%%%%%%%%%%%%%%%%%%%%%%%%%%%%%%
```

```
    if(s1_hat~=s1)
        errcnt=errcnt+1;
    end
    if(s2_hat~=s2)
        errcnt=errcnt+1;
    end
    symcnt=symcnt + 2;
  end

  BER(snri)=errcnt/symcnt;
  save alamouti_ML_BER.mat EbN0dB_vector BER
end
figure
semilogy( EbN0dB_vector, BER);
title('Alamouti coded, BPSK, Rayleigh fading');
xlabel('Eb/N0 [dB');ylabel('BER');
grid on
```

2.B-3   Explain why the part **'creation of the fading coefficients and the noise'** should not be placed outside of the **'while'** loop.

2.B-4   In an actual system, the streams of symbol pairs are continuously transmitted as illustrated in Table 28.3. However, in the m-file of 2.B-2, a loop is used to repeatedly transmit one symbol pair (**s1**, **s2**) at a time. Justify why we do not need to explicitly simulate other symbol pairs (**s3**, **s4**), (**s5**, **s6**), and so on.

2.B-5   Execute the completed m-file and capture the simulated BER plot.

2.B-6   Recall that the m-file completed in Section 4.C of Chapter 25 simulates the BER of MRC in Rayleigh fading channels. Open that m-file and set **L = 2**. Execute the m-file. Overlay the simulated BER of MRC with **L = 2** on the BER graph in 2.B-5. (The process to copy curves displayed in different figures generated in MATLAB and overlay them in one figure was discussed in 4.C-1 of Chapter 24.)

   (a) Capture the resulting BER graph.
   (b) Compare the two BER curves and check whether or not the BER of Alamouti code with MLD is equal to that of two-branch (**L**=2) receive diversity with MRC.

28.3   SIMPLE DETECTION OF ALAMOUTI CODE

3.A   One of the excellent features of Alamouti code is that detection based on a decision variable obtained by linearly combining the received signal pair (**r1**, **r2**) achieves the MLD performance. Let **z1** and **z2** denote the decision variables for

s1 and **s2**, respectively. They can be created by linearly combining **r1** and **r2** [1].

```
z1=conj(hA)*r1 - hB* conj(r2);
z2=conj(hB)*r1 + hA* conj(r2);
```

3.A-1 [T]Substitute the expressions for **r1** and **r2** completed in the m-file in 2.A into the expressions of **z1** and **z2** above.

(a) Rearrange the two expressions so that **s2** does not appear in the expression **z1** and **s1** does not appear in the expression **z2**.

(b) Show that the scaling terms for **s1** in **z1** and for **s2** in **z2** are both equal to '**(abs(hA)^2+abs(hB)^2)**'.

(c) Show that the noise terms in **z1** and **z2** have a zero mean and variance equal to '**(abs(hA)^2+abs(hB)^2)*N0**'.

(d) From (b) and (c), show that the instantaneous E_b/N_0 values in **z1** and **z2** are '**(abs(hA)^2+abs(hB)^2)*EbN0**'.

(e) Based on the instantaneous E_b/N_0 for **z1** and **z2**, show that the linear decoding process given above achieves the same performance as MRC with a diversity order **L** = 2.

(f) In 2.B-6(b), we showed that Alamouti code with MLD achieves the same performance as MRC with **L** = 2. From the results so far, does detection based on the decision variables obtained by linearly combining the received signals achieve the MLD performance? Justify your answer.

3.A-2 [T]Compare the instantaneous E_b/N_0 values of Alamouti code and the method described in Table 28.2, which were derived in 1.B-2(b)–1.B-2(e).

(a) Based on this comparison, assess the relative error performances of these two schemes.

(b) An Alamouti code is bandwidth more efficient than the scheme described in Table 28.2. Derive the relative bandwidth efficiency of the two schemes.

3.B [WWW]The m-file below simulates the BER with the simple linear combining method introduced in 3.A.

```
clear
EbN0dB_vector=[0 3 6 9 12 15]; % Set EbN0 values in dB.
Eb=2; %Total bit energy for two time slots T1 and T2. Refer to Table 28.3.

for snri=1:length(EbN0dB_vector)
  EbN0dB=EbN0dB_vector(snri);
  EbN0=10^(EbN0dB/10);
  N0=Eb/EbN0;
  errcnt=0; symcnt=0;
```

```
  while errcnt<500 %Decrease errcnt limit(=500) if the simulation takes long time but
the result will be less accurate.
    s1=sign(rand-0.5);
    s2=sign(rand-0.5);

    %%%Generating fading coefficient and noise %%%%%%%%%%%%%%
    hA=sqrt(1/2)*(randn+j*randn);
    hB=sqrt(1/2)*(randn+j*randn);
    n1=sqrt(N0/2)*(randn+j*randn); %Received noise at T1.
    n2=sqrt(N0/2)*(randn+j*randn); %Received noise at T2.
    %%%%%%%%%%%%%%%%%%%%%%%%%%%%%%%%%%%%%%%%%%%%%%%%

    %%%%%Generating Alamouti coded received signal%%%%%%%%%%%%
    r1=hA*s1+hB*s2+n1; %received signal at T1.
    r2=?;    %received signal at T2. Refer to Table 28.3
    %%%%%%%%%%%%%%%%%%%%%%%%%%%%%%%%%%%%%%%%%%%%%%%%

    %%%%%Simple linear combining for decoding!!!!%%%%%%%%%
    z1= ?; % Decision variable for s1
    z2= ?; % Decision variable for s2
    s1_hat=sign(real(z1));
    s2_hat=sign(real(z2));
    %%%%%%%%%%%%%%%%%%%%%%%%%%%%%%%%%%%%%%%%%%%%%%%%

    if(s1_hat~=s1)
        errcnt=errcnt+1;
    end
    if(s2_hat~=s2)
        errcnt=errcnt+1;
    end
    symcnt=symcnt + 2;
  end

  BER(snri)=errcnt/symcnt;
  save alamouti_BER.mat EbN0dB_vector BER
end
figure
semilogy( EbN0dB_vector, BER);
title('Alamouti coded, BPSK, Rayleigh fading');
xlabel('Eb/N0 [dB');ylabel('BER');
grid on
```

3.B-1   Complete the places marked by '?' and capture the completed m-file.

3.B-2   Compare the computational complexities of the exhaustive MLD implemented in the m-file of 2.B-2 and the simple linear combining method.

3.B-3   Execute the m-file and capture the simulated BER graph. Do not close the figure since it will be needed in 3.B-4.

3.B-4   Execute the m-file completed in 2.B-2 to generate the file **alamouti_ML_BER.mat** in the MATLAB work folder. After this, execute the three lines of code below to plot the BER of exhaustive MLD together with the BER of the simple linear combining scheme.

(a) Capture the resulting BER graph with the two BER curves.
(b) From the two BER curves, summarize the BER performances of MLD and the simple linear combining schemes for Alamouti code.

```
>>load alamouti_ML_BER.mat
>>hold on
>>semilogy( EbN0dB_vector, BER;'r')
```

3.B-5   Modify the m-file in 3.B-1 to extend the modulation method from BPSK to QPSK. (a) Capture the modified part. (b) Execute the modified m-file and capture the simulated BER graph.

3.B-6   [T]Other than the space time mapping given in Table 28.3, there are other mapping schemes that also maintain the Alamouti code properties.

(a) Design one or more of such mapping schemes and record them using the format shown in Table 28.4. Do not consider trivial variations to the mapping in Table 28.3 such as interchange the positions of **s1** and **s2** in time or in space.
(b) For each of the schemes designed in (a), derive its corresponding linear combining rule for the decision variables **z1** and **z2** and compare them with those for Alamouti code.

TABLE 28.4   Different Version of Alamouti Code.

Space	Time	
	T1	T2
Antenna A	?	?
Antenna B	?	?

28.4 [A]VARIOUS STBCs, THEIR DIVERSITY ORDERS, AND THEIR RATES

4.A [T]Alamouti code is a two-dimensional (space time) code. The received signal vector can be written in vector-matrix form as [1]

$$\begin{bmatrix} r_1 \\ r_2 \end{bmatrix} = \begin{bmatrix} ? & ? \\ ? & ? \end{bmatrix} \begin{bmatrix} h_A \\ h_B \end{bmatrix} + \begin{bmatrix} n_1 \\ n_2 \end{bmatrix}. \tag{28.1}$$

4.A-1 Complete the space time code matrix in equation (28.1) so that the expressions of **r1** and **r2** are the same as those implemented in the last two lines of the code fragment in 2.A.

4.A-2 Do the columns of the space time code matrix correspond to different time slots or different space elements (antennas)?

4.A-3 Define the rate R of the space time code as [2–5]

$$R = \frac{\text{Number of different symbols composing the space time code matrix}}{\substack{\text{Number of time slots used for transmitting one space time code} \\ (= \text{Number of rows in the space time code matrix})}}. \tag{28.2}$$

For the single transmit-antenna system, this rate equals 1. Determine the R of Alamouti code.

4.B [T]Now consider an STBC with the following received signal vector:

$$\begin{bmatrix} r_1 \\ r_2 \\ r_3 \\ r_4 \end{bmatrix} = \begin{bmatrix} s_1 & s_2 & s_3 \\ -s_2^* & s_1^* & 0 \\ s_3^* & 0 & -s_1^* \\ 0 & s_3^* & -s_2^* \end{bmatrix} \begin{bmatrix} h_A \\ h_B \\ h_C \end{bmatrix} + \begin{bmatrix} n_1 \\ n_2 \\ n_3 \\ n_4 \end{bmatrix}. \tag{28.3}$$

4.B-1 Determine the number of transmission antennas and the number of time slots in a block.

4.B-2 Determine the rate R of this code.

4.B-3 Determine the diversity order of this code.

4.C [A,WWW]Modify the m-file in 2.B-2 to simulate the BER of the STBC given in equation (28.3) by using exhaustive MLD.

4.C-1 Capture the modified m-file and the simulated BER plot.

4.C-2 Recall that the m-file completed in Section 4.C of Chapter 25 simulates the BER of MRC in Rayleigh fading channels. Open that m-file and set **L** = 3. Execute the m-file. Overlay the simulated BER curve of MRC with **L** = 3 on the BER graph generated in 4.C-1.

(a) Capture the resulting BER graph with both BER curves.

(b) Compare the two BER curves, one for the code expressed in equation (28.3) with MLD and one for MRC with L = 3, and summarize the relative BER performance of these two systems. Then check whether your answer to 4.B-3 is correct.

REFERENCES

[1] S. M. Alamouti, "A Simple Transmit Diversity Technique for Wireless Communications," *IEEE Journal on Selected Areas in Communications*, Vol. 16, No. 8, 1998, pp. 1451–1458.

[2] V. Tarokh, N. Seshadri, and A. R. Calderbank, "Space–Time codes for High Data Rate Wireless Communication: Performance Analysis and Code Construction," *IEEE Transactions on Information Theory*, Vol. 44, No. 2, 1998, pp. 744–765.

[3] V. Tarokh, A. Naguib, N. Seshadri, and A. R. Calderbank, "Space-Time Codes for High Data Rate Wireless Communication: Performance Criteria in the Presence of Channel Estimation Errors, Mobility and Multiple Paths," *IEEE Transactions Communications*, Vol. 47, 1999, pp. 199–207.

[4] V. Tarokh, H. Jafarkhani, and A. R. Calderbank, "Space–Time Block Coding for Wireless Communications: Performance Results," *IEEE Journal on Selected Areas in Communications*, Vol. 17, No. 3, 1999, pp. 451–460.

[5] V. Tarokh, H. Jafarkhani, and A. R. Calderbank, "Space–Time Block Codes from Orthogonal Designs," *IEEE Transactions on Information Theory*, Vol. 45, No. 5, 1999, pp. 744–765.

29

MIMO SYSTEM—PART II: SPATIAL MULTIPLEXING

- Implement detection processes for spatial multiplexing MIMO systems.
- Compare the performances of MIMO system with various detection methods.
- Investigate the performances of MIMO system with different system parameters.

29.1 MIMO FOR SPATIAL MULTIPLEXING

1.A System model.

MIMO systems may be broadly classified into two categories: spatial diversity (SD) systems and spatial multiplexing (SM) systems [1–3]. SD MIMO systems achieve diversity. Some simple SD MIMO systems were introduced in Chapter 28. SM MIMO systems exploit the multiple transmit and receive antennas to transmit multiple data streams simultaneously in the same frequency band to increase the spectral efficiency. Some of the existing literature defines only SM MIMO systems as MIMO systems.

In this chapter we study SM MIMO systems. Since the multiple streams of data are transmitted simultaneously in the same frequency band, there exists mutual interference among these data streams.

Let N_T and N_R denote the number of transmit (TX) and receive (RX) antennas, respectively. The signals received by the N_R RX antennas, written as an $N_R \times 1$ column vector $\mathbf{r}$, can be expressed as

$$\mathbf{r} = \mathbf{Hs} + \mathbf{n}, \tag{29.1}$$

Problem-Based Learning in Communication Systems Using MATLAB and Simulink, First Edition.
Kwonhue Choi and Huaping Liu.
© 2016 The Institute of Electrical and Electronics Engineers, Inc. Published 2016 by John Wiley & Sons, Inc.
Companion website: www.wiley.com/go/choi_problembasedlearning

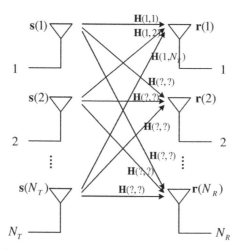

FIGURE 29.1   Fading coefficient diagram for SM MIMO.

where **s** is the $N_T \times 1$ data symbol vector whose kth element, **s**(k), denotes the data symbol transmitted from the kth TX antenna, **n** is an $N_R \times 1$ noise vector whose elements are complex Gaussian random variables, and **H** is the $N_R \times N_T$ fading coefficient (or channel gain) matrix. The model of the elements of **H** will be discussed next.

1.A-1   [T]From equation (29.1), the received signal at the kth RX antenna, **r**(k), can be expressed as

$$\mathbf{r}(k) = \mathbf{H}(?,?)\mathbf{s}(1) + (?,?)\mathbf{s}(?) + \cdots + (?,?)\mathbf{s}(N_T) + \mathbf{n}(?). \qquad (29.2)$$

Complete equation (29.2).

1.A-2   [T]From the model given by equation (29.1), the (n,m)th element of the channel matrix **H**, **H**(n,m), is the fading coefficient for the link from the ?-th TX antenna to the ?-th RX. Complete the two quantities marked by '**?**'.

1.A-3   [T]Fig. 29.1 shows the fading coefficient diagram from the TX antennas to the RX antennas. Complete all the places marked by '**?**' on the basis of equation (29.2).

29.2   MLD BASED ON EXHAUSTIVE SEARCH FOR SM MIMO

In this section, for convenience, whenever it does not cause a confusion, the variables/expressions we use in MATLAB and in the texts such as the number of RX antennas (N_R ,**Nr**), the received signal vector (**r**, r), the noise vector (**n**,n), and so forth, will be used interchangeably.

2.A [WWW]The m-file **ml_Nt4_bpsk.m** below is a user-defined MATLAB function that performs MLD (exhaustive search based) of an SM MIMO BPSK system with $N_T = 4$. The number of RX antennas N_R could be any nonnegative integer. This function takes the $N_R \times 1$ received signal vector **r** and the $N_R \times 4$ complex channel coefficient matrix **H** as its inputs and outputs **s_hat**, the estimate of the $N_T \times 1$ transmitted symbol vector s.

```
function s_hat=ml_Nt4_bpsk(r,H)

s_candidate_set=[];
dist_set=[];

for b1_candidate=[-1 1]
   for b2_candidate=[-1 1]
      for b3_candidate=[-1 1]
         for b4_candidate=[-1 1]
            s_candidate=[b1_candidate b2_candidate b3_candidate b4_candidate]';
            r_candidate=H*s_candidate;
            dist= sum(abs(? - r_candidate).^2);

            s_candidate_set=[s_candidate_set s_candidate];
            %Add s_candidate as a new column of s_candidate_set
            dist_set=[dist_set dist];
            %Add dist as an new element of dist_set
         end
      end
   end
end

[A B]=min(dist_set);
s_hat=s_candidate_set(:,B);
```

2.A-1   Determine the variable names that should replace '?' left in the code and justify it.

2.A-2   For each of the lines that contain '=', add a comment to explain what the variable on the left-hand side represents and justifies how the right-hand side expression is formulated accordingly. Capture the completed m-file with the comments.

2.B   Extension to the different number of TX antennas.

2.B-1   [WWW]We can extend the m-file in 2.A to the case of $N_T = 3$ as shown below. Complete the m-file and save it as **ml_Nt3_bpsk.m**. Compare **ml_Nt3_bpsk.m** with **ml_Nt4_bpsk.m** to find all modified or deleted lines and explain why these lines should be modified or deleted.

```
function s_hat=ml_Nt3_bpsk(r,H)

s_candidate_set=[];
dist_set=[];

for b1_candidate=[-1 1]
   for b2_candidate=[-1 1]
      for b3_candidate=[-1 1]
            s_candidate=[b1_candidate b2_candidate b3_candidate]';
            r_candidate=H*s_candidate;
            dist= sum(abs(?-r_candidate).^2);

            s_candidate_set=[s_candidate_set s_candidate];
            %Add s_candidate as a new column of s_candidate_set
            dist_set=[dist_set dist];
            %Add dist as an new element of dist_set
      end
   end
end

[A B]=min(dist_set);
s_hat=s_candidate_set(:,B);
```

2.B-2   Similarly, modify the m-file for the case of $N_T = 2$. Save the modified m-file as **ml_Nt2_bpsk.m**. Capture **ml_Nt2_bpsk.m**.

2.C [WWW]Modify **ml_Nt4_bpsk()** as follows in order to extend to the case of quadrature phase shift keying (QPSK). Save the modified m-file as **ml_Nt4_qpsk()**. Explain why the modified lines should be modified so.

```
function s_hat=ml_Nt4_qpsk(r,H)
...
for b1_candidate=[-1-j, -1+j, 1-j, 1+j]
   for b2_candidate=[-1-j, -1+j, 1-j, 1+j]
      for b3_candidate=[-1-j, -1+j, 1-j, 1+j]
         for b4_candidate=[-1-j, -1+j, 1-j, 1+j]
            ...
         end
      end
   end
end
...
```

2.D Complexity of maximum likelihood detection (MLD).

2.D-1 In the nested 'for' loops of **ml_Nt4_bpsk.m** in 2.A, count the total number of iterations needed to generate a new candidate (hypothesis) of the received signal and to calculate the distance between the candidate and the received signal.

2.D-2 In the nested 'for' loops of **ml_Nt4_qpsk.m** in 2.C, count the total number of iterations needed to generate a new candidate (hypothesis) of the received signal and to calculate the distance between the candidate and the received signal.

2.D-3 Generalize the results in 2.D-1 and 2.D-3 into an $N_R \times N_T$ MIMO system that employs a modulation with an alphabet size M.

2.D-4 Summarize the problems using MLD based on exhaustive search.

29.3 ZERO FORCING DETECTION

For systems with $N_R \geq N_T$, zero forcing (ZF) detection is a method to multiply the received signal $\mathbf{r}$ by the pseudo-inverse of the channel matrix $\mathbf{H}$ to generate the decision variables in a vector $\mathbf{z}$ for the N_T simultaneously transmitted symbols:

$$\begin{aligned}
\mathbf{z} &= (\mathbf{H}^T\mathbf{H})^{-1}\mathbf{H}^T \times \mathbf{r} \\
&= (\mathbf{H}^T\mathbf{H})^{-1}\mathbf{H}^T \times (\mathbf{H}\mathbf{s} + \mathbf{n}) \\
&= \mathbf{s} + (\mathbf{H}^T\mathbf{H})^{-1}\mathbf{H}^T\mathbf{n},
\end{aligned} \tag{29.3}$$

where $(\cdot)^T$ denotes transpose and $(\mathbf{H}^T\mathbf{H})^{-1}\mathbf{H}^T$ is the pseudo-inverse of $\mathbf{H}$; if $\mathbf{H}$ is a full-rank square matrix, then it equals the inverse of $\mathbf{H}$. Note that the signal term in $\mathbf{z}$ is equal to $\mathbf{s}$, which shows that ZF detection completely eliminates the mutual interference among the simultaneously transmitted symbols in forming the decision variables.

3.A [WWW]The user-defined MATLAB function **zf_bpsk( )** below implements ZF detection for BPSK MIMO systems. It takes the received signal vector r and channel matrix $\mathbf{H}$ as its inputs. The function **pinv(·)** computes the pseudo-inverse of a matrix. Complete the second line and save the m-file as **zf_bpsk.m**. Capture the completed m-file.

```
function s_hat=zf_bpsk(r,H)
z=pinv(?)*r;
s_hat=sign(real(z));
```

3.B Explain why **real()** and **sign()** are used in the last line.

3.C [WWW]Modify **zf_bpsk.m** for QPSK simulation as shown below. Complete the second and third lines and save the m-file as **zf_qpsk.m**. Capture the completed m-file.

```
function s_hat=zf_qpsk(r,H)
z=pinv(?)*r;
s_hat=sign(real(z))+ j * ??;
```

29.4   NOISE ENHANCEMENT OF ZF DETECTION

4.A   [WWW]In the ZF detection expressed by equation (29.3), the data symbol vector s is recovered exactly for any invertible channel matrix **H**. However, the noise term after ZF becomes pinv(**H**)**n**. Therefore the performance of ZF detection will be determined by the variance of the elements of the vector pinv(**H**)**n**.

The variance of the elements of pinv(**H**)**n** depends on the condition of **H**. Let us compare the variances of the elements of pinv(**H**)**n** for the following two examples of **H**, expressed as Ha and Hb.

```
>> Ha=[2, 1; 1, 2]
Ha =
   2   1
   1   2

>> Hb=[1, 1; 1, -1]
Hb =
   1   1
   1  -1
```

4.A-1   Since the channel matrices **Ha** and **Hb** are 2 by 2 matrices, the received vector **n** is a 2(=**Nr**) × 1 vector and the noise vector after ZF is also a 2 by 1 vector. Substituting **Ha** given above into **pinv(H)*n**, we can express the two elements of **pinv(H)*n** as the linear combination of **n(1)** and **n(2)**. For example, because **pinv(Ha) =** $(\text{Ha})^{-1} = \begin{bmatrix} 2/3 & -1/3 \\ -1/3 & 2/3 \end{bmatrix}$, the first element of the vector **pinv(Ha)*n** is expressed as **(2/3)*n(1) +(-1/3)*n(2)**. Similarly, express the second element of the vector **pinv(Ha)*n**, the first and second elements of the vector **pinv(Hb)*n**, as the linear combination of **n(1)** and **n(2)**.

4.A-2   The two elements **n(1)** and **n(2)** are i.i.d. complex Gaussian random variables. Let **v_n** denote the variance of **n(1)** and **n(2)**. Based on the expressions obtained in 4.A-1 and using the formula in equation (15.6), show that the variance of elements of **pinv(Ha)*n** is equal to **5/9*v_n** and the variance of elements of **pinv(Ha)*n** is equal to **1/2*v_n**.

4.A-3   According to 4.A-2, for ZF detection, does channel **Ha** or **Hb** result in a lower BER? Justify the answer.

4.A-4   According to 4.A-3, for ZF detection, the channel **Ha** results in a higher BER compared with the channel **Hb** although the elements of **Ha** are greater than the elements of **Hb**. This phenomenon is called "noise enhancement." Explain what kind of channel matrices cause a high noise enhancement.

4.A-5   When the channel matrix is ill-conditioned (like **Ha**), ZF will result in a high noise. Give another example of an ill-conditioned channel matrix that causes an even higher noise enhancement than the channel matrix **Ha**, that is, a channel matrix whose elements' magnitudes are greater than those of **Ha** but will result in a worse BER performance if ZF is used. Mathematically justify your example.

4.B   [WWW]The m-file below simulates the BERs of ZF detection and MLD assuming that the channel matrix **H** is equal to the **Ha** given in 4.A.

```
clear
Nr=2;
Ha=[2, 1;1, 2];
Hb=[1, 1;1, -1];

EsN0dB=3;
EsN0=10^(EsN0dB/10);
Es=Nr*1;
N0=Es/EsN0;
ErrCntZF=0; ErrCntML=0;SymCnt=0;
while ErrCntML<500
%Change 500 to a smaller value to speed up simulation. However, the simula-
tion error will increase instead.
    s=sign(rand(2,1)-0.5);
    H=Ha; % Change to H=Hb in case of simulating the BER of the channel Hb
    n=sqrt(N0/2)*(randn(2,1)+j*randn(2,1));
    r=H*s+n;

    s_hat_zf=zf_bpsk(r,H); % zf_bpsk() should be in the same folder.
    ErrCntZF=ErrCntZF+sum(s~=s_hat_zf);

    s_hat_ml=ml_Nt2_bpsk(r,H); % ml_Nt2_bpsk() should be in the same folder.
    ErrCntML=ErrCntML+sum(s~=s_hat_ml);

    SymCnt=SymCnt+2;
end
ber_zf=ErrCntZF/SymCnt
ber_ml=ErrCntML/SymCnt
```

4.B-1   For each of the lines in bold, explain what the variable on the left-hand side represents and justify how the right-hand side expression is properly formulated accordingly.

4.B-2   (a) Execute the m-file and capture the BER results.
   (b) Change the line '**H=Ha**' in the m-file into '**H=Hb**'. Execute the m-file and capture the results.

4.C   If the columns of the channel matrix are orthogonal, then ZF achieves the same performance as MLD. (a) Do the results in 4.B-2 confirm this fact? (b) [A]Explain why ZF achieves the same performance as MLD if the columns of the channel matrix are mutually orthogonal.

4.D   Based on the results obtained in 4.B-2, compare the BERs of ZF detection for the two example channel matrices **Ha** and **Hb**. Are the BER results consistent with the noise variance analysis result in 4.A-2 and 4.A-3?

4.E   In the m-file, set the channel matrix **H** to the one selected in 4.A-5 and execute the modified m-file. Capture the results and determine whether your answer to 4.A-5 is correct.

29.5   SUCCESSIVE INTERFERENCE CANCELLATION DETECTION

The successive interference cancellation (SIC) detection scheme for SM MIMO typically involves the following steps [1, 2].

Step 1: Multiply the received signal vector by a linear detection matrix to create the decision variable vector **z**. For example, if the linear ZF scheme is applied, then the linear detection matrix would be **pinv(H)**.

```
z=pinv(H)*r;
```

Step 2: From **z**, detect only the N_T-th symbol (the corresponding MATLAB variable is **s(Nt)**) and store the detection result.

```
s_hat(Nt) =sign(real(z(Nt))); % for BPSK
s_hat(Nt) =sign(real(z(Nt)))+j*sign(imag(z(Nt))); % for QPSK
```

Step 3: Assuming a correct decision in Step 2, that is, assuming that **s_hat(Nt)** is equal to **s(Nt)**,

Step 3-1. Reconstruct the signal portion for **s(Nt)** in the received signal vector **r**. Refer to the note below for this step.

```
H(:,Nt)*s_hat(Nt);;
```

NOTE: The desired signal in the received vector **r** given by equation (29.1) can be decomposed as

$$\mathbf{r} = \mathbf{H}(:,1) \times s(1) + \mathbf{H}(:,2) \times s(2) + \cdots + \mathbf{H}(:,N_T) \times s(N_T), \qquad (29.4)$$

where $\mathbf{H}(:,k)$ denotes the kth column vector of $\mathbf{H}$. From this decomposition, we note that the contribution of $s(k)$ in r is $\mathbf{H}(:,k) \times s(k)$.

Step 3-2. Since the signal portion that carries **s(Nt)** in the received signal vector **r** acts as the interference to the remaining **Nt**-1 symbols, we cancel this interference from the received signal. Hence we replace the received vector **r** as follows.

```
r=r-H(:,Nt)*s_hat(Nt);
```

Step 4: At this point, to the other **Nt**-1 data streams the updated **r** in Step 3-2 is equivalent to the received signal as if the **Nt**-th TX antenna did not transmit any signal, effectively reducing the system to an **Nr** × (**Nt**-1) MIMO system. Now we replace **H** and **Nt** as follows.

```
H=H(:, 1:(Nt-1));
Nt=Nt-1;
```

Step 5: Repeat Steps 1–4 with r updated in Step 3-2, and **H** and **Nt** updated in Step 4 until all **Nt** symbols are detected.

Step 6: Return **s_hat** whose elements are successively stored in Step 2.

5.A [WWW]The m-file **sic_zf_bpsk.m** below is a user-defined MATLAB function that performs SIC for the received signal vector **r** and the channel matrix **H** assuming BPSK modulation.

```
function s_hat=sic_zf_bpsk(r,H)

[Nr Nt]=size(H);
s_hat=zeros(Nt,1); %To initialize s_hat.

while Nt>0
   z=pinv(H)*r;
   s_hat(Nt)=sign(real(z(Nt)));
   r=r-H(:,Nt)*s_hat(Nt);   % or r = r-H*[0 0 0....s_hat(Nt)]';
   H=H(:,1:(Nt-1));
   Nt=Nt-1;
end
```

5.A-1 Add a detailed explanation to each line as a comment. Refer to the steps explained above.

5.A-2 [WWW]The function below extends the function **sic_zf_bpsk.m** so that it works with QPSK modulation, for which the data symbols candidates are $\{-1\text{-}j, -1\text{+}j, 1\text{-}j, 1\text{+}j\}$. For this extension, only the line that computes **s_hat** (the 6th line) needs to be changed. Complete this line and save the m-file as **sic_zf_qpsk()**. Capture the completed m-file.

```
function s_hat=sic_zf_qpsk(r,H)

[Nr Nt]=size(H);
s_hat=zeros(Nt,1);

while Nt>0
  z=pinv(H)*r;
  s_hat(Nt)=sign(real(z(Nt))) + j* ? ;
  r=r- H(:,Nt)*s_hat(Nt); %or r=r-H*[zeros(Nt-1, 1); s_hat(Nt)];
  H=H(:,1:(Nt-1));
  Nt=Nt-1;
end
```

5.B Ordered SIC.

5.B-1 Note that inside the '**while**' loop in the m-file, we substitute **r=H*s+n** into **z=pinv(H)*r** to generate the ZF decision variable vector **z**, resulting in **z=s+pinv(H)*n**, where **pinv(H)*n** is the noise vector. Let **Hinv** denote the matrix **pinv(H)** and **v_n** denote the variance of elements of the received noise vector **n**. Each element of the noise vector after ZF, **Hinv*n**, is a linear combination of the elements of **n**, where the weights depend on the channel matrix. Specifically, the **k**-th element of **Hinv*n** is a linear combination of the elements of **n** with weights equal to **Hinv(k,:)**, the **k**-th row of **Hinv**. Analyzing it a bit further, we can show that the variance of the **k**-th element of **Hinv*n** is equal to **sum(Hinv(k,:).^2) *v_n**. Prove this relationship.

5.B-2 According to 5.B-1, after ZF, the instantaneous variances of the effective noise to the Nt simultaneously transmitted symbols, **s(k)**, $\cdots$, **s(Nt)**, are **sum(Hinv(k,:).^2)*v_n**, k=1, $\cdots$, **Nt**, respectively. Apparently, the instantaneous noise variances for the **Nt** data symbols are different, depending on the instantaneous channel matrix. Therefore it is possible to improve the performance of the SIC scheme described above by switching the antenna indexes between the symbol with the smallest noise component and the original symbol at the **Nt**-th antenna, allowing the detection to start with the data symbol that has the smallest noise component. Such a scheme is called an ordered successive interference cancellation (OSIC) scheme [1–3].

The code fragment below implements this reordering process. Explain in detail how the variable **B** is set to the index of the element with the minimum variance among all the elements of the noise vector **pinv(H)*n**.

```
sym_index=1:Nt;
Hinv=pinv(H);
T=sum(abs(Hinv').^2);%sum(A) is the column vector which takes the sum of each row
of the matrix A as its elements.
[A B]=min(T);

C=H(:,Nt);  H(:,Nt)=H(:,B); H(:,B)=C;
D=sym_index(Nt); sym_index(Nt)=sym_index(B);sym_index(B)=D;
```

5.B-3   The last line '**D=sym_index(Nt); sym_index(Nt)=sym_index(B);sym_index (B)=D;**' is for storing the original symbol index in the rearranged symbol order. Let us assume that **Nt** = 4 and the second symbol in **z** has the minimum noise variance before we execute the code lines in 5.B-2. Determine **sym_index** after executing these lines.

5.B-4   Equation (29.4) shows that if we rearrange the order of the symbols in vector s on the basis of the variances of the noise component associated with each symbol, then the columns of the channel matrix **H** must be rearranged accordingly. The line '**C=H(:,Nt); H(:,Nt)=H(:,B); H(:,B)=C;**' accomplishes this. Explain how this line works in detail.

5.B-5   [WWW]The m-file **osic_zf_bpsk.m** below is a user-defined MATLAB function that implements OSIC given the received signal vector r assuming BPSK modulation and the channel matrix **H**. Compare this m-file with **sic_zf_bpsk.m** completed in 5.A for nonordered SIC. For the two lines in bold, explain what they do and in what way they differ from the two corresponding lines in **sic_zf_bpsk.m** completed in 5.A.

```
function s_hat=osic_zf_bpsk(r,H)

[Nr Nt]=size(H);
sym_index=1:Nt;

while Nt>0
    Hinv=pinv(H);
    T=sum(abs(Hinv').^2);
    [A B]=min(T);

    C=H(:,Nt); H(:,Nt)=H(:,B); H(:,B)=C;
    D=sym_index(Nt); sym_index(Nt)=sym_index(B);sym_index(B)=D;

    z=pinv(H)*r;
    s_hat(sym_index(Nt))=sign(real(z(Nt)));
```

```
    r=r-H(:,Nt)*s_hat(sym_index(Nt));
    H=H(:,1:(Nt-1));
    Nt=Nt-1;
 end
```

5.B-6   [WWW]The function below extends the function **sic_zf_bpsk.m** so that it works with QPSK modulation, for which the data symbols candidates are {-1-j, -1+j, 1-j, 1+j}. Only the line that computes **s_hat** needs to be changed. Complete this line and save the modified m-file as **osic_zf_qpsk()**. Capture the completed m-file.

```
function s_hat=osic_zf_qpsk(r,H)
...
while Nt>0
  ...
  s_hat(sym_index(Nt))=sign(real(z(Nt)))+j*??;
  ...
End
```

5.B-7   We will see in the next section that OSIC outperforms SIC. Justify why identifying the most reliable symbol and canceling it first at each SIC iteration improves the performance of SIC.

29.6   BER SIMULATION OF ZF, SIC, OSIC, AND ML DETECTION SCHEMES

6.A   [WWW]The m-file below simulates the BER performances of ZF, SIC, OSIC, and ML detection methods for an SM system with **Nr** = 4, **Nt** = 4, and BPSK modulation. Examine the two scaling factors in bold in the two lines that generate **H** and **n**, and explain why they should be set as they are.

```
clear
EbN0dB_vector=0:5:30; %Lower the limit of EbN0dB(currently 30) to speed up the sim-
ulation.
Eb=1; %Because we set the symbol by s=sign(rand(Nt,1)-0.5) below.
Nr=4;Nt=4;

for snr_i=1:length(EbN0dB_vector)
  EbN0dB=EbN0dB_vector(snr_i);
  EbN0=10^(EbN0dB/10);
  N0=Eb/EbN0;

  Nerrs_zf=0;  Nerrs_sic=0;  Nerrs_osic=0;  Nerrs_ml=0;
  Nbits=0;
```

```
    Nerrs_stop=100;%Decrease Nerrs_stop if the simulation takes long time but the
    simulation error will increase.
    while Nerrs_osic < Nerrs_stop

      s=sign(rand(Nt,1)-0.5);

      H=sqrt(0.5/Nr)*(randn(Nr,Nt)+j*randn(Nr,Nt));
      n=sqrt(N0/2)*(randn(Nr,1)+j*randn(Nr,1));

      r=H*s+n;

      shat_zf=zf_bpsk(r,H);  shat_sic=sic_zf_bpsk(r,H);
      shat_osic=osic_zf_bpsk(r,H);  shat_ml=ml_Nt4_bpsk(r,H);

      Nerrs_zf=Nerrs_zf+sum(shat_zf~=s);
      Nerrs_sic=Nerrs_sic+sum(shat_sic~=s);
      Nerrs_osic=Nerrs_osic+sum(shat_osic~=s);
      Nerrs_ml=Nerrs_ml+sum(shat_ml~=s);

      Nbits=Nbits+Nt;
    end

    BER_vector(snr_i,:)=[Nerrs_zf Nerrs_sic Nerrs_osic  Nerrs_ml]/Nbits

end
figure
semilogy(EbN0dB_vector, BER_vector)
legend('ZF', 'SIC','OSIC', 'ML')
xlabel('E_b/N_0');ylabel('BER');grid
```

6.B   Execute the m-file. (a) Capture the simulated BER graph. (b) Order the detection schemes according to their BER performances from highest to lowest.

6.C   [WWW]Modify the m-file in 6.A as shown below to simulate the case with **Nr** = 2 and **Nt** = 2.

```
...
Nr=2;Nt=2;
...
for snr_i=1:length(EbN0dB_vector)
  ...
  while Nerrs_osic < Nerrs_stop
    ...
    shat_ml=ml_Nt2_bpsk(r,H);
    ...
```

TABLE 29.1   BER Comparison of Various Detection Schemes.

Detection scheme	Does the BER increase or decrease as Nr and Nt increase? (Yes/No)	Justification
ZF		
SIC		
OSIC		
ML		

6.C-1   Execute the modified m-file and capture the simulated BER graph.

6.C-2   Answer the following questions:

(a) Is the order of the detection schemes according to their performances with **Nr** = 2 and **Nt** = 2 the same as that in 6.B(b), that is, the case with **Nr** = 4 and **Nt** = 4?

(b) Let us assume that the order of the four schemes according to their error performance from highest to lowest is DS1, DS1, DS3, and DS4. At BER = 10e-3, measure (from the figure) the relative performance gap in dB between DS1 and DS2, and between DS2 and DS3, and between DS3 and DS4 for the case of (**Nr, Nt**) = (2,2). Also do the same for the case of (**Nr, Nt**) = (4,4). How do the gaps change (increase or decrease) as the number of antennas changes from (2,2) to (4,4)?

6.C-3   As the number of antennas increases from (2,2) to (4,4), does the BER of each detection scheme increase or decrease? Justify your answer in Table 29.1.

6.D   [WWW]The m-file in 6.A has been modified to simulate the QPSK system with **Nr** = 4 and **Nt** = 4. The modified parts are highlighted in bold below.

```
clear
...
...
for snr_i=1:length(EbN0dB_vector)
...
    while Nerrs_osic < Nerrs_stop
        s=sign(rand(Nt,1)-0.5)+j*sign(rand(Nt,1)-0.5);
        ...
        shat_zf=zf_qpsk(r,H); shat_sic=sic_zf_qpsk(r,H);
        shat_osic=osic_zf_qpsk(r,H); shat_ml=ml_Nt4_qpsk(r,H);

        Nerrs_zf=Nerrs_zf+sum(real(shat_zf)~=real(s))+sum(imag(shat_zf)~=
        imag(s));
```

```
Nerrs_sic=Nerrs_sic+sum(real(shat_sic)~=real(s))+sum(imag(shat_sic)~=
imag(s));
Nerrs_osic=Nerrs_osic+sum(real(shat_osic)~=real(s))+sum(imag(shat_
osic)~=imag(s));
Nerrs_ml=Nerrs_ml+sum(real(shat_ml)~=real(s))+sum(imag(shat_ml)~=
imag(s));
...
Nbits=Nbits+2*Nt;
...
```

6.D-1   Justify why the line **'Nbits=Nbits+Nt'** should be changed into **'Nbits=Nbits+2*Nt'**.

6.D-2   Execute the modified m-file and capture the simulated BER curves.

6.D-3   Compared with the BPSK system, have the relative BER performances of the four detection schemes changed?

29.7   RELATIONSHIP AMONG THE NUMBER OF ANTENNAS, DIVERSITY, AND DATA RATE

[WWW]The m-file below simulates the BER performance of MLD for each of the following cases of (**Nr, Nt**): (2, 2), (3, 2), and (2, 3).

```
clear
EbN0dB_vector=0:5:15;
Eb=1;

for snr_i=1:length(EbN0dB_vector)
   EbN0dB=EbN0dB_vector(snr_i);  EbN0=10^(EbN0dB/10);  N0=Eb/EbN0;

   Nerrs_2by2=0;  Nerrs_2by3=0;  Nerrs_3by2=0;
   NsymsNt2=0;  NsymsNt3=0;
   Nerrs_stop=200;
   while Nerrs_2by3 < Nerrs_stop
      %%%%%%%%%%%%%%%%%%%%%%%%%%%%%%%%%%%%%%%%%%
      Nr=2;Nt=2;
      s=sign(rand(Nt,1)-0.5);
      H=sqrt(1/Nr)*(randn(Nr,Nt)/sqrt(2)+j*randn(Nr,Nt)/sqrt(2));
      n=sqrt(N0/2)*(randn(Nr,1)+j*randn(Nr,1));
      r=H*s+n;
      shat=ml_Nt2_bpsk(r,H);
      Nerrs_2by2=Nerrs_2by2+sum(shat~=s);
```

```
%%%%%%%%%%%%%%%%%%%%%%%%%%%%%%%%%%%%%%%%%
Nr=3;Nt=2;
s=sign(rand(Nt,1)-0.5);
H=sqrt(1/Nr)*(randn(Nr,Nt)/sqrt(2)+j*randn(Nr,Nt)/sqrt(2));
n=sqrt(N0/2)*(randn(Nr,1)+j*randn(Nr,1));
r=H*s+n;
shat=ml_Nt2_bpsk(r,H);
Nerrs_3by2=Nerrs_3by2+sum(shat~=s);

NsymsNt2=NsymsNt2+2;
%%%%%%%%%%%%%%%%%%%%%%%%%%%%%%%%%%%%%%%%%
Nr=2;Nt=3;
s=sign(rand(Nt,1)-0.5);
H=sqrt(1/Nr)*(randn(Nr,Nt)/sqrt(2)+j*randn(Nr,Nt)/sqrt(2));
n=sqrt(N0/2)*(randn(Nr,1)+j*randn(Nr,1));
r=H*s+n;
shat=ml_Nt3_bpsk(r,H);
Nerrs_2by3=Nerrs_2by3+sum(shat~=s);

    NsymsNt3=NsymsNt3+3;
  end

  BER_vector(snr_i,:)=[Nerrs_2by2/NsymsNt2 Nerrs_3by2/NsymsNt2 Nerrs_2by3/
  NsymsNt3]
end
figure
semilogy(EbN0dB_vector(1:snr_i), BER_vector)
legend('Nr=2, Nt=2', 'Nr=3, Nt=2','Nr=2, Nt=3')
xlabel('E_b/N_0');ylabel('BER');grid
```

7.A Execute the m-file above and capture the simulated BER curves.

7.B Compare the slope of the BER curves for the cases of (**Nr, Nt**) = (2, 2) and (3, 2).

(a) For which case does the BER curve have a larger slope?

(b) The relationship between the slope of the BER curve and diversity order was investigated in Chapter 25. Justify why the slope of the BER curve increases as Nr increases.

7.C Compare the BER curves for the cases of (**Nr, Nt**) = (2, 2) and (2, 3).

(a) Which case has a worse BER performance?

(b) Justify why the BER performance is worse for the case of Nt = 3.

7.D Although the BER performance decreases as Nt increases, a larger Nt is beneficial to the system in a different perspective. Summarize the benefits.

7.E Find scenarios where a system with Nr > Nt is preferable to a system with Nr < Nt. Also find scenarios where the latter is preferable to the former.

REFERENCES

[1] G. J. Foschini, "Layered Space–Time Architecture for Wireless Communication in a Fading Environment When Using Multiple Antennas," *Bell Labs Technical Journal,* Vol. **1**, No. 2, 1996, pp. 41–59.

[2] G. D. Golden, G. J. Foschini, R. A. Valenzuela, and P. W. Wolniansky, "Detection Algorithm and Initial Laboratory Results Using V-BLAST Space–Time Communication Architecture," *Electronics Letters,* Vol. **35**, No. 1, 1999, pp.14–16. .

[3] D. Tse and P. Viswanath, *Fundamentals of Wireless Communication*, Cambridge, UK: Cambridge University Press, 2005.

30

NEAR-ULTRASONIC WIRELESS ORTHOGONAL FREQUENCY DIVISION MULTIPLEXING MODEM DESIGN

- Transmit an image file through a near-ultrasonic (NUS) wireless channel.
- Investigate transmit and receive algorithms for NUS orthogonal frequency division multiplexing systems.
- Observe and analyze the waveforms and spectra of NUS systems at major processing stages.

30.1 IMAGE FILE TRANSMISSION OVER A NEAR-ULTRASONIC WIRELESS CHANNEL

In this section we transmit an image file over a near-ultrasonic (NUS) wireless channel. The image file is orthogonal frequency division multiplexing (OFDM) modulated and transmitted from the speaker of a phone over an NUS wireless channel. The microphone in a PC samples the received signal and demodulates it to restore the image data.

1.A Prepare a handheld audio device (e.g., a smartphone) that is capable of playing .wav files through its internal speaker. Also prepare a laptop or desktop PC with an internal microphone (MIC) that has installed MATLAB on it.

Step 1. [WWW]Copy the following three files into the MATLAB work folder in your laptop or desktop PC.

Problem-Based Learning in Communication Systems Using MATLAB and Simulink, First Edition.
Kwonhue Choi and Huaping Liu.
© 2016 The Institute of Electrical and Electronics Engineers, Inc. Published 2016 by John Wiley & Sons, Inc.
Companion website: www.wiley.com/go/choi_problembasedlearning

- **NUS_AOFDM_TX.m** and **NUS_AOFDM_DEM.m** from the companion website.
- Your photo image file. If you do not have one in your PC, take one with your phone and transfer it to the MATLAB work folder via email or a USB cable.

Step 2. Change the filename of your photo to **myphoto** and maintain the file extension.

Step 3. Execute **NUS_AOFDM_TX.m**. NOTE: If your photo file extension is not 'jpg', then open **NUS_AOFDM_TX.m** and replace the argument jpg in the fourth line '**A = imread('myphoto' , 'jpg')** with your photo file extension.

Step 4. Make sure that **NUS_AOFDM_TX.wav** is created in the MATLAB work folder. Move it to your phone memory or disk via email or USB cable. Do not accept any file format conversion or encoding if you are asked during the process of transferring the .wav file to your phone. For iPhones, the Gmail (Google mail) attachment and outlook attachment are two ways confirmed working.

Step 5. In the audio setting of your laptop, for recording, select the internal MIC as the default MIC. To avoid distortion in recording, disable the sound effects if there are any.

Step 6. Get ready so that you can start to run **NUS_AOFM_DEM.m** immediately and at any time needed. For example, you may type in **NUS_AOFM_DEM** in the command window but do not press the return key to run it; when you need to run this file, just need to press the return key. Alternatively, you may open **NUS_AOFM_DEM.m** and get ready to click the green play (run) button.

Step 7. Set the phone speaker volume to the maximum level and place the phone within 5 cm from the PC's internal MIC. Orient the speaker of the phone (not the receiver speaker that will be placed near your ear while you are making a phone call) toward the PC's MIC.

Step 8. Play **NUS_AOFDM_TX.wav** in the phone. After the wave file starts playing, run **NUS_AOFDM_DEM.m** which has been set ready to run in the PC in Step 6. Be sure not to stop or pause playing the wav file until **NUS_AOFDM_DEM.m** finishes running.

1.A-1   When all these steps are finished, capture the following.

(a) **Figure 1.** Be sure to maximize the window size prior to capturing. Save the **Figure 1** in .fig format, since it will be needed in Section 30.3.

(b) **Figure 3.**

(c) The value of **fo_pilot** returned in the command window. This value will be used later in 3.D-3.

1.A-2   Increase the distance between the phone and the PC bit by bit and repeat Step 7 at each distance. No need to save **Figure 1**, nor to record the value of **fo_pilot**. Record the maximum distance where the demodulated image data in **Figure 3** is recognizable.

30.2 ANALYSIS OF OFDM TRANSMITTER ALGORITHMS AND THE TRANSMITTED SIGNALS

2.A In the part **'OFDM signal parameter'** of **NUS_AOFDM_TX.m**, all system parameters to generate the NUS OFDM signal are set and their definitions are explained using comments.

2.A-1 Tabulate the parameter names, definitions, and values.

2.A-2 From the values tabulated in 2.A-1, calculate the following.

(a) Subcarrier frequency spacing in [Hz].
(b) The total signal bandwidth in [Hz], that is, the highest subcarrier frequency minus the lowest subcarrier frequency.
(c) The guard interval (cyclic prefix) length in [seconds].

2.B In the part **'OFDM modulated image data packet generation'** of the file **NUS_AOFDM_TX.m**, the image data are represented by the elements of the vector **Pixel**. This data stream is then OFDM modulated, and the complex baseband OFDM signal is stored in vector **x**.

In this part of the m-file, in the inner **'for'** loop, a certain number of image pixels are modulated into one OFDM symbol. Then a certain number of ODFM symbols are concatenated to generate one OFDM frame. In the outer **'for'** loop, the multiple OFDM frames for the entire image are concatenated to form one transmission packet. That is, **k** in the outer **'for'** loop is the OFDM frame index in the transmission packet and kk is the OFDM symbol index in each OFDM frame.

2.B-1 Fig. 30.1 shows the data packet structure as a frequency (subcarrier)–time (OFDM symbol) matrix for the transmission of one image. Each cell denotes one subcarrier of an OFDM symbol. From the definition of parameters listed in 2.A-1 and the part **'OFDM modulated image data packet generation'** of the m-file, determine the values of the three variables, A, B, and C in the diagram.

2.B-2 In part **'OFDM modulated image data packet generation'** of the m-file, the input to the IFFT for generating an OFDM symbol is contained in the vector **D**. According to the m-file, a certain subcarrier is allocated as the pilot subcarrier. (a) Determine the pilot subcarrier index. (b) Shade all the cells for the pilot subcarriers in Fig. 30.1.

2.B-3 The outer and inner **'for'** loops show that each OFDM frame has one fixed OFDM symbol, which is called the "pilot OFDM symbol." The pilot OFDM symbol is known to the receiver for frame synchronization and the channel estimation, whereas the pilot subcarrier described in 2.B-2 is used for carrier recovery in the receiver. Analyze the m-file, determine the OFDM symbol position in each OFDM frame, and mark all shaded OFDM symbols in Fig. 30.1.

2.B-4 Examine the m-file and determine (a) How many image data symbols (elements of the vector **Pixel**) are carried by one subcarrier of one OFDM symbol? (b) If

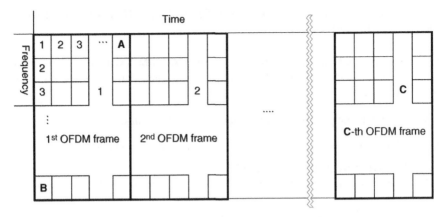

FIGURE 30.1 Packet structure to transmit one image.

one subcarrier carries more than one image data symbol, explain how it is possible to transmit multiple symbols on a single subcarrier.

2.C In part **'up conversion'** of the file **NUS_AOFDM_TX.m**, the complex baseband signal vector **x** is first up-converted into NUS frequencies. Then the sampled vector of the real-valued passband OFDM signal `tx` is generated.

2.C-1 After up-conversion, the frequency of the first (leftmost) subcarrier equals **fc**. Explain this on the basis of the m-file.

2.C-2 Based on the pilot subcarrier location determined in 2.B-2(a) and the frequency of the first subcarrier, determine the pilot subcarrier frequency in the passband.

2.D Execute **NUS_AOFDM_TX.m** and capture **Figure 2**.

2.D-1 According to the m-file, what are plotted in **Figure 2**?

2.D-2 Based on the plots in **Figure 2**, summarize all evidences observed that verify that the OFDM signal is generated correctly in the NUS band.

2.D-3 In the top subplot of **Figure 2**, zoom in the area of the OFDM spectrum and capture it.

2.D-4 From the figure in 2.D-3, (a) measure the bandwidth of the OFDM signal and record it. (b) Is the measured bandwidth consistent with the answer to 2.A-2(b)? If not, revisit 2.A-2(b).

2.D-5 From the captured figure in 2.D-3, (a) measure the pilot subcarrier frequency and record it. (b) Is the measured bandwidth consistent with the answer to 2.C-2? If not, revisit 2.C-2.

2.E In the first line of the part **'Packet repetition and side lobe suppression'** of the m-file, OFDM modulated image data packet **tx** is repeated one more time in order to avoid a packet loss in the receiver even if we start executing the m-file late after the .wav file is being played.

In the second line, the side lobes (spectrum skirt) of the OFDM signal are suppressed by passing the OFDM signal through a bandpass filter. To check the necessity of this bandpass filtering, comment out the second line '**tx=conv(tx,bpf)**' and run **NUS_AOFDM_TX.m** again.

2.E-1   (a) Capture **Figure 2**. (b) Compare the captured figure with the one in 2.D. Assess the difference of the OFDM spectra with and without bandpass filtering.

2.E-2   Execute **soundsc(tx,fs)** in the command line. You may or may not hear any sound. Explain why it should be the case.

2.E-3   Based on 2.E-1 and 2.E-2, document why bandpass filtering for the OFDM signal is required prior to it being transmitted.

2.E-4   Uncomment the line '**tx=conv(tx,bpf)**' and run **NUS_AOFDM_TX.m** again. Then repeat 2.E-2.

30.3   ANALYSIS OF OFDM RECEIVER ALGORITHMS AND THE RECEIVED SIGNALS

3.A   In the part '**Near ultrasonic sound sampling**' of the file **NUS_AOFDM_DEM.m**, the received signal by the MIC in the PC is sampled and converted into a MATLAB vector.

3.A-1   Open **Figure 1** saved in 1.A-1. According to the m-file, what are plotted in the first and second subplots of **Figure 1**?

3.A-2   Compare the first and second subplots with the captured plots in 2.D. (a) Document the main differences (b). Explain what causes such differences.

3.B   In the part '**Front end BPF stage**' of the file **NUS_AOFDM_DEM.m**, the OFDM spectrum located in the NUS band is extracted by bandbass filtering.

3.B-1   According to the m-file, what are plotted in the third and fourth subplots of **Figure 1**?

3.B-2   In **Figure 1**, compare the first subplot with the third one; also compare the second subplot with the fourth one. Document the reasons that bandpass filtering is needed at this stage.

3.B-3   For all the lines in the part '**Front end BPF stage**', explain what the variable of the left-hand side represents and how the right-hand side expression is formulated accordingly.

3.B-4   In the fourth subplot, zoom in the area of OFDM spectrum and capture it.

3.B-5   According to the captured spectrum in 3.B-4, determine whether or not the wireless channel is frequency selective for the signal being transmitted. Justify your answer.

3.C   In the part **'Down conversion'** of the file **NUS_AOFDM_DEM.m**, the output of the BPF is down-converted into the baseband. The vector **cbb** is the sampled version of the complex baseband signal.

For all the lines in the part **'Down conversion'**, explain what the variable of the left-hand side represents and how the right-hand side expression is formulated accordingly.

3.D   In the part **'Carrier recovery'** of the file **NUS_AOFDM_DEM.m**, the residual frequency offset in the complex baseband signal after down-conversion is measured and compensated. The vector **rcbb** is the sampled version of the frequency-offset-compensated complex baseband signal.

Since a carrier signal of the same frequency **fc** is required for up-conversion in the TX and for down-conversion in the RX (see the part **'Down conversion'**), ideally there should not be a carrier frequency offset in the down-converted complex baseband signal. In practice, however, the D/A converter clock for the phone audio device that plays the .wav file and the A/D converter clock for the PC audio device cannot be identical. This results in a residual frequency error in the down-converted complex baseband signal vector cbb, which should be compensated for before demodulation.

If the residual frequency offset is zero, then the pilot subcarrier frequency in the complex baseband signal rcbb is zero, and thus the pilot subcarrier spectrum is located at zero frequency. On the contrary, if the frequency offset is not zero, then the pilot subcarrier spectrum is located at the frequency offset. Thus, in the part 'Carrier recovery' of NUS_AOFDM_DEM.m, the pilot subcarrier in cbb is extracted by a narrowband low pass filter, and then its center frequency, which is equal to the residual frequency offset, is measured by using FFT. In the first three lines, 'cbb' is down-sampled 100:1 to lower the computational complexity of FFT to a practically reasonable level.

3.D-1   For the fourth line to the last line of the part **'Carrier recovery'**, explain what the variable of the left-hand side represents and how the right-hand side expression is formulated accordingly.

3.D-2   The fifth subplot in **Figure 1** shows the spectrum of the pilot subcarrier extracted from rcbb by using a low pass filter. Zoom in the spectrum peak to measure the residual frequency error and record it.

3.D-3   The value of **fo_pilot** recorded in 1.A-1(c) is the estimated residual frequency error. Is the measured frequency error in 3.D-2 approximately equal to **fo_pilot**?

3.E   To demodulate each OFDM symbol, we need to separately perform FFT on the properly partitioned portions of **rcbb**. Proper partitioning of **rcbb** requires locating the starting sample of each OFDM symbol. The part **'Frame synchronization'** in **NUS_AOFDM_DEM.m** locates the starting point of the frame first. Then the starting point of each OFDM symbols is determined by using the frame starting point.

First, the complex baseband signal vector **rcbb** passes through the matched filter, which is matched to the pilot OFDM symbol. The filter output is then partitioned into parts, each with a length equal to OFDM frame length. After this, the parts are aligned up in parallel and summed up. Because the same pilot OFDM symbol is repeatedly and periodically inserted in every OFDM frame, the pilot OFDM symbols at the matched filter output throughout the packet are coherently added up while the data OFDM symbols are added up randomly.

Consequently, as the accumulation goes on, the peak value of the matched filter filtered pilot OFDM symbol gets significantly larger than the off-peak value in the accumulator output. The starting sample of the pilot OFDM symbol is one OFDM symbol ahead of the peak sample point of the filtered (by the matched filter) pilot OFDM symbol.

In the first three lines, in order to lower the computational complexities of frame synchronization, the complex baseband signal sample vector **rcbb** is down-sampled to generate its down-sampled version **rcc_dec**.

3.E-1   From the fifth line to the last line of the part '**Frame synchronization**', the output of the matched filter, which is matched to the pilot OFDM symbol, is partitioned into parts, each with a length equal to the frame length and then accumulated to find the peak point. The starting sample index of the pilot OFDM symbol is then determined and stored in the variable **ST**.

Document what the variable on the left-hand side of the equal sign represents and how the right-hand side expression is properly formulated accordingly.

3.E-2   The fourth line '**rcbb_dec=abs(conv(rcbb_dec, conj(Pf_GI_dec(end:-1:1))));**' realizes the filtering process by a filter whose impulse response is equal to '**conj(Pf_GI_dec(end:-1:1))**' and then takes the absolute value of the filter output.

(a) The variable **Pf_GI_dec** is the down-sampled version of the vector **Pf_GI** that is defined in the transmitter m-file **NUS_AOFDM_TX.m**. The impulse response of the filter '**conj(Pf_GI_dec(end:-1:1))**' is the conjugate and time-reflected (left-right) version of **Pf_GI_dec**.

Explain what the filtering process in the fifth line is supposed to do.

(b) Based on the purpose of the filtering, explain why the impulse response of the filter is set as '**conj(Pf_GI_dec(end:-1:1))**', that is, why **Pf_GI_dec** is time-reflected and conjugated.

(c) Assuming imperfect carrier recovery, explain why **abs( )** is performed after this filtering process.

3.E-3   The sixth subplot of **Figure 1** shows the filtered and accumulated OFDM frame. Is this result shown in this plot what you expected to see? Justify the answer.

3.F   In the part '**OFDM demodulation with Channel compensation**', the OFDM symbols in the packet are demodulated one by one by using FFT. The channel gains at each subcarrier are estimated by using the FFT output of the periodically inserted pilot OFDM symbols. The FFT output vector of the data OFDM symbols is compensated for by using the estimated channel fading gains. Finally, the image data stream `Pixel` is recovered.

3.F-1   The OFDM symbols are demodulated through the outer '**for**' loop one by one by increasing the OFDM symbol index **k**. (a) In the '**for**' loop, the first line '**T=fft(rcbb(ST+GI+ NpGI*(k-1)+(1:Np)));**' performs FFT of the up-sampled vector of the current (**k**-th) OFDM symbol. Explain why the index of the vector **rcbb** is set as '**ST+GI+ NpGI*(k-1)+(1:Np)**'.

(b) The second line '**D=T(1:N);**' extracts the first **N** samples of the FFT output to get the modulated symbol vector at **N** subcarriers. Explain why **D**, that is, the first **N** samples of the FFT output, corresponds to the modulated symbol vector at **N** subcarriers.

3.F-2   Inside the outer '**for**' loop, the line '**if mod(k,Ns)==1**' checks whether the current (**k**-th) OFDM symbol is pilot or data. Explain how it does this.

3.F-3   For the two lines '**ch:=D./P;**' and '**ch_tbl(frcnt,:)=ch;**' below the line '**if mod(k,Ns)==1**', document what the variable on the left-hand side of the equal sign represents and why it is set or computed as the right-hand side. The index **frcnt** is the frame counter. The vector **P** is defined in the part '**OFDM signal parameter**' of **NUS_AOFDM_TX.m**. Section 2.B of Chapter 27 will help answer this question.

3.F-4   In the line '**D_buffer=[D_buffer ; D];**' below '**else**', the data symbol vector **D** for the current OFDM symbol in the current OFDM frame is accumulated and stored as an additional row of the matrix **D_buffer**. If the condition checking '**if mod(k,Ns)==1**' returns 'true' after each iteration, then a new frame will start. Thus, at this point, all the data symbol vectors of the current frame have been stored in **D_buffer**. Then the innermost '**for**' loop performs channel compensation for the data symbols in the current OFDM frame in **D_buffer** to generate the final demodulation values of the data symbols and recover the **Pixel** stream in the current OFDM frame.

According to the frame structure investigated in 2.B-3, the data OFDM symbols in each frame are packed between two pilot OFDM symbols of two adjacent frames. Inside the innermost '**for**' loop, Pch and Cch are the estimated channel gain vectors from the pilot OFDM symbols of the current frame and the next frame, respectively. The channel gain vectors for the data OFDM symbols in the middle are estimated through linear interpolation using the two end values, that is, **Pch** and **Cch**.

For the three lines inside the innermost '**for**' loop, document what the variable on the left-hand side of the equal sign represents and why it is set or computed as the right-hand side.

3.F-5   Recall that the fourth subplot in **Figure 1** shows the received signal spectrum. Zoom into its spectrum portion. The seventh subplot shows the 3-D surface of the estimated channel gain magnitude over the time and frequency matrix. Compare these two subplots. Does the estimated channel gain magnitude match well the actual one? Justify your answer.

3.F-6   Based on the seventh subplot in **Figure 1**, is the current wireless channel more selective in time or in frequency? Justify your answer.

3.G   The last part '**Image packet synchronization**' first locates the start of the image data packet from the demodulated pixel steam **DemPixel**. Then the vector **DemPixel** is cyclically shifted so that the first element of it becomes the beginning of the packet.

3.G-1   For all lines in the part '**Image packet synchronization**', document what the variable on the left-hand side of the equal sign represents and why it should be set or computed as the right-hand side.

3.G-2   What does the ninth subplot in **Figure 1** show? Is it what you expected to see?

30.4   EFFECTS OF SYSTEM PARAMETERS ON THE PERFORMANCE

4.A   Insert the following two lines right after the line '**RX_time=TX_time+1;**' at the beginning part of **NUS_AOFM_DEM.m** and save the m-file.

- **load NUS_AOFDM_TX_tx.mat;**
- **soundsc(tx,fs)**

Now, running **NUS_AOFM_DEM.m** will transmit the NUS OFDM signal from the PC's speaker, which is then received and demodulated. Therefore, without playing the .wav file in the phone, transmission and reception are executed within the m-file.

4.A-1   Maximize the PC audio volume and then run **NUS_AOFM_DEM.m**. Capture the received image in **Figure 3**.

4.A-2   Open **NUS_AOFDM_TX.m**. and change the line '**GI=ceil(Np*1/8);**' into '**GI=0**'. Run **NUS_AOFDM_TX.m** and then run **NUS_AOFDM_DEM.m**. Capture the received image in **Figure 3**.

4.A-3   Compare the quality between the received images in 4.A-1 and 4.A-2. Justify the comparison result.

4.B   Answer the following questions:

4.B-1   In **NUS_AOFDM_TX.m**, **Np** is set proportional to **N**. This maintains a constant total data speed regardless of **N**. Prove this assuming **GI** = 0.

4.B-2   Recall that the total bandwidth has been calculated from **N**, **Np**, and **fs**. Show that with **Np** proportional to **N**, the total bandwidth is also constant, regardless of **N**.

4.B-3   Run **NUS_AOFDM_TX.m** with **N** = 16 and 256 and capture the received images in **Figure 3** for both cases.

4.B-4 Compare the transmitted signal waveforms in terms of the peak amplitude for $N = 16$, 64 (captured in 2.D), and 256. Which case has the highest peak power and why?

4.C Restore $N = 64$ in **NUS_AOFDM_TX.m**. The line **'Pixel=Pixel(ILindex)'** in **NUS_AOFDM_TX.m** interleaves the image pixel elements in **Pixel**. Comment out this line and run **NUS_AOFDM_TX.m**.

4.C-1 Capture **Figure 2**. Compare the transmitted signal waveform with the one captured in 2.D. Which one has a higher peak?

4.C-2 Considering the typical characteristics of adjacent pixels of a photo, explain why the transmitted waveform has a high peak power if the pixel stream is only OFDM-modulated as is.

4.C-3 Explain why interleaving of the pixel elements decreases the peak power of the OFDM modulated waveform.

4.D The algorithms and the parameter settings in the two m-files provided could be improved in terms of performance and complexity.

4.D-1 Revise the two m-files to improve the received image quality. (a) Explain where and why you revise so. (b) Capture the received images before and after the revision and confirm that the image quality did improve after the revision. For fair comparison, be sure to maintain the same signal bandwidth and the same channel environment, for example, noise level and audio volume.

4.D-2 Revise the two m-files to achieve a higher spectral efficiency, for example, a lower bandwidth while the packet transmission time is not changed, or a shorter packet transmission time while the bandwidth is not changed. (a) Explain where and why you revise so. (b) Capture the results that show a higher spectral efficiency, for example, smaller bandwidth or shorter transmission time. (c) Capture the received images before and after the revision to confirm that the image quality is not reduced due to the revision.

4.D-3 Revise the two m-files to reduce the computational complexity without lowering the received image quality. (a) Explain where and why you revise so. (b) Calculate the computation reduction ratio for the algorithm part revised. (c) Capture the received images before and after the revision to confirm that the image quality is not reduced due to the revision.

INDEX

Problem-Based Learning in Communication Systems Using MATLAB and Simulink, First Edition.
Kwonhue Choi and Huaping Liu.
© 2016 The Institute of Electrical and Electronics Engineers, Inc. Published 2016 by John Wiley & Sons, Inc.
Companion website: www.wiley.com/go/choi_problembasedlearning

IEEE PRESS SERIES ON
DIGITAL AND MOBILE COMMUNICATION

John B. Anderson, *Series Editor*
University of Lund

Printed and bound by CPI Group (UK) Ltd, Croydon, CR0 4YY

16/04/2025

14658579-0004